Aerothermodynamics and Jet Propulsion

Get up to speed with this robust introduction to the aerothermodynamics principles underpinning jet propulsion, and learn how to apply these principles to jet engine components. This book is suitable for undergraduate students in aerospace and mechanical engineering, and for professional engineers working in jet propulsion. This textbook includes consistent emphasis on fundamental phenomena and key governing equations, providing students with a solid theoretical grounding on which to build practical understanding; clear derivations from first principles, enabling students to follow the reasoning behind key assumptions and decisions, and successfully apply these approaches to new problems; practical examples grounded in real-world jet propulsion scenarios illustrate new concepts throughout the book giving students an early introduction to jet and rocket engine considerations; and online materials for course instructors, including solutions, figures, and software resources, to enhance student teaching.

Paul G. A. Cizmas is Professor of Aerospace Engineering at Texas A&M University. Over the past 25 years, he has conducted research into numerical simulations of transport phenomena in propulsion systems covering a large range of topics, including reduced-order modeling, aeroelasticity, combustion, and computational fluid dynamics. He is a Fellow of American Society of Mechanical Engineers and an Associate Fellow of American Institute of Aeronautics and Astronautics.

Aerothermodynamics and Jet Propulsion

Paul G. A. Cizmas

Texas A&M University

CAMBRIDGE
UNIVERSITY PRESS

CAMBRIDGE
UNIVERSITY PRESS

University Printing House, Cambridge CB2 8BS, United Kingdom

One Liberty Plaza, 20th Floor, New York, NY 10006, USA

477 Williamstown Road, Port Melbourne, VIC 3207, Australia

314-321, 3rd Floor, Plot 3, Splendor Forum, Jasola District Centre, New Delhi – 110025, India

103 Penang Road, #05–06/07, Visioncrest Commercial, Singapore 238467

Cambridge University Press is part of the University of Cambridge.

It furthers the University's mission by disseminating knowledge in the pursuit of
education, learning, and research at the highest international levels of excellence.

www.cambridge.org
Information on this title: www.cambridge.org/9781108480758
DOI: 10.1017/9781108691055

First published 2022

Printed in the United Kingdom by TJ Books Limited, Padstow Cornwall

A catalogue record for this publication is available from the British Library.

Library of Congress Cataloging-in-Publication Data
Names: Paul G. A. Cizmas, author.
Title: Aerothermodynamics and jet propulsion / Paul G. A. Cizmas.
Other titles: Aerothermodynamics and jet propulsion Description: New York :
    Cambridge University Press, [2021] | Includes bibliographical references and index.
Identifiers: LCCN 2021029819 (print) | LCCN 2021029820 (ebook) |
    ISBN 9781108480758 (hardback) | ISBN 9781108691055 (epub)
Subjects: LCSH: Aerothermodynamics. | Jet propulsion. |
    BISAC: SCIENCE / Mechanics / Fluids
Classification: LCC TL574.A45 C57 2021 (print) | LCC TL574.A45 (ebook) |
    DDC 629.132/3–dc23
LC record available at https://lccn.loc.gov/2021029819
LC ebook record available at https://lccn.loc.gov/2021029820

ISBN 978-1-108-48075-8 Hardback

Brief Contents

Contents

Preface

This textbook was developed from the course notes put together for the second-semester junior class, Aerothermodynamics and Propulsion, taught at Texas A&M University. This class is followed by two senior courses: Aerospace Propulsion, a jet engine design class, and Rocket Propulsion, a rocket engine design class. Although this textbook was not conceived for these design courses, it provides essential foundational knowledge for design. Fundamentally, the purpose of the book is to enhance the aerodynamics and thermodynamics background of students, and to enable them to apply this knowledge to understanding jet propulsion.

Aims of the Text

This text is written primarily for undergraduate students in their third year of study, and it also serves as a self-study for students and engineers interested in the field. It emphasizes the fundamental phenomena of aerothermodynamics and their governing equations, as well as the simplifying assumptions used when applying these governing equations to solving propulsion-related problems. Derivations of the governing equations from first principles are included, so that students can follow the reasoning and the assumptions made during this process. It is important to include these derivations because if students do not understand how these equations were derived, it is quite probable that they may apply them incorrectly. Furthermore, if students do not understand the reasoning process, they might not be able to apply these principles when solving new problems.

The concepts presented in the book are always followed by examples relevant to propulsion. In this way, the student is exposed to jet engines and rocket engines well before reaching the chapters that describe these engines.

Structure of the Book

The book is split into three parts. Part I presents the basic fluid mechanics and thermo-dynamics laws and derives the governing equations for different levels of approximation. Part II considers the specific aspects of aerodynamics and thermodynamics that apply to air-breathing engines, and describes and examines the jet engine components. Part III presents a classification of rocket engines and describes the fundamentals of rocket performance.

Both Part II and Part III rely on the material covered in Part I. Part II and Part III are independent of each other.

Part I begins with a classification of propulsion systems and a brief overview of the history of jet propulsion (Chapter 1). The jet propulsion principle is then introduced using an empirical description. Chapter 2 presents a review of aerothermodynamics. The Reynolds Transport Theorem is used to derive the conservation equations; the thermodynamic laws are established, and their expressions derived for a control volume. In Chapter 3, dimensional and dimensionless reference speeds are introduced, along with a discussion of isentropic and nonisentropic flows. Flows through nozzles are also included. Normal and oblique shock waves are then examined, and solution methods are presented for air and combustion products.

Chapter 4 is concerned with both viscous and thermal boundary layers, and applies them to propulsion problems. Chapter 5 is a brief introduction to combustion. First a classification of fuels is presented, followed by the thermochemical laws. The heats of formation and reaction needed to calculate the adiabatic flame temperature are presented next. It ends with a description of standard fuel.

Part II of the book discusses air-breathing engines. Chapter 6 derives the thrust equation and establishes the engine performance parameters. After introducing the Brayton cycle, the real cycles of the turbojet, turbofan, turboprop, and ramjet engines are analyzed. Chapter 7 presents the jet engine components: inlet diffusers, compressors and fans, combustors, turbines, and exhaust nozzles. The performance of these engine components is then connected to the real cycles covered in Chapter 6. Chapter 8 is devoted to thrust augmentation and such topics as water injection, afterburning, and intraturbine combustion.

Part III concludes with rocket engines and offers in Chapter 9 a classification and succinct presentation of the essential features and general equations related to rocket engines. Chapter 10 presents the performance of chemical rocket engines.

Teaching with this Book

The content of this textbook typically exceeds what can be covered in one semester. The instructor has the option to tailor the material to accommodate a variety of course syllabi based on the specifics of the program at their school.

A large number of examples and problems are included to help the reader understand the concepts and practice the methods introduced in the textbook. The difficulty of the problems varies, so the instructor can tailor homework assignments, tests, and exams accordingly. Solutions are offered to the instructor for all problems. The instructor is also offered web access to several codes that calculate (1) the thermodynamic properties of air and combustion products; (2) normal and oblique shock waves; (3) the Fanno line; (4) the Rayleigh line; (5) adiabatic temperature for different fuels; (6) real cycles analysis for all jet engines; (7) rocket thrust, nozzle exit velocity, and mass flow rate; (8) radial velocity variation in the axial compressor stage; and (9) radial velocity variation in the axial turbine stage.

Finally, I would like to thank Professors Adrian Bejan and John Slattery, who guided me before and during the genesis of this book. Several chapters of the book were finalized during the summer of 2019 when I visited the DLR Institute of Aeroelasticity in Göttingen, Germany. I am grateful to my host at the institute, Professor Dr. Holger Hennings, and his colleagues, Drs. Virginie Chenaux, Jens Nitzsche, and David Quero Martin, with whom I had numerous interesting technical discussions as well as many relaxing moments. I would also like to express appreciation for the feedback received from my former colleague Dr. Gabriel Marinescu. Last but not least, I am grateful for the suggestions and comments I received from the reviewers of the manuscript.

Nomenclature

Roman

A	–	Area
A_{cr}	–	Critical area
a	–	Speed of sound
a_0	–	Speed of sound at stagnation temperature
c	–	Chord length or Circumference length
a_{cr}	–	Critical speed of sound
C_f	–	Skin friction drag coefficient
c_p	–	Specific heat at constant pressure
c_v	–	Specific heat at constant volume
E	–	Energy
e	–	Specific energy per unit mass
F	–	Impulse function
$\vec{F}_\mathcal{V}$	–	Body force
$\vec{F}_e$	–	External force
$\vec{F}_i$	–	Inertia force
$\vec{F}_p$	–	Pressure force
$\vec{F}_\mathcal{S}$	–	Surface force
$\vec{F}_{vis}$	–	Viscous force
F_{cr}	–	Critical impulse function
$\vec{f}$	–	External force per unit mass
f	–	Fuel-air mass ratio
g	–	Gravitational acceleration
H	–	Shape factor
h	–	Enthalpy
h_0	–	Stagnation enthalpy
$\vec{I}$	–	Linear momentum
I	–	Rothalpy or Impulse
$\vec{K}$	–	Angular momentum
$\mathcal{M}$	–	Molecular mass
M	–	Mach number
$\vec{M}_{F\mathcal{V}}$	–	Momentum due to body forces
$\vec{M}_{F\mathcal{S}}$	–	Momentum due to surface forces
m	–	Mass
$\dot{m}$	–	Mass flow rate

n	–	Number of degrees of freedom of the molecule
$\hat{n}$	–	Normal unit vector
$\mathcal{P}$	–	Power
$\mathcal{P}_{\text{shaft}}$	–	Shaft power
$\mathcal{P}_{\text{shear}}$	–	Shear power
p	–	Static pressure
p_0	–	Stagnation pressure
Pr	–	Prandtl number
Q	–	Heat transfer
$\dot{Q}$	–	Heat transfer rate
$\mathcal{R}$	–	Universal gas constant
R	–	Gas constant
R'	–	Degree of reaction
$\vec{r}$	–	Point vector
Re	–	Reynolds number
$\mathcal{T}$	–	Thrust
T	–	Static temperature or Thwaites parameter
T_0	–	Stagnation temperature
t	–	Time
S	–	Entropy
s	–	Specific entropy per unit mass
St	–	Stanton number
U	–	Internal energy or Transport velocity
u	–	Specific internal energy per unit mass
u^*	–	Friction velocity, $u^* = \sqrt{\tau_{\text{wall}}/\rho}$
$\mathcal{V}$	–	Volume
$\vec{V}$	–	Velocity
$\vec{V}^*$	–	Control volume velocity
V_{max}	–	Maximum velocity for a given stagnation temperature
v	–	Specific volume per unit mass
W	–	Work transfer or Relative velocity
W_n	–	Work transfer due to normal stresses
w	–	Specific work transfer per unit mass
(x, y, z)	–	Cartesian coordinates
y_1	–	Height of the element adjacent to the airfoil
y^+	–	Non-dimensional number, $y^+ = u^* y_1 / v$

Greek

α	–	Angle of absolute velocity
β	–	Angle of relative velocity
γ	–	Ratio of specific heats or Stagger angle

δ^*	–	Displacement thickness
η_p	–	Propulsion efficiency
η_th	–	Thermal efficiency
θ	–	Momentum thickness or Turning angle
λ	–	Mean free path or Lambda number or Boundary layer parameter of Excess air
μ	–	Dynamic viscosity
ν	–	Kinematic viscosity
ρ	–	Density
σ	–	System surface or Solidity
σ^*	–	Surface of control volume τ^*
τ	–	System volume
τ^*	–	Control volume
τ_wall	–	Shear stress at wall
Φ	–	Equivalence ratio
ϕ	–	Flow coefficient
Ψ	–	Work coefficient
ω	–	Angular velocity

Hebrew

א	–	Generic variable

Subscripts

0	–	Stagnation
1	–	Initial or upstream the shock wave
2	–	Final or Downstream the shock wave
$-\infty$	–	Upstream infinity
a	–	Atmospheric or Air or Axial component
cr	–	Critical
cm	–	Control mass
cv	–	Control volume
e	–	Exit
i	–	Inlet
n	–	Normal component
stoich	–	Stoichiometric
t	–	Tangential component
u	–	Component in the direction of the transport velocity, U

Part I

Basic Fluid Mechanics and Thermodynamics for Propulsion

1   Jet Propulsion Principle

1.1   Introduction

This introductory chapter starts with a classification of propulsion systems. This allows us to get familiar with some of the nomenclature used in this text. A brief history of jet propulsion is presented in order to understand the evolution of propulsion systems. The jet propulsion principle is then presented, and the expression of jet engine thrust is introduced using elementary arguments. A rigorous derivation of the thrust expression will be presented in Chapter 6.

1.2   Propulsion Systems Classification

Propulsion, according to the *Merriam-Webster Dictionary*, is the action or process of driving a body forward or onward. A *propulsion system* is the device that accomplishes this task. The force generated by the propulsion system that produces the locomotion is called *thrust*.

There are numerous ways to classify propulsion systems because there are numerous criteria that can be used. One option is to classify propulsion systems based on the means used for *thrust generation*. Several options are possible for thrust generation: jet, propeller, fan, and combinations of these, as shown in Table 1.1. The entire thrust generated by turbojet and ramjet engines is produced by the jet. In a turboprop engine, most of the thrust is generated by the propeller and a small fraction by the jet. In a turbofan engine, most of the thrust is generated by the fan and a small fraction by the jet. The entire thrust generated by the piston engine used on airplanes is produced by the propeller.

A second way of classifying propulsion systems is based on the *source of energy* used to produce the thrust. The most common source of energy used by today's propulsion systems is *chemical energy*. Chemical energy is used by all piston engines and jet engines (turbojets and ramjets), by most chemical rocket engines, and by some electric rocket engines. The propellant for turbojets and ramjets is a mixture of fuel and air. The propellant for chemical rockets and electric rockets is stored propellant.

3

Table 1.1 Several ways of classifying propulsion systems.

Criterion	Type	Propulsion system
A. Cycle type	Intermittent cycle	Piston engine
	Continuous cycle	Jet engine
		Rocket engine
B. Air usage	Air-breathing	Piston engine
		Jet engine
	Nonair-breathing	Rocket engine
C. Thrust generation	Propeller	Piston engine
	Jet	Rocket engine
		Ramjet
		Turbojet
	Propeller & jet	Turboprop
	Fan & jet	Turbofan
D. Power source	Chemical energy	
	Nuclear energy	
	Solar energy	
E. Combined	Turbo compound piston engine	
	Turbo-ramjet	
	Turbo-rocket	
	Ramjet-rocket	
F. Gas turbine	Turbojet	
	Turboprop and turboshaft	
	Turbofan	
	Unducted fan (or propfan)	

Another source of energy for propulsion is *nuclear energy*. The main advantage of nuclear energy is the fact that it has the largest density of energy per unit mass of fuel. A drawback of using nuclear energy is the danger posed by the radioactive material. Nuclear energy can be used in combination with chemical energy, whereby the chemical energy is used to provide the thrust at takeoff and the nuclear energy is used for deep space missions. Technical feasibility of using nuclear energy has been demonstrated for turbojets, ramjets, chemical rockets, electrical rockets, and nuclear fission rockets.

A third source of energy for propulsion is *solar energy*. Electric rockets, solar heated rockets, and solar sails can use solar energy. The propellant of the solar heated rocket is stored hydrogen, and the propellant for solar sails is reflected photons (not stored in the propulsion system).

Propulsion systems can either use air from the atmosphere to generate thrust or carry the oxidizer on board. The former propulsion systems are air-breathing engines and the latter are nonair-breathing engines. Air-breathing engines include piston engines and jet engines. Electric, nuclear fission, and most chemical rockets are nonair-breathing engines.

Another criterion for classifying the propulsion systems is the *cycle type*. An intermittent cycle is used by piston engines, while a continuous cycle is used by turbojets, ramjets, and rocket engines.

Jet engines can be classified into gas turbine engines and ramjet engines. Gas turbine engines can be further classified into turbojet, turbofan, turboprop, and turboshaft engines. A detailed presentation of gas turbine and ramjet engines is made in Chapter 6.

1.3 Brief History of Jet Propulsion

Although jet propulsion is a technology that revolutionized transportation in the last seven decades, its beginnings date back more than one thousand years. This section briefly presents some of the most important moments in the history of jet propulsion.

1.3.1 Rocket Engines

The rocket engine is the oldest propulsion system based on jet propulsion. The development of the rocket engine came as a result of the discovery of gunpowder, around year 850 AD in China. Gunpowder emerged as an accidental outcome of the work of alchemists, who were experimenting with potassium nitrite, sulfur, and charcoal. The first flying rockets were used around 1150 at fireworks displays. These rocket engines used solid propellant based on gunpowder.

The rockets were further developed by trial and error, mainly for military applications. They were used by Great Britain in the war with the USA in 1812. The flying range of these rockets was approximately 3,000 yards, but their accuracy was rather poor. In the last part of the nineteenth century, solid propellant rockets were also used for nonmilitary applications, such as whaling, signaling, and transfer of lifelines between ships.

Further developments of rocket engines were due to the quest of exploring space. This led in the beginning of the twentieth century to a more systematic development, both analytical and experimental, of rockets and rocket engines. This development was greatly affected by the contribution of three pioneers: Konstantin Tsiolkovsky, Robert Goddard, and Hermann Oberth (Fig. 1.1) [Clark, 1972].

Tsiolkovsky was a Russian high school teacher interested in how to send vehicles into space [Ley, 1954]. In 1903, he published the paper "Exploration of Space with Reactive Devices," in which he described how one can escape Earth's gravitational field with a rocket. He calculated, using what is now called the Tsiolkovsky equation, that the horizontal speed needed for a minimal orbit around Earth is 8 km/s. Based on his calculations, he argued the need to use a multistage rocket. Tsiolkovsky also proposed the use of liquid oxygen and hydrogen as propellants.

(a)   (b)   (c)

Figure 1.1 Pioneers of rockets and rocket engines. (a) Konstantin Tsiolkovsky (1857–1935) (courtesy of Universal History Archive/Universal Images Group/Getty Images), (b) Robert Goddard (1882–1945) (courtesy of Hulton Archive/Getty Images), and (c) Hermann Oberth (1894–1989) (courtesy of Mondadori/Getty Images).

Goddard was a professor of physics at Clark University in Worcester, Massachusetts. In his 1919 publication *A Method for Reaching Extreme Altitudes*, he presented a mathematical analysis of methods for reaching high altitudes, including traveling to the moon [Archer and Saarlas, 1996, p. 25]. Goddard was also an experimentalist. He worked with solid propellant (black and smokeless powder fuel), but he also appreciated the advantages of liquid propellants, such as liquid oxygen and hydrogen. He was granted approximately 200 patents, including patents for the design of a rocket combustion-chamber nozzle, a propellant feed system, and a multistage rocket (1914).

Hermann Oberth was a physicist born in Hermannstadt (Sibiu), a Transylvanian town in Austro-Hungaria (now Romania). After the First World War, Oberth studied physics in Germany, first in Munich and then at the University of Heidelberg [Teodorescu, 2004]. His proposed doctoral dissertation on rocket science was rejected, being considered unrealistic. Oberth, however, published his work privately in 1923 in a 92-page book titled *The Rocket into Planetary Space*. Disappointed by the German system of education, Oberth stated "Our educational system is like an automobile which has strong rear lights, brightly illuminating the past. But looking forward, things are barely discernible" [Teodorescu, 2004].

In his book, Oberth presented the possibility of rocket journeys into space, and also included an assessment and design of hydrogen and alcohol fuels. Oberth received his license in physics from the University of Cluj, Romania, in 1923, for the same rocketry paper that was rejected in Germany. In 1929, Oberth published an expanded presentation of his work in a 429-page book titled *Ways to Spaceflight*. Unlike other experts in engineering and physical sciences, Oberth did not have the chance to teach at the college or university level. To support his family, he ended up teaching physics and mathematics at a high school in

Figure 1.2 Wernher von Braun (1912–1977) (courtesy of NASA/Interim Archives/Getty Images).

Mediaş, Romania, between 1924 and 1938. In the fall of 1929, Oberth fired his first liquid rocket engine, supported by students of the Technical University of Berlin. One of the students was the eighteen-year-old Wernher von Braun, a leading figure in the development of rocket technology in Germany and the United States. In 1938, Oberth left Romania and went first to Vienna, Austria, which was part of Germany at the time. Although a Romanian citizen, Oberth started working in 1941 at Peenemünde, where his former student Wernher von Braun (1912–1977) had begun building the world's first rocket (Fig. 1.2).

Some of the most important dates in the history of space exploration are as follows: (1) October 4, 1957, when the first artificial satellite of the Earth, Sputnik, was launched by USSR; (2) April 12, 1961, when Yuri Gagarin (USSR) was the first human to orbit the Earth; (3) July 20, 1969, when Neil Alden Armstrong and Edwin Eugene "Buzz" Aldrin, Jr. became the first humans to land on the moon, fulfilling the goal that John F. Kennedy proposed in his speech of May 25, 1961: "I believe that this nation should commit itself to achieving the goal, before this decade is out, of landing a man on the Moon and returning him safely to the Earth."

1.3.2 Jet Engines

Before talking about the history of jet engines, let's recall briefly the status of piston engines in the first part of the twentieth century. The propulsion system used by the Wright brothers in their first flight was a piston engine that delivered 12 hp at the shaft. The power-to-weight ratio of this engine was only 0.05 hp/lb.

By the end of World War I, the power of piston engines used on airplanes had reached 400 hp. At the end of World War II, the most powerful piston engines delivered approximately 5,000 hp. The power-to-weight ratio had increased to 0.8 hp/lb. The overall efficiency of the propulsion system (engine and propeller) reached $\eta = 29\%$. The increase of power, power-to-weight ratio, and propulsion system efficiency resulted in a constant increase in airplane speed, which reached 720 km/h for piston engine–powered airplanes during World War II, as shown in Table 1.2.

Table 1.2 Evolution of engine power and airplane speed.

	1903	World War I	World War II	Today
		⟵ Piston Engine		Jet Engine ⟶
Power	12 hp	400 hp	5,000 hp	100,000 hp (equivalent)
Airplane Speed	50 km/h	240 km/h	720 km/h	more than Mach 3

(a)   (b)

Figure 1.3 Pioneers of jet engines. (a) Frank Whittle (1907–1996) (courtesy of Central Press/Getty Images) and (b) Hans Joachim Pabst von Ohain (1911–1998).

Although the first jet engines were developed in the 1930s, early jet engine patents were filed as early as 1908. A patent for a jet engine based on piston machinery was introduced by Lorin in 1908, followed by a ramjet patent in 1913. Guillaume filed a patent for a jet engine based on turbomachinery in 1921. Probably the most important patents for the development of jet engines were filed by Frank Whittle in 1930, when he was only twenty-three years old. These patents covered the turbojet engine, the axial and radial compressors, the combustor, the axial turbine, and the exhaust nozzle.

The first jet engine was developed by Dr. Hans Joachim Pabst von Ohain, who obtained his doctorate in physics at the University of Göttingen in Germany (Fig. 1.3). Dr. von Ohain was only twenty-two years old when he conceived the idea of a continuous cycle

(a)   (b)

Figure 1.4 First airplanes powered by jet engines. (a) Heinkel He-178 (courtesy of Apic/Getty Images) and (b) Gloster E28/39 (courtesy of SSPL/Getty Images).

combustion engine in 1933. In 1934, he patented a jet engine design similar to that of Frank Whittle but different in internal arrangement. Dr. von Ohain's engine ran in February 1937 using hydrogen. The engine was developed with the support of Ernst Heinkel, the owner of the Heinkel Aircraft Factory [Bathie, 1996, p. 6]. The engine had a radial (or centrifugal) compressor and a radial turbine. A third engine design, called HeS 3, which used diesel fuel and produced 4.4 kN thrust, was used on the Heinkel 178 airplane. This airplane was designed around Dr. von Ohain's engine. The first successful flight took place on August 27, 1939.

The first jet engine running with liquid fuel was developed by Frank Whittle in April 1937. The engine produced 6.3 kN thrust. The engine was installed on the Gloster Meteor E28/29 airplane, shown in Fig. 1.4, that flew for the first time on May 15, 1941.

The first jet engine designed in the USA was the Westinghouse J30, in 1943. This engine was designed and built in eight months. In 1943, General Electric built Whittle's engine, W2B, under license. From the very beginning, the power-to-weight ratio of jet engines (25 hp/lb) was significantly larger than that of reciprocating engines (5 hp/lb).

In the early 1950s the turboprops were developed, followed by high-bypass-ratio turbofans in the 1960s. One should note, however, that the first turbofan development was initiated in 1936.

The evolution of the jet engine can be illustrated by the development of its component parts. Table 1.3 shows the variation of compressor pressure ratio, stage pressure ratio, and compressor efficiency between 1943 and today.

1.3.2.1 Evolution of Airplanes Powered by Jet Engines

The development of jet engines has influenced the evolution of airplanes. This section briefly presents the evolution of airplanes powered by jet engines. Interestingly, the first airplane powered by jet did not use a jet engine but a piston engine. This airplane was designed and built by the Romanian Henry Coandă in 1910, well before the first jet engine was built.

Table 1.3 Evolution of compressor pressure ratio and efficiency.

Year	Compressor pressure ratio, π^*	Stage pressure ratio, π^*_{stage}	Compressor efficiency, $\eta_{polytropic}$
1943	3:1	1.15:1	79%
2000	50:1	1.4:1	92%

The thrust produced by the engine was 2.1 kN. The concept was later used by the Italian aircraft manufacturer Caproni on the experimental jet aircraft Caproni-Campini N.1, which first flew in 1940. The performance of the N.1 airplane powered by a "motorjet" engine, which consisted of a compressor driven by a piston engine, was underwhelming.

The first jet engine airplane was the Heinkel 178 and used von Ohain's HeS 3 jet engine described in the previous section. Only one Heinkel 178 airplane was built, but the lessons learned on the 178 were used for the development of the Heinkel 280. The Heinkel 280 was the first jet engine fighter. Its maiden flight was on March 30, 1941. Soon after that, the Messerschmitt Me 262 was introduced. The maximum speed of the Me 262 airplane was 880 km/h. The Me 262 was equipped with two Jumo 004 engines. The designer of the Jumo 004 was the Austrian Anselm Franz (1900–1994), who after World War II was brought to the United States as part of Operation Paperclip.[1] After working for the US Air Force, he was hired in 1951 by Lycoming where he developed several turboshaft engines. The engines designed by Amselm Franz at Lycoming include the T53, the world's first helicopter turboshaft engine, which powers the Bell Aircraft UH-1 Huey and AH-1 Cobra helicopters and the OV-1 Mohawk ground attack aircraft, and the AGT-1500 turboshaft that powers the M1 Abrams tank. Anselm Franz rose through the ranks at Lycoming to become the vice president of the company in 1963.

The first supersonic jet bomber was the B-58 Hustler, developed by Convair in the early 1950s. The first flight of the B-58 was on November 11, 1956. The Valkyrie XB-70 airplane of North American reached Mach 3 on October 14, 1965 (Fig. 1.5). Unfortunately, the second XB-70 was hit by a Starfighter F-104 during a formation flight with four other GE-powered aircraft. The last flight of the Valkyrie was on February 4, 1969. Currently, the speed record is held by the Lockheed SR-71 Blackbird, whose maximum speed exceeds Mach 3.

The development of jet engines influenced the evolution of commercial airplanes as well. Three generations of commercial jet airplanes can be identified. The first generation includes the De Havilland Comet, Sud-Aviation Caravelle, and Boeing 707. The second generation includes the Boeing 747, Lockheed L1011 Tristar, and Douglas DC-10. This is the generation of wide-body aircraft and high-bypass-ratio turbofan engines. The third generation includes the supersonic commercial airplanes Concorde (airplane speed of 2400 km/h (or Mach 2.2)) and the Tupolev 144.

---

[1] Operation Paperclip was a secret program of the US Joint Intelligence Objectives Agency that brought to the United States from Germany more than 1,600 German scientists, engineers, and technicians to work for the US government, primarily between 1945 and 1959.

(a)   (b)

Figure 1.5 First Mach 2 and 3 supersonic airplanes. (a) Convair B-58 Hustler at the National Museum of the United States Air Force, Dayton, Ohio (US Air Force photo) and (b) Valkyrie XB-70 (courtesy of Ralph Crane/The LIFE Picture Collection/Getty Images).

Figure 1.6 Closed cylinder with internal pressure p_0 and external pressure p_a.

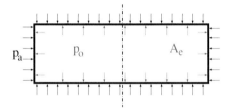

Figure 1.7 Half cylinder with internal pressure p_0 and external pressure p_a.

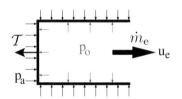

1.4   Jet Propulsion Principle

Jet propulsion is a type of locomotion in which a reaction force is applied to a device by the momentum of ejected matter. This section will present the jet propulsion principle, using a very simple experiment [Hill and Peterson, 1992, p. 5]. Let us consider the closed cylinder with internal pressure p_0 and external pressure p_a shown in Fig. 1.6. The side area of the cylinder is A_e.

Let us suppose that the cylinder can be instantly divided into two parts, following the vertical axis shown in Fig. 1.6. As shown in Fig. 1.7, the thrust acting on the left part of the cylinder before the fluid leaves the half cylinder is equal to the area A_e times the pressure difference $p_0 - p_a$:

$$\mathcal{T} = A_e(p_0 - p_a). \tag{1.1}$$

Figure 1.8 Half cylinder with inlet (and outlet) mass flow.

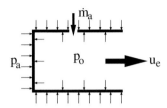

As the fluid leaves the half cylinder with velocity u_e, the pressure p_0 decreases to p, and the thrust is proportional to

$$\mathcal{T} \propto A_e(p - p_a).$$

When the pressure p reaches p_a, the thrust $\mathcal{T}$ approaches zero. To maintain a pressure that is higher than the atmospheric pressure p_a, a mass flow rate $\dot{m}_a$ needs to be fed into the half cylinder, as shown in Fig. 1.8.

Let us assume that the rate at which the fluid flows out of the cylinder is $\dot{m}_e$ and that the mass-averaged exhaust velocity is u_e. The thrust acting on the cylinder, or the force acting on the fluid inside the cylinder, shown in Fig. 1.8, is proportional to $\dot{m}_e$ and u_e:

$$\mathcal{T} \propto \dot{m}_e u_e. \tag{1.2}$$

The arguments made for establishing (1.1) and (1.2) were such that these equations are dimensionally correct. We do not know, however, whether some multiplying constants, such as a 2 or 1/2 factor, should be included or not. To be certain, the thrust must be derived in a rigorous manner. This will be done in Chapter 6, which will show that this simple reasoning is correct and that the expression of the thrust in a jet engine is

$$\mathcal{T} = \dot{m}_e u_e - \dot{m}_a u + A_e(p_e - p_a), \tag{1.3}$$

where $\dot{m}_a$ is the mass flow rate of air at the inlet in the jet engine, and u is the airplane speed.

Bibliography

R. D. Archer and M. Saarlas. *An Introduction to Aerospace Propulsion*. Prentice Hall, 1996. 6

W. W. Bathie. *Fundamentals of Gas Turbines*. John Wiley & Sons, Inc., second edition, 1996. 9

J. Clark. *Ignition! An Informal History of Liquid Rocket Propellants*. Rutgers University Press, 1972. 5

P. Hill and C. Peterson. *Mechanics and Thermodynamics of Propulsion*. Addison-Wesley, second edition, 1992. 11

W. Ley. *Rockets, Missiles and Space Travel*. Chapman & Hall, 1954. 5

H. N. Teodorescu. Hermann Oberth and his professional geography in the European context of the XXth century. In *EMC'04 Conference*, Iaşi, Romania, 2004. 6

2 Aerothermodynamics Review

2.1 Introduction

This chapter presents a review of the laws that govern the aerodynamics and thermodynamics of gases in jet engines. Mastery of these laws is crucial for understanding why and how propulsion systems work. This chapter is divided into two parts: conservation laws and thermodynamics laws. This split is somewhat arbitrary since conservation laws and thermodynamics laws overlap. For example, the energy conservation law is also known as the first law of thermodynamics.

This chapter starts by introducing the Reynolds transport theorem [Reynolds, 1903], which allows us to calculate the time derivative of an integral that is time-dependent. The Reynolds transport theorem is then used to derive the mass, momentum, and angular momentum conservation equations. The energy conservation law, also known as the first law of thermodynamics, is subsequently introduced, followed by the second law of thermodynamics. The first and second laws of thermodynamics are also presented for a control volume, since this formulation is useful for applications. A summary of these fundamental equations closes this chapter.

2.2 Reynolds Transport Theorem

The Reynolds transport theorem is a useful mathematical theorem because it allows us to calculate the time derivative of an integral whose argument is a function of space and time. This theorem was introduced in 1903 by Osborne Reynolds, an Irish-born British professor of engineering at the University of Manchester. The proof of the Reynolds transport theorem is given in Appendix D. The Reynolds transport theorem will be used herein to derive the conservation equations for a control volume or control mass.

There are two things we have to consider before introducing the Reynolds transport theorem: the open system and the closed system. An *open system*, or flow system, is a system whose defining boundaries can be crossed by the flow of mass. Such a system is referred to as a *control volume*, while the boundary of the open system is referred to as a *control surface*. The control volume is defined as a region of constant shape and size, that has no resistance to the passage of mass. The control volume is often assumed to be fixed in space relative to the observer [Hill and Peterson, 1992, p. 24], although this assumption is not necessary.

A *closed system* is defined by a boundary impermeable to mass flow. Such a system is referred to as a control mass. The *control mass* is a collection of matter of fixed identity enclosed by an invisible, massless, flexible surface through which no matter can pass.

Both open and closed systems, or control volumes and control masses, will be used for the derivation of the conservation equations. The Reynolds transport theorem can be applied to either control volumes or control masses. The following sections will introduce the formulations of the theorem applied to both systems.

2.2.1 Time Derivatives: Partial, Total, and Material

Because the Reynolds transport theorem will be used here to calculate time derivatives of integrated quantities, let us pause briefly to discuss three types of time derivative that we will encounter: (1) the partial time derivative, (2) the total time derivative, and (3) the material (or substantial) time derivative. The example given by Bird et al. [1960, p. 73], namely the problem of reporting the concentration c of fish in the Kickapoo River, will be used to illustrate the differences between these time derivatives. From the very beginning, let us recognize that since fish move, the fish concentration is a function of position (x, y, z) and time, t.

2.2.1.1 Partial Time Derivative, $\partial c / \partial t$

Let's assume the observer is on a bridge and observes how the concentration of fish below him changes over time. Because the observer has a fixed position in space, he observes the concentration changes at a constant location (x, y, z), that is,

$$\frac{\partial c}{\partial t} = \left(\frac{\partial c}{\partial t} \right)_{(x, y, z)}.$$

For this reason, the partial time derivative is also called the local time derivative.

2.2.1.2 Total Time Derivative, dc/dt

Suppose the observer is now traveling in a motor boat on the river, going sometimes upstream, sometimes downstream, and sometimes across the current. When the observer now reports the change of fish concentration with respect to time, the values he reports must reflect the motion of the boat. Consequently, the variation of the concentration in time, that is, the total time derivative, is

$$\frac{dc}{dt} = \frac{\partial c}{\partial t} + \frac{\partial c}{\partial x}\frac{dx}{dt} + \frac{\partial c}{\partial y}\frac{dy}{dt} + \frac{\partial c}{\partial z}\frac{dz}{dt} = \frac{\partial c}{\partial t} + (\vec{V}_b \cdot \nabla)c$$

where dx/dt, dy/dt, and dz/dt are the components of the velocity of the boat with respect to a fixed point, $\vec{V}_b$, that is,

$$\vec{V}_b = \hat{i}\frac{dx}{dt} + \hat{j}\frac{dy}{dt} + \hat{k}\frac{dz}{dt},$$

and where $\hat{i}, \hat{j}$, and $\hat{k}$ are the unit vectors in the x-, y-, and z-directions.

The nabla operator, ∇[1], is defined as

$$\nabla \bullet = \hat{i}\frac{\partial \bullet}{\partial x} + \hat{j}\frac{\partial \bullet}{\partial y} + \hat{k}\frac{\partial \bullet}{\partial z}.$$

The $\bullet$ is a place holder for the variable on which the nabla operator gets applied. Using the dot product rule, we see that the operator $(\vec{V}_b \cdot \nabla)$ is equal to

$$(\vec{V}_b \cdot \nabla)\bullet = \frac{\partial \bullet}{\partial x}\frac{dx}{dt} + \frac{\partial \bullet}{\partial y}\frac{dy}{dt} + \frac{\partial \bullet}{\partial z}\frac{dz}{dt}.$$

2.2.1.3 Material (or Substantial) Time Derivative, Dc/Dt

Suppose now that the observer turns off the engine of the motor boat. The boat starts drifting along with the river current. The observer reports how the fish concentration changes in time while following the water (the material). This derivative is a particular kind of total time derivative, in which the boat velocity is equal to the local stream velocity, $\vec{V}$:

$$\frac{Dc}{Dt} = \frac{\partial c}{\partial t} + (\vec{V} \cdot \nabla)c.$$

For this reason, the material time derivative is also called the "derivative following the motion."

2.2.2   Reynolds Transport Theorem for a Control Volume

Let us consider a variable $\aleph = \aleph(x, y, z, t)$ that can be any scalar-, vector-, or tensor-valued function, such as density, ρ, density multiplied by the velocity vector, $\rho\vec{V}$, etc. The symbol aleph, $\aleph$, the first letter of the Hebrew alphabet, is used here to underline the generality of the argument. Further consider that τ^* is the control volume and σ^* is the surface of τ^*. Then, the Reynolds transport theorem states that[2] [Slattery, 1999, p. 18]

$$\frac{d}{dt}\int_{\tau^*} \aleph(x, y, z, t)\, d\tau = \int_{\tau^*} \frac{\partial \aleph}{\partial t}\, d\tau + \int_{\sigma^* \equiv \partial \tau^*} \aleph\, (\vec{V}^* \cdot \hat{n})\, d\sigma, \tag{2.1}$$

where $\vec{V}^*$ is the velocity of the control volume. The proof of the Reynolds transport theorem is included in Appendix D.

---

[1] A more precise notation is $\vec{\nabla}$, which that reminds us that this is a vectorial operator.
[2] Additional reading: Frank White, *Fluid Mechanics*, McGraw-Hill, Forth Edition, 1999, pp. 133-141

2.2.3   Reynolds Transport Theorem for a Control Mass

Using the same variable $\aleph = \aleph(x, y, z, t)$, and with τ as the volume of the closed system and σ as the system surface, that is, the boundary of the volume, the Reynolds transport theorem states that

$$\frac{d}{dt} \int_{\tau} \aleph(x, y, z, t) \, d\tau = \int_{\tau} \frac{\partial \aleph}{\partial t} \, d\tau + \int_{\sigma \equiv \partial \tau} \aleph \, (\vec{V} \cdot \hat{n}) \, d\sigma. \tag{2.2}$$

Note that τ^* and σ^* denote the volume and surface of a control volume (or open system), while τ and σ denote the volume and surface of a control mass (or closed system).

The Reynolds transport theorem can be used to derive the mass, momentum, and energy conservation equations. An example that shows how to use the mass conservation equation is presented in the following section, as an alternative to the traditional approach.

2.3   Conservation Laws

Throughout this book it will be assumed that the average distance the molecules move between successive collisions, λ, aka the mean free path, is much smaller than a characteristic dimension of the body L. Considering that at a pressure of 1 atmosphere and temperature of 288 K, the mean free path is $\lambda = 80$ nm, it is clear that $\lambda / L \ll 1$[3] for any component of a jet engine with a characteristic dimension L. Consequently, one can ignore the atomic structure of the fluid and assume that it is a continuous medium (or continuum). Therefore, one can assume that the fluid can be subdivided into infinitesimal pieces of identical structure. As a result, the density, pressure, and velocities are point properties.

This section will derive the mass, momentum, and angular momentum conservation laws. In addition, the equation of motion will be derived, and its relationship to the momentum conservation equation will be determined.

2.3.1   Mass Conservation Law

The mass conservation law applied to a closed system (or control mass) states that the mass, m_{cm}, of the system does not vary in time,

$$\frac{d}{dt} m_{cm} = 0, \tag{2.3}$$

that is, mass is not being created or destroyed in a closed system. Although this form of the mass conservation law is straightforward, its usefulness is rather limited. A more useful form of the mass conservation law is obtained by using an open system (or control volume) as opposed to a closed system. Let us consider the control volume shown in Fig. 2.1, where $\hat{n}$ is the unit normal to the control surface, pointing outwards, $d\sigma$ is the area of a small

---

[3] The ratio between the mean free path and a characteristic dimension of the body is the Knudsen number, $\mathrm{Kn} := \lambda / L$.

Figure 2.1 Control volume.

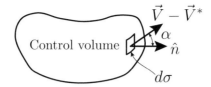

element on the surface of the control volume, and $\vec{V}$ is the absolute velocity vector of the fluid penetrating the control surface that moves with velocity $\vec{V}^*$. Note that the velocity $\vec{V} - \vec{V}^*$ is relative to the control surface. The symbol α denotes the angle between the velocity vector and the vector normal to the surface. The mass conservation law applied to a control volume states that the difference between the mass flow rate in, $\dot{m}_{\text{in}}$, and the mass flow rate out, $\dot{m}_{\text{out}}$, is equal to the time variation of mass flow rate in the control volume,

$$\frac{d}{dt} m_{\text{cv}} = \dot{m}_{\text{in}} - \dot{m}_{\text{out}},$$

where the dot denotes time derivative. The mass of fluid inside the control volume τ^* is

$$m_{\text{cv}} = \int_{\tau^*} \rho \, d\tau.$$

The mass flow rate through the surface $d\sigma$ is

$$d\dot{m} = \rho(\vec{V} - \vec{V}^*) \cdot \hat{n} \, d\sigma,$$

where ρ is the fluid density.

Combining the above three equations, one obtains the mass conservation equation for a control volume:

$$\boxed{\frac{d}{dt} \int_{\tau^*} \rho \, d\tau + \int_{\sigma^* \equiv \partial \tau^*} \rho(\vec{V} - \vec{V}^*) \cdot \hat{n} \, d\sigma = 0}. \tag{2.4}$$

When α is less than 90 deg on a portion of surface σ^*, then the corresponding part of the second integral of equation (2.4) is positive and represents the mass flow rate out. Similarly, the part of the surface σ^* where the angle α is larger than 90 deg corresponds to the mass flow rate in.

The Reynolds transport theorem can be used to derive the mass, momentum, and energy conservation equations. In the following example, the mass conservation equation is derived, as an alternative to the methodology presented above.

Example 2.3.1 Derive the mass conservation equation using the Reynolds transport theorem.

Solution

Let us choose $\aleph = \rho$, deduct equation (2.1) from (2.2), take into account that for a closed system (or control mass), $\frac{d}{dt} \int_{\tau} \rho \, d\tau = 0$, and assume that at moment t, $\tau = \tau^*$ and therefore

$\sigma = \sigma^*$. The volume integrals are equal, $\int_\tau \frac{\partial \rho}{\partial t}\, d\tau = \int_{\tau^*} \frac{\partial \rho}{\partial t}\, d\tau$, because the volumes τ and τ^* are the same and the function under the integral is the same. One obtains.

$$\frac{d}{dt} \int_{\tau^*} \rho\, d\tau = \int_{\sigma^* \equiv \partial \tau^*} \rho(\vec{V}^* - \vec{V}) \cdot \hat{n}\, d\sigma. \tag{2.5}$$

If the velocity of the control volume is $\vec{V}^* = 0$, that is, the control volume is fixed, equation (2.5) reduces to

$$\boxed{\frac{d}{dt} \int_{\tau^*} \rho\, d\tau + \int_{\sigma^* \equiv \partial \tau^*} \rho\vec{V} \cdot \hat{n}\, d\sigma = 0} \tag{2.4}$$

As expected, (2.5) derived using the Reynolds transport theorem is identical to the mass conservation equation (2.4) derived in the traditional way.

In the following paragraphs, the integral formulation of the mass conservation equation will be transformed into a differential formulation. This process requires several steps. First, assuming $\aleph = \rho$ and $\vec{V}^* = 0$ in (2.1), one obtains

$$\frac{d}{dt} \int_{\tau^*} \rho d\tau = \int_{\tau^*} \frac{\partial \rho}{\partial t}\, d\tau. \tag{2.6}$$

Second, combining equations (2.4) and (2.6) gives

$$\int_{\tau^*} \frac{\partial \rho}{\partial t}\, d\tau + \int_{\sigma^* \equiv \partial \tau^*} \rho\vec{V} \cdot \hat{n}\, d\sigma = 0. \tag{2.7}$$

Third, using Gauss' theorem (or the divergence theorem):

$$\int_\tau \nabla \cdot \vec{a}\, d\tau = \int_{\sigma \equiv \partial \tau} \vec{a} \cdot \hat{n}\, d\sigma, \quad \forall \quad \vec{a}, \tau, \tag{2.8}$$

and assuming that $\vec{a} = \rho\vec{V}$, equation (2.7) becomes

$$\int_{\tau^*} \frac{\partial \rho}{\partial t}\, d\tau + \int_{\tau^*} \nabla \cdot (\rho\vec{V})\, d\tau = 0$$

or

$$\int_{\tau^*} \left[\frac{\partial \rho}{\partial t} + \nabla \cdot (\rho\vec{V}) \right] d\tau = 0.$$

Since the volume τ^* is arbitrary, the quantity under the integral sign must be zero, that is,

$$\boxed{\frac{\partial \rho}{\partial t} + \nabla \cdot (\rho\vec{V}) = 0}. \tag{2.9}$$

Recall that this equation was derived assuming $\vec{V}^* = 0$.

There are two important particular cases of equation (2.9). For steady flows, equation (2.9) becomes

$$\nabla \cdot (\rho \vec{V}) = 0$$

or

$$\frac{\partial(\rho u)}{\partial x} + \frac{\partial(\rho v)}{\partial y} + \frac{\partial(\rho w)}{\partial z} = 0.$$

If the flow is incompressible, then equation (2.9) becomes

$$\nabla \cdot \vec{V} = 0 \qquad\qquad (2.10)$$

or

$$\frac{\partial u}{\partial x} + \frac{\partial v}{\partial y} + \frac{\partial w}{\partial z} = 0.$$

2.3.2 Equation of Motion

Before introducing the momentum conservation equation, let us derive the equation of motion. The equation of motion is the relationship equating the rate of change of momentum of a portion of fluid and the sum of all forces acting on that portion of fluid, excluding inertial forces [Batchelor, 1967, p. 137]. We will then demonstrate in the following section that the equation of motion is a particular case of the momentum conservation equation. The equation of motion can also be determined by stating that the sum of all forces acting on the portion of fluid, including the inertial forces, equals zero. The second option will be used herein. Therefore, to derive the equation of motion, all forces acting on a selected portion of the fluid must be estimated.

Let us now evaluate the forces acting on the control mass $\tau(t)$ bounded by a surface $\sigma(t)$, shown in Fig. 2.2. The forces acting on the control mass are inertial forces, volume forces, and surface forces. The volume and surface forces make up the external forces.

2.3.2.1 Inertial Forces

The inertial forces are proportional to the mass and acceleration of the control volume,

$$\vec{F}_i = - \int_\tau \rho \frac{D\vec{V}}{Dt} \, d\tau,$$

Figure 2.2 Forces acting on a control volume (viscous forces are neglected).

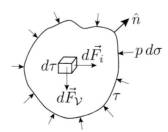

where $\vec{V}$ is the velocity of the volume $d\tau$. The material derivative $D\vec{V}/Dt$ represents the acceleration of the material volume $d\tau$.

2.3.2.2 Volume (or Body) Forces

Volume forces are long-range forces, that is, forces that decrease slowly as the distance between interacting elements increases. Examples of volume forces are gravity, electromagnetic forces, and fictitious forces, such as centrifugal forces, which act on the mass elements when their motion is described by an accelerating set of axes [Batchelor, 1967, p. 7]. Because of the slow variation of long-range forces with distance, these forces can be assumed to act equally on all the matter within a small element of volume. As a result, the total force acting on the element of volume is proportional to the volume size. The volume forces, which are also called *body forces*, can be written as

$$\vec{F_V} = \int_\tau \rho \vec{f} \, d\tau. \tag{2.11}$$

The force per unit mass, $\vec{f}$, has the dimensions of acceleration. A common example for $\vec{f}$ is $\vec{f} = -\hat{k} \cdot g$, where g is the gravitational acceleration and $\hat{k}$ is the unit vector in the vertical direction, pointing upward.

2.3.2.3 Surface Forces

Surface forces are short-range forces. These forces have a direct molecular origin, decrease rapidly with increasing distance between interacting elements, and are significant only when the distance is of the order of the separation of molecules of the fluid [Batchelor, 1967, p. 7]. The surface forces, $\vec{F_S}$, consist of pressure forces, $\vec{F_p}$, and viscous forces, $\vec{F_{vis}}$, that is,

$$\vec{F_S} = \vec{F_p} + \vec{F_{vis}}.$$

The pressure forces are

$$\vec{F_p} = - \int_{\sigma \equiv \partial \tau} \hat{n} p \, d\sigma, \tag{2.12}$$

where p is the pressure on the control mass, and $\hat{n}$ is the unit vector normal to the surface σ and points outward. The negative sign indicates that the force $\vec{F_p}$ is in the direction opposite to the normal $\hat{n}$, that is, it is acting on the control mass. The viscous forces are due to the shear and normal viscous stresses.

2.3.2.4 Derivation of the Equation of Motion

According to Newton's first law, the sum of all forces acting on a system must be zero for the system to be in equilibrium, such that

$$\int_\tau \rho \frac{D\vec{V}}{Dt}\, d\tau = \vec{F}_\mathcal{V} + \vec{F}_\mathcal{S}.$$ (2.13)

If the viscous forces are small and can be neglected compared to the pressure forces, then using (2.11) and (2.12) yields

$$\int_\tau \rho \frac{D\vec{V}}{Dt}\, d\tau = \int_\tau \rho \vec{f}\, d\tau - \int_{\sigma \equiv \partial\tau} p\hat{n}\, d\sigma.$$

Using Gauss' theorem on the last term of the above equation, one obtains

$$\int_\tau \rho \frac{D\vec{V}}{Dt}\, d\tau = \int_\tau \rho \vec{f}\, d\tau - \int_\tau \nabla p\, d\tau.$$ (2.14)

The equilibrium equation (2.14) can be rearranged as

$$\int_\tau \left(\rho \frac{D\vec{V}}{Dt} - \rho \vec{f} + \nabla p \right) d\tau = 0.$$ (2.15)

The volume τ is arbitrary because no restriction was placed on it. Therefore, (2.15) is satisfied for any volume τ. This holds true only if the quantity under the integral sign is zero,

$$\rho \frac{D\vec{V}}{Dt} - \rho \vec{f} + \nabla p = 0,$$

which can be rearranged as

$$\boxed{\rho \frac{D\vec{V}}{Dt} = \rho \vec{f} - \nabla p}.$$ (2.16)

Equation (2.16) is the equation of motion for inviscid flows.

Example 2.3.2 Derive the unsteady Bernoulli equation using the equation of motion for inviscid flows (2.16).

Solution
Equation (2.16) can be used to derive the *unsteady Bernoulli equation*. Let us rewrite (2.16) using the definition of the material derivative

$$\frac{\partial \vec{V}}{\partial t} + (\vec{V} \cdot \nabla)\vec{V} = -\frac{\nabla p}{\rho} + \vec{f}.$$ (2.17)

Using the identity

$$\nabla(\vec{a} \cdot \vec{a}) = 2(\vec{a} \cdot \nabla)\vec{a} + 2\vec{a} \times (\nabla \times \vec{a}),$$

the equation (2.17) can be written as

$$\frac{\partial \vec{V}}{\partial t} + \frac{1}{2}\nabla(V^2) - \vec{V} \times (\nabla \times \vec{V}) = -\frac{\nabla p}{\rho} + \vec{f}. \tag{2.18}$$

Let us now integrate equation (2.18) along a stream line $d\vec{s}$:

$$\int \frac{\partial \vec{V}}{\partial t}\, d\vec{s} + \frac{1}{2}\int \nabla(V^2)\, d\vec{s} - \int \vec{V} \times (\nabla \times \vec{V})\, d\vec{s} = -\int \frac{\nabla p}{\rho}\, d\vec{s} + \int \vec{f}\, d\vec{s}. \tag{2.19}$$

The third integral of the right-hand side is zero because the vector $\vec{V} \times (\nabla \times \vec{V})$ is orthogonal to the vector $d\vec{s}$ (or $\vec{V}$). Rearranging the definition of the gradient,

$$\nabla \phi := \frac{d\phi}{ds} \cdot \vec{s},$$

where ϕ is any scalar, as

$$\frac{d\phi}{ds} = \nabla \phi \cdot \vec{s}$$

and using the identity

$$d\vec{s} = \vec{s}\, ds,$$

Eq. (2.19) becomes

$$\int \frac{\partial \vec{V}}{\partial t}\, d\vec{s} + \frac{1}{2}\int \frac{d(V^2)}{ds}\, ds = -\int \frac{1}{\rho}\frac{dp}{ds}\, ds + \int \vec{f}\, d\vec{s}. \tag{2.20}$$

Assuming the external force is $\vec{f} = -g\,\hat{k}$, and since $d\vec{s} = \hat{i}dx + \hat{j}dy + \hat{k}dz$, (2.20) becomes the unsteady Bernoulli equation

$$\boxed{\frac{V^2}{2} + gz + \int \frac{dp}{\rho} = -\int \frac{\partial \vec{V}}{\partial t}\, d\vec{s}.} \tag{2.21}$$

The unsteady Bernoulli equation is a useful relationship for studying unsteady flows.

2.3.3　Momentum Conservation Law

This section uses Newton's second law to derive a general form of the momentum conservation equation. The momentum conservation law states that the sum of the external forces $\vec{F}$ acting on a system is equal to the time derivative of the system impulse, $\vec{I}$:

$$\frac{d}{dt}\vec{I} = \vec{F}, \tag{2.22}$$

where the impulse is defined as

$$\vec{I} := \int_{\tau} \rho \vec{V} d\tau$$

and the force acting on the system consists of the volume and surface forces

$$\vec{F} = \vec{F}_V + \vec{F}_S \tag{2.23}$$

such that (2.22) becomes

$$\frac{d}{dt} \int_\tau \rho \vec{V} d\tau = \vec{F}_V + \vec{F}_S. \tag{2.24}$$

The time derivative of the impulse can be written using the Reynolds transport theorem:

$$\frac{d\vec{I}}{dt} = \frac{d}{dt} \int_\tau \rho \vec{V} d\tau = \int_\tau \frac{\partial}{\partial t} (\rho \vec{V}) d\tau + \int_{\sigma \equiv \partial \tau} \rho \vec{V} (\vec{V} \cdot \hat{n}) d\sigma. \tag{2.25}$$

Combining (2.22), (2.23), and (2.25) yields

$$\int_\tau \frac{\partial}{\partial t} (\rho \vec{V}) d\tau + \int_{\sigma \equiv \partial \tau} \rho \vec{V} (\vec{V} \cdot \hat{n}) d\sigma = \vec{F}_V + \vec{F}_S. \tag{2.26}$$

Using $\aleph = \rho \vec{V}$ in the Reynolds transport theorem for a control volume (2.1) yields:

$$\int_{\tau^*} \frac{\partial (\rho \vec{V})}{\partial t} d\tau + \int_{\sigma^* \equiv \partial \tau^*} \rho \vec{V} (\vec{V}^* \cdot \hat{n}) d\sigma = \frac{d}{dt} \int_{\tau^*} \rho \vec{V} d\tau. \tag{2.27}$$

Let us choose the control volume such that at time t, the volume τ is identical to τ^*. Consequently,

$$\int_{\tau^*} \frac{\partial (\rho \vec{V})}{\partial t} d\tau = \int_\tau \frac{\partial (\rho \vec{V})}{\partial t} d\tau. \tag{2.28}$$

Deduct (2.27) from (2.26) and use (2.28) to obtain:

$$\boxed{\frac{d}{dt} \int_{\tau^*} \rho \vec{V} d\tau + \int_{\sigma^* \equiv \partial \tau^*} \rho \vec{V} \left[(\vec{V} - \vec{V}^*) \cdot \hat{n} \right] d\sigma = \vec{F}_V + \vec{F}_S} . \tag{2.29}$$

If one chooses that the control volume is fixed, *i.e.*, $\vec{V}^* = 0$, then the particular form of (2.29) is

$$\frac{d}{dt} \int_{\tau^*} \rho \vec{V} d\tau + \int_{\sigma^* \equiv \partial \tau^*} \rho \vec{V} (\vec{V} \cdot \hat{n}) d\sigma = \vec{F}_V + \vec{F}_S. \tag{2.30}$$

For steady flows, (2.30) reduces to

$$\int_{\sigma^* \equiv \partial \tau^*} \rho \vec{V} (\vec{V} \cdot \hat{n}) d\sigma = \vec{F}_V + \vec{F}_S. \tag{2.31}$$

If the viscous forces $\vec{F}_{vis}$ are much smaller than the pressure forces $\vec{F}_p$, and can be neglected compared to the pressure forces $\vec{F}_p$, then (2.31) reduces to

$$\int_{\sigma^* \equiv \partial \tau^*} \rho \vec{V} (\vec{V} \cdot \hat{n}) d\sigma = \vec{F}_V + \vec{F}_p. \tag{2.32}$$

2.3.3.1   Equation of Motion vs. Momentum Conservation Equation

To determine whether there is a connection between the equation of motion (2.13) and the momentum conservation equation (2.24), let us start by proving the following equality:

$$\frac{d}{dt} \int_{\tau} \rho \aleph \, d\tau = \int_{\tau} \rho \frac{D\aleph}{Dt} \, d\tau. \tag{2.33}$$

Note that when $\aleph = 1$, (2.33) reduces to the mass conservation equation (2.3) because $m_{cm} = \int_{\tau} \rho \, d\tau$.

To prove (2.33), let us use the Reynolds transport theorem applied to a closed system, where $\aleph$ has been replaced by $\rho \aleph$:

$$\frac{d}{dt} \int_{\tau} \rho \aleph \, d\tau = \int_{\tau} \frac{\partial}{\partial t}(\rho \aleph) \, d\tau + \int_{\sigma \equiv \partial \tau} (\rho \aleph)(\vec{V} \cdot \hat{n}) \, d\sigma.$$

The second term of the right-hand side is rewritten using Gauss' theorem:

$$\frac{d}{dt} \int_{\tau} \rho \aleph \, d\tau = \int_{\tau} \frac{\partial}{\partial t}(\rho \aleph) \, d\tau + \int_{\tau} \nabla(\rho \aleph \vec{V}) \, d\tau,$$

and the $\partial / \partial t$ and ∇ operators are applied inside the integrals of the right-hand-side terms,

$$\frac{d}{dt} \int_{\tau} \rho \aleph \, d\tau = \int_{\tau} \left[\aleph \frac{\partial \rho}{\partial t} + \rho \frac{\partial \aleph}{\partial t} + \aleph \nabla(\rho \vec{V}) + \rho \vec{V} \nabla \aleph \right] d\tau.$$

By advantageously grouping the terms of the right-hand side and using the mass conservation equation (2.9), one obtains

$$\frac{d}{dt} \int_{\tau} \rho \aleph \, d\tau = \int_{\tau} \left\{ \aleph \underbrace{\left[\frac{\partial \rho}{\partial t} + \nabla(\rho \vec{V}) \right]}_{0} + \rho \left(\frac{\partial \aleph}{\partial t} + \vec{V} \nabla \aleph \right) \right\} d\tau.$$

Using the definition of the material derivative in the right-hand side, the identity (2.33) is proven. Note that (2.33) was proven formally, without specifying whether $\aleph$ is a scalar or vector.

If one chooses $\aleph = \vec{V}$ in equation (2.33), one obtains:

$$\frac{d}{dt} \int_{\tau} \rho \vec{V} \, d\tau = \int_{\tau} \rho \frac{D\vec{V}}{Dt} \, d\tau. \tag{2.34}$$

Substituting (2.34) into (2.24) yields

$$\int_{\tau} \rho \frac{D\vec{V}}{Dt} \, d\tau = \vec{F}_V + \vec{F}_S,$$

which is identical to (2.13).

Example 2.3.3 Calculate the drag on a symmetric airfoil shown in Fig. 2.3 assuming the flow incidence to be zero. The velocity upstream of the airfoil is $U = $ constant, and the variation of the velocity downstream of the airfoil is known and equal to $u(y)$. One assumes the flow is steady and incompressible.

Figure 2.3 Airfoil in wind tunnel.

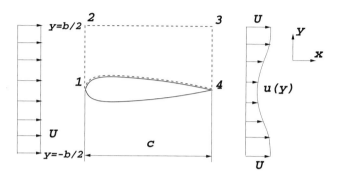

Solution

Consider the control volume 1-2-3-4-1 shown in Fig. 2.3 by the dashed line. Let us first write the mass conservation equation for a control volume:

$$\frac{d}{dt} \int_{\tau^*} \rho \, d\tau + \int_{\sigma^*} \rho \vec{V} \cdot \hat{n} \, d\sigma = 0 \tag{2.4}$$

where the term $\frac{d}{dt} \int_{\tau^*} \rho d\tau = 0$ because the flow is steady. The second term of the mass conservation equation becomes

$$\int_0^{b/2} \rho \, U \hat{i} \, (-\hat{i}) \, dy + \int_0^{c} \rho \left(U \hat{i} + v_e(x) \hat{j} \right) \hat{j} \, dx + \int_0^{b/2} \rho \, u \, (y) \, \hat{i} \cdot \hat{i} \, dy = 0 \tag{2.35}$$

where $v_e(x)$ is the velocity normal to the boundary 23. It is assumed that the dimension normal to the plane xy is unity. Note that the integral along the airfoil surface is zero because the velocity $\vec{V}$ and the normal $\hat{n}$ are orthogonal. After simple algebra, (2.35) becomes

$$-\rho \, U \frac{b}{2} + \rho \int_0^c v_e(x) \, dx + \rho \int_0^{b/2} u(y) dy = 0,$$

which yields the expression of the integral of the v_e velocity along the boundary 23

$$\int_0^c v_e(x) \, dx = U \frac{b}{2} - \int_0^{b/2} u(y) dy. \tag{2.36}$$

The momentum conservation equation (2.29) can be simplified by neglecting: (i) the volume force $\vec{F}_V$ due to gravitational force, which is small compared to the other terms, and (ii) the viscous forces, $\vec{F}_{vis}$. In addition, the flow is steady, and consequently the first term of the right-hand side is zero. We will also assume that the control volume is fixed, that is, $\vec{V}^* = 0$, such that (2.29) becomes

$$\vec{F}_p = \int_{\sigma^*} \rho\vec{V}\left(\vec{V}\cdot\hat{n}\right).d\sigma.$$ (2.37)

The pressure forces $\vec{F}_p$ are obtained by integrating over the boundary surface. Note that the pressure along surfaces 12, 23, and 34 is equal to atmospheric pressure, p_a. The pressure along surface 41 will be written as $p = p_a + \Delta p$. The force due to pressure becomes

$$\vec{F}_p = -\oint_{\sigma^*} p\hat{n}d\sigma = -\left(\int_{\sigma_{12}} p_a\hat{n}d\sigma + \int_{\sigma_{23}} p_a\hat{n}d\sigma + \int_{\sigma_{34}} p_a\hat{n}d\sigma + \int_{\sigma_{41}} (p_a+\Delta p)\hat{n}d\sigma\right) =$$

$$= -\left(\oint_{\sigma^*} p_a\hat{n}d\sigma + \int_{\sigma_{41}} \Delta p\hat{n}d\sigma\right).$$

Since the atmospheric pressure, p_a is constant on the closed surface σ^*, the force due to pressure acting on the control volume 1234 reduces to

$$\vec{F}_p = -\int_{\sigma_{41}} \Delta p\hat{n}d\sigma.$$

The force $\vec{F}_p$ represents the force with which the airfoil, or more precisely, the upper half of the airfoil, acts on the control volume 1234. According to Newton's third law of motion, this force $\vec{F}_p$ is equal (and has the opposite sign) to the force with which the fluid acts on the upper half of the airfoil. The force acting on the upper side of the symmetrical airfoil at zero angle of attack will generate half of the airfoil drag and possibly a component, F_y, in the y direction:

$$\vec{F}_p = -\left(\frac{1}{2}D\hat{i} + F_y\hat{j}\right).$$ (2.38)

Note that the force in the y direction on the upper side of the airfoil will be canceled by the force in the y direction on the pressure side, because the airfoil is symmetric and has a zero angle of attack. The momentum conservation equation (2.37) becomes

$$-\frac{D}{2}\hat{i} - F_y\hat{j} = \int_0^{b/2} \rho U\hat{i}\left[U\hat{i}\left(-\hat{i}\right)\right]dy+$$

$$+ \int_0^c \rho\left(U\hat{i} + v_e(x)\hat{j}\right)\left[\left(U\hat{i} + v_e(x)\hat{j}\right)\hat{j}\right]dx + \int_0^{b/2} \rho u(y)\cdot\hat{i}\left[u(y)\hat{i}\cdot\hat{i}\right]dy$$ (2.39)

and then simplifies to

$$-\frac{D}{2}\hat{i} - F_y\hat{j} = -\rho U^2\frac{b}{2}\hat{i} + \rho\int_0^c \left(U\hat{i} + v_e\hat{j}\right)v_edx + \rho\int_0^{b/2} u^2(y)\hat{i}dy.$$

The two components of the vectorial equation (2.39) are

$$-\frac{D}{2} = -\rho U^2 \frac{b}{2} + \rho U \int_0^c v_e dx + \rho \int_0^{b/2} u^2(y) dy \tag{2.40}$$

$$-F_y = \rho \int_0^c v_e^2 \, dx. \tag{2.41}$$

The force F_y acting on the upper part of the airfoil cannot be determined since neither the variation of velocity v_e nor the integral of v_e^2 are known. Since the airfoil is symmetrical and the angle of attack is zero, the sum of forces F_y acting on the upper and lower side of the airfoil is zero. The airfoil drag, however, can be calculated by substituting (2.36) into (2.40):

$$-\frac{D}{2} = -\rho U^2 \frac{b}{2} + \rho U \left(U\frac{b}{2} - \int_0^{b/2} u(y) \, dy \right) + \rho \int_0^{\frac{b}{2}} u^2(y) \, dy,$$

so that

$$D = 2\rho \left(U \int_0^{b/2} u(y) dy - \int_0^{b/2} u^2(y) dy \right).$$

2.3.4 Angular Momentum (or Moment of Momentum) Conservation Law

The angular momentum, or moment of momentum, $\vec{K}$, is defined as

$$\vec{K} := \sum_j m_j \vec{r}_j \times \vec{V}_j = \int_\tau \rho(\vec{r} \times \vec{V}) \, d\tau.$$

The angular momentum conservation law states that the time rate of change of angular momentum is equal to the sum of moments of external forces, $\sum_j \vec{r}_j \times \vec{F}_j$, where the summation is done over all the particles j:

$$\frac{d}{dt}\vec{K} = \sum_j \vec{r}_j \times \vec{F}_j.$$

One option for deriving the angular momentum conservation equation is to apply the same approach used to derive the momentum conservation equation. Consequently, for the angular momentum one substitutes $\aleph$ in the Reynolds transport theorem (2.2) by $\rho(\vec{r} \times \vec{V})$,

$$\frac{d\vec{K}}{dt} = \frac{d}{dt} \int_\tau \rho(\vec{r} \times \vec{V}) \, d\tau = \int_\tau \frac{\partial \left[\rho(\vec{r} \times \vec{V}) \right]}{\partial t} \, d\tau + \int_{\sigma \equiv \partial\tau} \rho(\vec{r} \times \vec{V})(\vec{V} \cdot \hat{n}) \, d\sigma,$$

and then does the algebra using a method similar to that of Section 2.3.3.

A simpler way to derive the angular momentum equation is to define a vector $\vec{W}$

$$\vec{W} := \vec{r} \times \vec{V}$$

and replace $\vec{V}$ in (2.29) by $\vec{W}$. In addition, the forces of the right-hand-side term of (2.29) are replaced by their moments. Consequently, the angular momentum equation can be formally derived as

$$\frac{d}{dt} \int_{\tau^*} \rho(\vec{r} \times \vec{V}) \, d\tau + \int_{\sigma^* \equiv \partial\tau^*} \rho(\vec{r} \times \vec{V}) \left[(\vec{V} - \vec{V}^*) \cdot \hat{n} \right] d\sigma = \vec{M}_{F_V} + \vec{M}_{F_S}, \qquad (2.42)$$

where $\vec{M}_{F_V}$ is the momentum of the external forces

$$\vec{M}_{F_V} := \int_{\tau} \rho(\vec{r} \times \vec{f}) \, d\tau,$$

and $\vec{M}_{F_S}$ is the momentum of the pressure forces (if viscous forces are neglected)

$$\vec{M}_{F_S} := - \int_{\sigma \equiv \partial\tau} p(\vec{r} \times \hat{n}) \, d\sigma.$$

Example 2.3.4 What is the angular velocity of the sprinkler shown in Fig. 2.4? Assume that the mass flow rate through each nozzle is $\dot{m}$, the area of each nozzle is A, the water density is ρ, $r_1 = r$, $r_2 = 2r$, and the friction in the bearing is neglected. Compute how many rotations per second the sprinkler does if $r = 0.1$ m, $\dot{m} = 0.1$ kg/s, $A = 25$ mm^2, and $\rho = 1000$ kg/m^3.

Figure 2.4 Sprinkler.

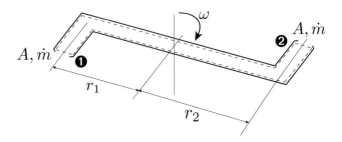

Solution
Consider the control volume shown in Fig. 2.4 by the dashed line. Neglect viscous forces and assume the flow is steady and velocity $V^* = 0$ such that the angular momentum equation (2.42) becomes

$$\int_{\sigma^* \equiv \partial\tau^*} \rho(\vec{r} \times \vec{V}) \left(\vec{V} \cdot \hat{n} \right) d\sigma = \vec{M}_{F_V} + \vec{M}_{F_S}. \qquad (2.43)$$

Neglect gravitational effects, that is, $\vec{M}_{F_V} = 0$. Assume constant pressure in the sprinkler, such that $\vec{M}_{F_S} = 0$. The angular momentum equation (2.43) reduces to

$$(r_1 V_{a1}\dot{m} + r_2 V_{a2}\dot{m})\hat{k} = 0. \tag{2.44}$$

where V_{a1} and V_{a2} are the absolute velocities, and $\hat{k}$ is the unit vector parallel to the axis of rotation and points upwards. The relative velocity is $\dot{m}/(\rho A)$, and the transport velocity is $r_i\omega$, where i is 1 or 2. Consequently, the absolute velocities are $V_{a1} = \dot{m}/(\rho A) - r_1\omega$ and $V_{a2} = \dot{m}/(\rho A) - r_2\omega$.

Substituting the absolute velocities in (2.44) yields

$$r_1\left(\dot{m}/(\rho A) - r_1\omega\right) + r_2\left(\dot{m}/(\rho A) - r_2\omega\right) = 0$$

such that the angular velocity is

$$\omega = \frac{r_1 + r_2}{r_1^2 + r_2^2}\frac{\dot{m}}{\rho A} = \frac{3}{5r}\frac{\dot{m}}{\rho A}.$$

Substituting the numerical values of the variables yields

$$\omega = \frac{3}{5 \cdot 0.1}\frac{0.1}{1000 \cdot 25 \cdot 10^{-6}} = 24 \text{ rad/s,}$$

and the frequency is

$$f = \omega/(2\pi) = 3.82 \text{ Hz.}$$

2.4   Thermodynamics Laws

This section presents the four thermodynamics laws, from the zeroth law to the third law of thermodynamics. The section starts by introducing the terminology needed to describe the thermodynamics laws. The thermodynamics laws are then presented in chronological order, from the first introduced, the first and second laws, to the last introduced, the zeroth law of thermodynamics.

2.4.1   Elements of Thermodynamics Terminology

This section presents a minimal set of the nomenclature used for describing the thermodynamics laws. This nomenclature includes the basic notions briefly described in the following paragraphs [Shapiro, 1953, p. 24], [Bejan, 1988, p. 4].

Fundamental law A fundamental law is a generalization of experimental results beyond the region covered by the experiments themselves. The first law of thermodynamics and the second law of thermodynamics are examples of fundamental laws.

Theorem A theorem is a statement whose validity depends upon the validity of a given set of laws. Carnot's theorem (1824), Clausius' theorem (1854), and Reynolds' transport theorem (1903) are examples of theorems.

Corollary A corollary is a more or less self-evident statement following a definition, law, or theorem. The corollaries of the second law of thermodynamics are examples of this definition.

Thermodynamic property A thermodynamic property is any observable characteristic of a system: for example, temperature, T; volume, τ; pressure, p; velocity, $\bar{V}$; entropy, S; or internal energy, U. Some properties can be measured, such as temperature, pressure, velocity; others cannot be measured directly but can be derived based on measurements, such as entropy, internal energy.

Extensive and intensive properties Some properties depend on the size of the system, such as volume, mass, internal energy, U. These are called extensive properties. Other properties do not depend on the size of the system, such as density, temperature, pressure, mass specific internal energy, u. These are called intensive properties.

State In thermodynamics, the state defines the condition of a system, identified through the properties of the system. When the condition of a system is identified through the intensive properties of the system, the state is an intensive state.

Work transfer interaction, W The interaction between two systems as a result of a boundary force between the two systems displacing the common boundary through a distance is called a work transfer interaction.

Heat transfer interaction, Q The interaction between two systems as a result of a temperature difference between the two systems is called a heat transfer interaction. Note that a heat transfer interaction, as opposed to a work transfer interaction, does not require displacement of the common boundary.

Process A process is a change of state, which describes how a system changes from one state to another. Examples of processes include:

1 An adiabatic process is a process in which there are no heat transfer interactions.
2 A reversible process is a process that can be reversed by using infinitesimal changes in some property of the system. The process must occur without loss or dissipation of energy. Because of the infinitesimal changes, the system is in thermodynamic equilibrium during the entire process. In addition, because of the infinitesimal changes, the time required for the process to finish would be infinite, that is, perfectly reversible processes are impossible. For practical purposes, however, if the system undergoing changes responds much faster than the applied change, the deviation from reversibility may be negligible.
3 An isobaric process is a process that occurs at constant pressure.
4 An isochoric process is a process that occurs at constant volume.

Cycle A process that has identical initial and final states is a cycle.

Path A succession of states followed by a system from the initial to the final state, as shown in Fig. 2.5, is called path.

Phase A phase is a collection of all the parts of a system throughout which all physical properties of a material are essentially uniform.

Figure 2.5 States and path on a $T - s$ diagram.

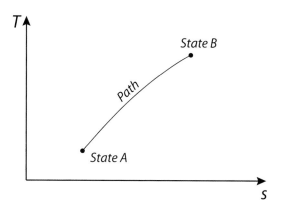

Figure 2.6 Heat engine sign convention: Q_{1-2} and W_{1-2} are both positive.

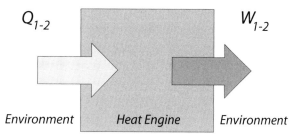

Quantities of state (properties) Quantities of state, or properties, are quantities that do not depend on the path of the process. Pressure, temperature, internal energy, and entropy are examples of quantities of state.

Quantities of process (non-properties) Quantities of process, or nonproperties, are quantities that depend on the path of the process. Work transfer interaction, W, and heat transfer interaction, Q are examples of quantities of process.

Heat engine A system of fixed identity that undergoes a cyclic process during which there are work and heat interactions (transfers) with the surroundings is called a heat engine. A heat engine converts thermal energy and chemical energy to mechanical energy that can be used to do mechanical work. Examples of heat engines include a steam power plant and a gas turbine.

Heat engine sign convention The heat engine sign convention states that heat transfer into the system and work transfer out of the system are positive, as shown in Fig. 2.6.

Heat source The part of the surroundings from which the heat engine receives heat during the cycle is called the heat source. The heat transfer from the heat source is usually denoted by Q_1.

Heat sink The part of the surroundings to which the heat engine gives, sends, or rejects heat during the cycle is called a heat sink. The heat transfer to the heat sink is usually denoted by Q_2.

Work of engine during a complete cycle The useful output of the engine is the work during a complete cycle, W_e. The work of an engine is

$$W_e := Q_1 - Q_2. \tag{2.45}$$

Thermal efficiency Thermal efficiency, η_{th}, is defined as the ratio between the useful output of the engine W_e and the heat transfer from the heat source Q_1:

$$\eta_{\text{th}} := \frac{W_e}{Q_1} = \frac{Q_1 - Q_2}{Q_1} = 1 - \frac{Q_2}{Q_1}. \tag{2.46}$$

Perpetual motion machine of the second kind A perpetual motion machine of the second kind is a heat engine with thermal efficiency $\eta_{\text{th}} = 100\%$, that is, a machine that spontaneously converts thermal energy into mechanical work, therefore violating the second law of thermodynamics.[4]

2.4.2 First Law of Thermodynamics

The first law of thermodynamics, formulated in early 1850 by William John MacQuorn Rankine, Rudolph Clausius, and William Thomson (Lord Kelvin), states that the difference between heat transfer and work transfer equals the energy change:

$$\underbrace{Q_{1-2} - W_{1-2}}_{\text{energy interaction (non-properties)}} = \underbrace{E_2 - E_1,}_{\text{energy change (property)}} \tag{2.47}$$

where Q_{1-2} is the heat transfer and W_{1-2} is the work transfer between states 1 and 2. Both Q_{1-2} and W_{1-2} are nonproperties, while the energy E is a property.

The first law of thermodynamics can also be written for an incremental change of state as:

$$\delta Q - \delta W = dE. \tag{2.48}$$

Note that different infinitesimal increments δ and d are used to make the distinction between nonproperties and properties. The first law of thermodynamics for a cycle becomes

$$\oint \delta Q = \oint \delta W,$$

since $E_1 = E_2$.

The energy term of the right-hand side of (2.47) can be divided into terms that correspond or not to macroscopically identifiable forms of energy storage [Bejan, 1988, p. 18–21]:

$$\underbrace{E_2 - E_1}_{\text{Energy}} = \underbrace{U_2 - U_1}_{\text{Internal energy}} + \underbrace{\frac{1}{2}mV_2^2 - \frac{1}{2}mV_1^2}_{\text{Kinetic energy}} + \underbrace{mgz_2 - mgz_1}_{\text{Gravitational potential energy}} + (E_2 - E_1)_{\text{other}}, \tag{2.49}$$

where $(E_2 - E_1)_{\text{other}}$ denotes other macroscopic forms of energy storage, such as, kinetic energy due to rotational motion, electrical capacitance, or electrical inductance. The internal energy, U, as a name suggests, is a form of energy that cannot be identified macroscopically. The term "internal energy" and notation, U, were introduced by Rudolf Julius Emanuel Clausius (1822–1888), a German physicist, and William John Macquorn Rankine (1820–1872), a Scottish mechanical engineer.

---

[4] A perpetual motion machine of the first kind produces work without input of energy, therefore violating the first law of thermodynamics, the law of conservation of energy.

2.4.3 Second Law of Thermodynamics

The second law of thermodynamics describes the direction of natural processes. The law was an empirical finding, accepted as an axiom of thermodynamics. Later on, statistical mechanics was used to explain the microscopic origin of the law. Nicolas Leonard Sadi Carnot (1796–1832), a French military engineer and physicist, is credited with the first formulation of the second law in 1824, when he showed that there is an upper limit to the efficiency of conversion of heat to work in a heat engine [Carnot, 1960]. His ideas were pursued after his death by Rudolf Clausius and William Thomson, Lord Kelvin (1824–1907), a British physicist, who introduced the most prominent classical statements of the second law.

2.4.3.1 Statement of the Second Law of Thermodynamics

The second law of thermodynamics states that *it is impossible to achieve a perpetual motion of the second kind.* Consequently, the thermal efficiency cannot be 100%. Using (2.46), this implies that the heat transfer to the heat sink must be different from zero. As a result, a system that delivers work must have both a heat source and a heat sink in order to function. An equivalent statement of the second law of thermodynamics is: *No system can pass through a complete cycle of states and deliver positive work to the surroundings while exchanging heat with only a single source of heat at uniform temperature.*

The second law of thermodynamics is connected with the fact that real processes are irreversible. Irreversibility, in propulsion, can be traced to four basic causes: (1) viscosity, (2) heat conduction, (3) mass diffusion, and (4) unrestrained (or free) expansion. The first three causes are manifestations of molecular action.

2.4.3.2 Corollaries of the Second Law of Thermodynamics

As mentioned in the thermodynamics terminology Section 2.4.1, a corollary is a self-evident statement following a definition, law, or theorem. This section presents the six corollaries of the second law that are seminal in introducing new concepts of theoretical and practical importance.

Corollary 1 No heat engine operating between two heat reservoirs of fixed and uniform temperature can have a greater thermal efficiency, η_{th}, than a reversible heat engine that operates between same two reservoirs, that is,

$$\eta_{th\,rev} > \eta_{th\,irrev}.$$

Corollary 2 All reversible heat engines operating between the same two reservoirs of fixed and uniform temperature have the same thermal efficiency.

Corollary 3 The ratio between the heat source and the heat sink is equal to the ratio between the temperature of the heat source and the temperature of the heat sink:

$$\frac{Q_1}{Q_2} = \frac{T_1}{T_2}.$$

Corollary 4 *Inequality of Clausius*: Perpetual motion of the second kind is possible unless

$$\oint \frac{\delta Q}{T} \le 0. \tag{2.50}$$

In other words, if $\oint \frac{\delta Q}{T} \le 0$ then perpetual motion of the second kind is impossible.

Corollary 5 In a cycle made up of reversible steps, $\oint (\frac{\delta Q}{T})_{\text{rev}} = 0$ or $\int (\frac{\delta Q}{T})_{\text{rev}}$ depends only on the end states and not on the intermediate series of states.

Corollary 5 implies that $(\frac{\delta Q}{T})_{\text{rev}}$ is an exact differential. Consequently, $(\frac{\delta Q}{T})_{\text{rev}}$ is a property, *i.e.*, a quantity of state. The variation of entropy dS was defined to be equal to the property $(\frac{\delta Q}{T})_{\text{rev}}$:

$$\boxed{dS := \left(\frac{\delta Q}{T}\right)_{\text{rev}}.} \tag{2.51}$$

The term "entropy" was coined by Rudolf Clausius in 1865, although the same property was described earlier by William Rankine, who called it "thermodynamic function" and denoted it by ϕ. The entropy is a property of matter that measures the degree of disorder at the microscopic level. The natural state of affairs is that entropy is produced by all processes. Entropy can be produced but not destroyed, according to the second law of thermodynamics. From (2.51) one concludes that a process is *isentropic* if it is both reversible and adiabatic. For an adiabatic system $(dS)_{\text{adiabatic}} \ge 0$. Using the temperature–entropy $(T - s)$ diagram, Fig. 2.7 compares an isentropic process with a reversible adiabatic process for two cases: compression and expansion. Note that an isentropic process is a limiting case for a real adiabatic process. Note also that s is the specific entropy, that is, $s = S/m$, where m denotes mass. Other specific variables will be defined later in Table 2.1.

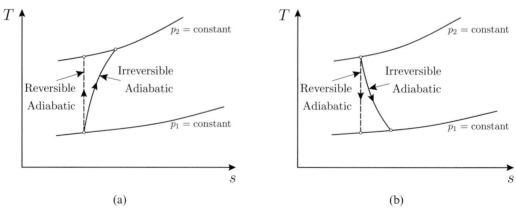

Figure 2.7 Reversible vs. irreversible processes in a temperature–entropy diagram. (a) Compression and (b) expansion.

Table 2.1 Definition of thermodynamics variables.

Specific energy	$e := E/m$
Specific internal energy	$u := U/m$
Specific entropy	$s := S/m$
Specific volume	$v := V/m = 1/\rho$
Enthalpy	$H := U + pV$
Specific enthalpy	$h := u + pv = u + p/\rho$
Specific heat capacity at constant pressure	$c_\mathrm{p} := (\partial h/\partial T)_\mathrm{p}$
Specific heat capacity at constant volume	$c_\mathrm{v} := (\partial u/\partial T)_\mathrm{v}$

Figure 2.8 Temperature-entropy diagram of a reversible process.

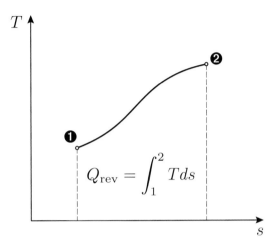

Using (2.51), the heat transfer of a reversible process is

$$Q_\mathrm{rev} = \int_1^2 T\, ds \qquad (2.52)$$

which is the area below the path from state 1 to state 2 in the $T - s$ diagram shown in Fig. 2.8. Although entropy is defined for a reversible process, the entropy is useful and will be used for irreversible processes, as will be shown in Corollary 6.

Corollary 6 The second law in analytical form results by combining Corollaries 4 and 5. Let us consider the cycle (1)-(2)-(1), in which the process (1)-(2) is arbitrary and the process (2)-(1) is reversible. According to Corollary 4,

$$\oint \frac{\delta Q}{T} \le 0 \qquad (2.50)$$

or

$$\int_1^2 \frac{\delta Q}{T} + \int_2^1 \left(\frac{\delta Q}{T}\right)_\mathrm{rev} \le 0. \qquad (2.53)$$

Using Eq. (2.51) of Corollary 5, Eq. (2.53) becomes

$$\int_1^2 \frac{\delta Q}{T} + S_1 - S_2 \leq 0$$

or

$$\underbrace{S_2 - S_1}_{\text{entropy change (property)}} \geq \underbrace{\int_1^2 \frac{\delta Q}{T}}_{\text{entropy transfer (non-property)}} . \tag{2.54}$$

Equation (2.54) can be written in differential form as

$$dS \geq \frac{\delta Q}{T}. \tag{2.55}$$

Equation (2.54) states that the entropy transfer never exceeds the entropy change. The difference between entropy change and entropy transfer is defined as the *entropy generation*, or *entropy production*:

$$S_{\text{gen}} = S_2 - S_1 - \int_1^2 \frac{\delta Q}{T} \geq 0. \tag{2.56}$$

Using the first law of thermodynamics, the entropy variation defined in (2.51) becomes:

$$dS = \left(\frac{\delta Q}{T}\right)_{\text{rev}} = \left(\frac{\delta W + dE}{T}\right)_{\text{rev}}.$$

Neglecting the kinetic and potential energy, and in the absence of electrical, magnetic, and capillary forces, the energy, E, is equal to the internal (or intrinsic) energy U. Assuming that the work is produced by normal stresses only, then $\delta W = p\delta V$. Consequently,

$$dS = \left(\frac{\delta W + dE}{T}\right)_{\text{rev}} = \left(\frac{p\delta V + dU}{T}\right)_{\text{rev}}.$$

Recall that the internal energy, U, is part of energy, E, which is independent of: (1) the system's motion, (2) gravitational forces, (3) electrical forces, (4) magnetic forces, and (5) capillary forces.

In light of the first law of thermodynamics, W and Q look equivalent. The second law of thermodynamics makes the fundamental distinction between W and Q: heat transfer is the energy interaction accompanied by entropy transfer, while work transfer is the energy interaction that takes place in the absence of entropy transfer. In addition to the heat transfer interaction, there is an entropy transfer interaction $\delta Q/T$.

2.4.4   Zeroth Law of Thermodynamics

The most recent law of thermodynamics is the zeroth law of thermodynamics. This law was introduced in 1931 by Sir Ralph Howard Fowler (1889–1944), a British physicist. The zeroth law of thermodynamics states that if systems (A) and (B) are in thermal equilibrium with a third system, then systems (A) and (B) are in equilibrium with each other.

The zeroth law of thermodynamics asserts that two systems are in thermal equilibrium when their temperatures are *identical*.

Each of the three laws of thermodynamics presented so far introduced a system property. The zeroth law introduced temperature, the first law introduced internal energy, and the second law introduced entropy.

2.4.5 Third Law of Thermodynamics

The third law of thermodynamics is presented for completeness, although it will not be applied herein to propulsion systems. The third law was introduced by the German chemist Walther Hermann Nernst (1864–1941) during the 1906–1912 period. For this reason, the third law is also known as the Nernst's theorem or postulate. The third law states that for a closed system in thermal equilibrium, the entropy of a system approaches a constant value as its temperature approaches absolute zero. A consequence of the third law is that it is impossible for any process, no matter how idealized, to reduce the entropy of a system to its absolute-zero value in a finite number of operations. While the third law was introduced as a *fundamental law*, justified by experiments, with the development of statistical mechanics, it changed to a *derived law*, derived from even more basic laws.

2.4.6 First Law of Thermodynamics for a Control Volume

The first law of thermodynamics presented in Section 2.4.2 is difficult to use "as is" in applications. For practical applications, it is much more useful to utilize the first law written for a control volume. This section presents the first law of thermodynamics for a control volume. Since in the reminder of this chapter we will be dealing with control volumes only, the notation will be simplified by removing the $*$ from the volume τ^* and surface σ^*.

2.4.6.1 Thermal Properties of Continuum

Before deriving the first law for a control volume, let us review the definitions of some of the basic variables shown in Table 2.1.

Note that specific energy can also be written as

$$e = u + \frac{V^2}{2} + gz + e_{\text{other}}, \tag{2.57}$$

where $u = u(T)$, $V^2/2$ is the kinetic energy term, gz is the potential energy term, and e_{other} includes other macroscopic forms of specific energy storage.

2.4.6.2 Derivation of the First Law of Thermodynamics for a Control Volume

Let us consider the control volume shown in Fig. 2.9, where the shaft penetrates the control volume and can allow work interaction with the surrounding environment. To derive the first law of thermodynamics for this control volume, one starts by dividing (2.48) by dt:

$$\frac{\delta Q}{dt} = \frac{dE}{dt} + \frac{\delta W}{dt}. \tag{2.58}$$

Figure 2.9 Schematic of a system that interacts with the environment.

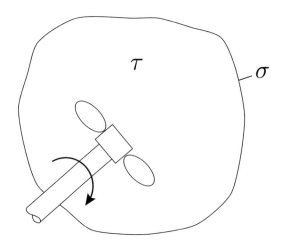

The first term of the right-hand side can be written using the Reynolds transport theorem as

$$\frac{dE}{dt} = \frac{d}{dt} \int_\tau (\rho e)\, d\tau = \int_\tau \frac{\partial(\rho e)}{\partial t}\, d\tau + \int_{\sigma \equiv \partial\tau} (\rho e)\, \vec{V} \cdot \hat{n}\, d\sigma.$$

Using the expression of energy e from (2.57) where e_{other} was neglected, one obtains

$$\frac{dE}{dt} = \int_\tau \frac{\partial(\rho e)}{\partial t}\, d\tau + \int_{\sigma \equiv \partial\tau} \rho \left(u + \frac{V^2}{2} + gz \right) \vec{V} \cdot \hat{n}\, d\sigma. \tag{2.59}$$

Let us find next the expression of the work due to normal stresses. The force acting on the surface $d\sigma$ is

$$\vec{F}_{d\sigma} = p\hat{n}d\sigma.$$

Assume that the displacement of the boundary due to the force $\vec{F}_{d\sigma}$ is $d\vec{\ell}$; then the work δW_n done by the normal stresses on the surface σ is

$$\delta W_n = \int_\sigma p\hat{n} \cdot d\vec{\ell}\, d\sigma. \tag{2.60}$$

Dividing (2.60) by dt yields

$$\frac{\delta W_n}{dt} = \frac{1}{dt} \int_\sigma p\hat{n} \cdot d\vec{\ell}\, d\sigma. \tag{2.61}$$

Note that the symbol "δ" was used instead of "d" because work is a non-property. Since $\vec{V} = d\vec{\ell}/dt$ and the specific volume $v = 1/\rho$, (2.61) becomes

$$\frac{\delta W_n}{dt} = \int_{\sigma \equiv \partial\tau} \rho(pv)\, \vec{V} \cdot \hat{n}\, d\sigma. \tag{2.62}$$

Because the time rate of work is power, (2.62) represents the power, $\mathcal{P}_n$, due to the normal stresses.

In addition to the work due to normal stresses, one must also evaluate the work due to shear stresses. The latter can be divided into shaft work and shear work. The shaft work is defined as the work done by the part of the shaft inside the system on the part outside the system. The shear work is defined as the work done at the boundaries of the system on adjacent fluid which is in motion. The time derivative of shaft work, W_{shaft}, is the shaft power, P_{shaft}, and the time derivative of the shear work, W_{shear}, is the shear power, P_{shear}.

Using (2.59), (2.62), and the notation $\dot{Q} := \delta Q/dt$ for the rate of heat transfer, (2.58) becomes

$$\dot{Q} = P_{shaft} + P_{shear} + \int_{\tau} \frac{\partial}{\partial t}(\rho e)\, d\tau + \int_{\sigma \equiv \partial \tau} \rho \left(h + \frac{V^2}{2} + gz \right) \vec{V} \cdot \hat{n}\, d\sigma \qquad (2.63)$$

where we used $h = u + pv$.

Example 2.4.1 Two jets shown in Fig. 2.10 are mixing downstream of a turbofan engine: a hot jet with a temperature of 800 K and a cold jet with a temperature of 273 K. The ratio of the cold mass flow rate and the hot mass flow rate is called the bypass ratio, BPR. What is the temperature of the mixed jets if the bypass ratio is (a) 1 and (b) 8? (Hint: neglect kinetic and potential energies).

Figure 2.10 Jets mixing at turbofan engine exit.

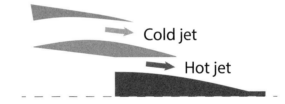

Cold jet

Hot jet

Solution
Let us use the following notation: $\dot{m}_c$ for the cold mass flow rate, $\dot{m}_h$ for the hot mass flow rate, and $\dot{m}_m$ for the mixed mass flow rate. Assuming steady flow, the mass conservation equation (2.4) yields:

$\dot{m}_c + \dot{m}_h = \dot{m}_m,$

and using the definition of the BPR,

$BPR + 1 = \dot{m}_m/\dot{m}_h.$

For steady flows in the absence of heat transfer and shaft and shear power, the energy conservation equation (2.63) yields

$\dot{m}_c h_c + \dot{m}_h h_h = \dot{m}_m h_m$

such that

$$h_m = \frac{BPR\, h_c + h_h}{BPR + 1}.$$ (2.64)

Assuming that the specific heat capacity is constant, (2.64) yields

$$T_m = \frac{BPR\, T_c + T_h}{BPR + 1}.$$

Therefore, $T_m = 536.5$ K for $BPR = 1$, and $T_m = 331.6$ K for $BPR = 8$.

2.4.7   Second Law of Thermodynamics for a Control Volume

Let us derive the expression of the second law of thermodynamics for a control volume by using Corollary 6 of the second law:

$$dS \geq \frac{\delta Q}{T}.$$ (2.53)

Writing the entropy, S, as a function of mass specific entropy, s, and dividing both sides of the inequality by dt, yields

$$\frac{d}{dt} \int_\tau \rho s\, d\tau \geq \frac{\delta Q}{dt} \cdot \frac{1}{T}.$$

Applying the Reynolds transport theorem to the left-hand side term yields

$$\boxed{\int_\tau \frac{\partial}{\partial t}(\rho s)\, d\tau + \int_{\sigma \equiv \partial \tau} \rho s \vec{V} \hat{n}\, d\sigma \geq \frac{\dot{Q}}{T} = \int_\sigma \frac{d\dot{Q}}{T},}$$ (2.65)

where the notation $\dot{Q} := \delta Q/dt$ was used.

   An example of how to apply the second law of thermodynamics for a control volume to an engineering problem is shown in Example 2.4.5, after introducing the variation of entropy formula in Section 2.4.8.2.

2.4.8   Thermodynamics of Single-Phase Gases

This section presents basic constitutive property relationships and models used to describe the macroscopic behavior of gases. The properties of mixtures will be treated separately in Chapter 5.

2.4.8.1   Pure Substance; Equation of State; Perfect Gas

A *pure substance* is a substance that has only two independent *static* properties, in the absence of electricity, magnetism, and capillarity effects. Most engineering fluids are pure substances.

Note that static properties, as opposed to dynamic properties, do not change with time. The static properties are not only equilibrium or thermodynamic properties, such as density and volume, but also transport coefficients, such as fluid viscosity.

The *equation of state* is an algebraic relationship between the properties of the substance. The simplest equation of state is the *ideal gas law*:

$$pv = RT \quad \text{or} \quad \frac{p}{\rho} = RT \quad \text{or} \quad pV = nRT, \tag{2.66}$$

where R is the gas constant $R = \mathcal{R}/\mathcal{M}$. $\mathcal{R}$ is the universal gas constant ($\mathcal{R} = 8.3143$ kJ/(kmol K)), $\mathcal{M}$ is the molar mass, n is the number of moles of gas, and V is the volume of gas. The ideal gas law was introduced in 1834 by Benoit Paul Emile Clapeyron (1799–1864), a French engineer and physicist. The *ideal gas* or *perfect gas* is a subset of pure substance that satisfies the ideal gas law. The ideal gas model assumes that the molecules are non-interacting point particles that have a random motion that obeys energy conservation. The ideal gas law, however, becomes increasingly inaccurate at higher pressures and lower temperatures. Several equations of state have been proposed to improve the shortcomings of (2.66), but none of them is general enough to accurately predict the properties of all substances under all conditions. One of the most well-known improvements of the ideal gas law is the van der Waals equation of state:

$$\left(p + \frac{a}{V_m^2}\right)(V_m - b) = \mathcal{R}T \tag{2.67}$$

where $V_m = V/n$ is the molar volume and a and b are positive constants that depend on the material. This equation was introduced in 1873 by Johannes Diderik van der Waals (1837–1923), a Dutch physicist.

Let us now find the variation of the internal energy, u, as a function of the temperature and specific heat capacity at constant volume, c_v. The internal energy, u, is only a function of temperature, T. Although u is only a function of T, one can formally write that for a pure substance

$$u = u(v, T)$$

and use the fact that $\left(\frac{\partial u}{\partial v}\right)_T = 0$. Differentiating $u(v, T)$ yields

$$du = \left(\frac{\partial u}{\partial v}\right)_T dv + \left(\frac{\partial u}{\partial T}\right)_v dT,$$

and using the definition of specific heat capacities at constant volume shown in Table 2.1, one obtains

$$du = c_v dT$$

Integrating this equations yields

$$u_2 - u_1 = \int_{T_1}^{T_2} c_v dT. \tag{2.68}$$

Having found the variation of internal energy as a function of temperature and specific heat capacity at constant volume, let us now use the same approach to obtain the variation of enthalpy as a function of temperature and specific heat capacity at constant pressure. Since the enthalpy is $h = u + pv = u + RT$, one concludes that the enthalpy is a function of temperature only, since the gas constant R is a constant, and the internal energy, u, is a function of temperature. For a pure substance, enthalpy can be written as $h = h(T,p)$ with a pressure variation at constant temperature $(\partial h/\partial p)_T = 0$.

Differentiating $h = h(T,p)$ yields

$$dh = \left(\frac{\partial h}{\partial p}\right)_T dp + \left(\frac{\partial h}{\partial T}\right)_p dT,$$

and using the definition of specific heat capacities at constant pressure and the fact that $(\partial h/\partial p)_T = 0$, one obtains

$$dh = \left(\frac{\partial h}{\partial T}\right)_p dT = c_p dT.$$

Integrating this equation yields

$$h_2 - h_1 = \int_{T_1}^{T_2} c_p dT. \tag{2.69}$$

At this point it is useful to find the relationship between the specific heat capacities c_p and c_v. According to the definition given in Table 2.1, the specific heat capacity at constant pressure is

$$c_p = \left(\frac{\partial h}{\partial T}\right)_p,$$

and since the enthalpy is a function of temperature only,

$$c_p = \frac{dh}{dT}. \tag{2.70}$$

Using the definition of enthalpy

$$\frac{dh}{dT} = \frac{d(u + RT)}{dT}$$

yields

$$c_p = c_v + R. \tag{2.71}$$

The ratio of specific heat capacities, γ (or κ), is defined as

$$\gamma := \frac{c_p}{c_v}. \tag{2.72}$$

From the kinetic energy theory of gases, $\gamma = (n+2)/n$, where n is the number of degrees of freedom in the molecule. For mono-atomic molecules, $n = 3$ such that $\gamma = 5/3$. For diatomic molecules, $n = 5$ such that $\gamma = 7/5$. This latter value is a good approximation for air at standard conditions, that is, at $T = 288.16$ K and $p = 1$ bar.

Example 2.4.2 Calculate the specific heats of a gas that has a gas constant, $R = 294$ J/(kg K) and a ratio of specific heat capacities, $\gamma = 1.35$.

Solution

From equations (2.71) and (2.72)

$$c_v = R/(\gamma - 1) = \tag{2.73}$$
$$= 294/(1.35 - 1) = 840 \text{ J/(kg K)}$$

and

$$c_p = \gamma R/(\gamma - 1) = \tag{2.74}$$
$$= 1.35 \times 294/(1.35 - 1) = 1134 \text{ J/(kg K)}.$$

Recall the perfect gas was defined as a subset of pure substance that satisfies the equation of state $pv = RT$. In addition, the perfect gas has a constant specific heat capacity.

For a perfect gas, the internal energy variation (2.68) becomes

$$u_2 - u_1 = c_v(T_2 - T_1), \tag{2.75}$$

and the enthalpy variation (2.69) becomes

$$h_2 - h_1 = c_p(T_2 - T_1). \tag{2.76}$$

If the specific heat capacity varies with temperature, then the gas is called a *semi-perfect gas*. For a semi-perfect gas, the internal energy variation becomes

$$u_2 - u_1 = \bar{c}_v(T_2 - T_1), \tag{2.77}$$

and the enthalpy variation becomes

$$h_2 - h_1 = \bar{c}_p(T_2 - T_1), \tag{2.78}$$

where $\bar{c}_v$ and $\bar{c}_p$ denote averaged values

$$\bar{c}_v = \frac{1}{T_2 - T_1} \int_{T_1}^{T_2} c_v dT \tag{2.79}$$

and

$$\bar{c}_p = \frac{1}{T_2 - T_1} \int_{T_1}^{T_2} c_p dT. \tag{2.80}$$

Example 2.4.3 Using the air thermodynamic properties tables of Appendix B, calculate the ratio of specific heat capacities of air at the inlet of a compressor where the temperature is 288.16 K and at the exit of the compressor where the temperature is 603.16 K.

Solution
The ratio of specific heat capacities is obtained from

$$a = \sqrt{\gamma R T} \quad \text{so that} \quad \gamma = a^2/(RT).$$

From Appendix B the speed of sound at temperature $T = 288.16$ K is $a = 340.4$ m/s, so that

$$\gamma = 340.4^2/(287.16 \times 288.16) = 1.4003,$$

and at temperature $T = 603.16$ K the speed of sound is $a = 488.1$ m/s, which yields

$$\gamma = 488.1^2/(287.16 \times 603.16) = 1.3755.$$

2.4.8.2 Entropy Variation

The first law of thermodynamics (2.48) can be written using intensive properties as

$$\delta q - \delta w = de. \tag{2.81}$$

Neglecting the kinetic and potential energies and assuming that the specific work transfer is $dw = p\,dv$ for a reversible process (and also for irreversible, [Hill and Peterson, 1992, p. 33]) (2.81) gives:

$$T ds = du + p dv$$

or

$$ds = \frac{du}{T} + \frac{p}{T} dv = \frac{du}{dT} \cdot \frac{dT}{T} + R\frac{dv}{v} = c_v \frac{dT}{T} + R\frac{dv}{v}.$$

Integrating the above equation from state 1 to state 2 yields

$$s_2 - s_1 = \int_{T_1}^{T_2} c_v(T)\frac{dT}{T} + R\ln\frac{v_2}{v_1}. \tag{2.82}$$

For a perfect gas, (2.82) becomes

$$s_2 - s_1 = c_v \ln\frac{T_2}{T_1} + R\ln\frac{v_2}{v_1}. \tag{2.83}$$

Using the relation (2.73) between R and c_v, the entropy variation (2.83) becomes

$$s_2 - s_1 = c_v \ln\frac{T_2}{T_1} + (\gamma - 1)c_v \ln\frac{v_2}{v_1} = c_v \ln\left[\frac{T_2}{T_1}\left(\frac{v_2}{v_1}\right)^{\gamma-1}\right]. \tag{2.84}$$

Using the equation of state for states 1 and 2,

$$p_1 v_1 = RT_1, \quad p_2 v_2 = RT_2,$$

yields $T_2/T_1 = p_2 v_2/(p_1 v_1)$ such that equation (2.84) can also be written as

$$s_2 - s_1 = c_v \ln \left[\frac{p_2}{p_1} \left(\frac{v_2}{v_1} \right)^\gamma \right] = c_v \ln \left[\frac{p_2}{p_1} \left(\frac{p_1}{p_2} \right)^\gamma \right]. \tag{2.85}$$

Similarly, by writing v_2/v_1 in equation (2.84) as a function of temperature and pressure ratios yields

$$s_2 - s_1 = c_v \ln \left[\frac{T_2}{T_1} \left(\frac{T_2}{T_1} \cdot \frac{p_1}{p_2} \right)^{\gamma-1} \right] = c_v \ln \left[\left(\frac{p_1}{p_2} \right)^{\gamma-1} \left(\frac{T_2}{T_1} \right)^\gamma \right]. \tag{2.86}$$

If the process that takes place between states 1 and 2 is isentropic, then $s_1 = s_2$. In this case, equations (2.84)–(2.86) yield

$$T v^{\gamma-1} = \text{const} \quad \text{or} \quad \frac{T}{\rho^{\gamma-1}} = \text{const} \quad \text{or} \quad \frac{\rho_1}{\rho_2} = \left(\frac{T_1}{T_2} \right)^{\frac{1}{\gamma-1}}, \tag{2.87}$$

$$p v^\gamma = \text{const} \quad \text{or} \quad \frac{p}{\rho^\gamma} = \text{const} \quad \text{or} \quad \frac{p_1}{p_2} = \left(\frac{\rho_1}{\rho_2} \right)^\gamma, \quad \text{and} \tag{2.88}$$

$$\frac{p^{\gamma-1}}{T^\gamma} = \text{const} \quad \text{or} \quad \frac{p}{T^{\frac{\gamma}{\gamma-1}}} = \text{const} \quad \text{or} \quad \frac{p_1}{p_2} = \left(\frac{T_1}{T_2} \right)^{\frac{\gamma}{\gamma-1}}. \tag{2.89}$$

Example 2.4.4 Calculate the temperature at the exit from a compressor knowing the temperature T_1 and pressure p_1 at the inlet of the compressor and the pressure p_2 at the exit of the compressor. The gas constant and the ratio of specific heat capacities are also known. Consider two cases: (a) compression is an isentropic process; (b) entropy increases in the compression process by Δs.

Solution
The $T - s$ diagram of the process is shown in Fig. 2.11.
(a) The process between states 1 and 2_i is isentropic; therefore using (2.89) yields

$$T_{2i} = T_1 \left(\frac{p_{2i}}{p_1} \right)^{\frac{\gamma-1}{\gamma}};$$

and since $p_{2i} = p_2$,

$$T_{2i} = T_1 \left(\frac{p_2}{p_1} \right)^{\frac{\gamma-1}{\gamma}}.$$

(b) The entropy variation

$$s_2 - s_1 = c_p \ln \frac{T_2}{T_1} - R \ln \frac{p_2}{p_1}$$

Figure 2.11 Temperature-entropy diagram of a compression process.

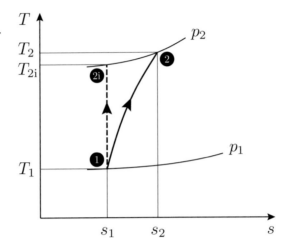

where $s_2 - s_1 = \Delta s$, yields

$$\Delta s = R \frac{\gamma}{\gamma - 1} \ln \frac{T_2}{T_1} - R \ln \frac{p_2}{p_1} = R \ln \left[\left(\frac{T_2}{T_1} \right)^{\frac{\gamma}{\gamma - 1}} \frac{p_1}{p_2} \right].$$

Therefore,

$$e^{\frac{\Delta s}{R}} = \left(\frac{T_2}{T_1} \right)^{\frac{\gamma}{\gamma - 1}} \frac{p_1}{p_2} \quad \rightarrow \quad T_2 = T_1 \left[\frac{p_2}{p_1} \cdot e^{\frac{\Delta s}{R}} \right]^{\frac{\gamma - 1}{\gamma}}.$$

The isentropic equation (2.88) can be used to obtain the variation of pressure as a function of the variation of density. By differentiating $p/\rho^\gamma = \text{const}$, one obtains

$$dp = \text{const}\ d\rho^\gamma = \frac{p}{\rho^\gamma} d\rho^\gamma = \frac{p}{\rho^\gamma} \gamma \rho^{\gamma - 1} d\rho = \gamma \frac{p}{\rho} d\rho,$$

which can be rearranged as

$$\frac{dp}{p} = \gamma \frac{d\rho}{\rho}. \tag{2.90}$$

An alternative derivation of (2.90) would be to take the logarithm of $p = \text{const}\ \rho^\gamma$ and then differentiate it.

Relations similar to (2.90) can be obtained between the variations of temperature, density, and pressure.

Problem 2.4.1 Using isentropic relations (2.87) and (2.89), show that

$$\frac{d\rho}{\rho} = \frac{1}{\gamma - 1} \frac{dT}{T} \tag{2.91}$$

$$\frac{dp}{p} = \frac{\gamma}{\gamma - 1} \frac{dT}{T}. \tag{2.92}$$

The entropy variation can also be written as a function of c_p instead of c_v. Let's once again start by writing the first law of thermodynamics:

$$T ds = du + p dv = du + d(pv) - v dp = dh - v dp,$$

and after dividing by temperature:

$$ds = \frac{dh}{T} - \frac{v \, dp}{T} = \frac{dh}{dT} \cdot \frac{dT}{T} - \frac{v \, dp}{T} = c_p \frac{dT}{T} - \frac{v \, dp}{T}.$$

Integrating this equation from state 1 to state 2 yields

$$s_2 - s_1 = \int_{T_1}^{T_2} c_p \frac{dT}{T} - R \int_{p_1}^{p_2} \frac{dp}{p} = \int_{T_1}^{T_2} c_p \frac{dT}{T} - R \ln \frac{p_2}{p_1}.$$

For a perfect gas this leads to

$$\boxed{s_2 - s_1 = c_p \ln \frac{T_2}{T_1} - R \ln \frac{p_2}{p_1}} \tag{2.93}$$

and for a semi-perfect gas

$$s_2 - s_1 = c_{p_{avg}} \ln \frac{T_2}{T_1} - R \ln \frac{p_2}{p_1},$$

where

$$c_{p_{avg}} = \frac{1}{\ln(T_2/T_1)} \int_{T_1}^{T_2} c_p \frac{dT}{T} \tag{2.94}$$

is different from the average value of (2.80).

For a perfect gas, the enthalpy variation during an isentropic process is

$$(\Delta h)_s = c_p (T_2 - T_1)_s = c_p T_1 \left[\left(\frac{T_2}{T_1} \right)_s - 1 \right] = c_p T_1 \left[\left(\frac{p_2}{p_1} \right)^{\frac{\gamma-1}{\gamma}} - 1 \right].$$

Example 2.4.5 The device shown in Fig. 2.12 can take two airstreams and produce a single exhaust stream. At inlet 1, the mass flow rate is 5 kg/s, temperature is 288 K, and pressure is 0.101 MPa. At inlet 2, the mass flow rate is 0.5 kg/s, temperature is 1500 K, and pressure is 5 MPa. The ratio of specific heats of the working fluid is $\gamma = 1.4$. Mass, momentum, and energy exchange mechanisms occur inside the device. Assuming the device operates in steady state and the kinetic and potential energies at all stations are insignificant, what is the maximum pressure at the outlet 3?

Figure 2.12 Pressure increasing device.

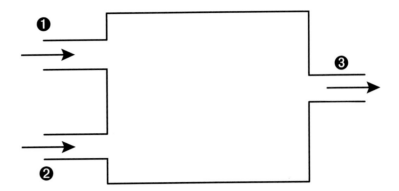

Solution

Mass conservation will be used to find out the mass flow rate at outlet 3:

$$\dot{m}_1 + \dot{m}_2 - \dot{m}_3 = 0. \tag{2.95}$$

Energy conservation will allow us to calculate the enthalpy at outlet 3:

$$\int_{\tau^*} \frac{\partial}{\partial t}(\rho e)d\tau + \int_{\sigma^*} \rho\left(h + \frac{V^2}{2} + gz\right)(\vec{V} \cdot \vec{n})d\sigma = \dot{Q} - \mathcal{P}. \tag{2.96}$$

With the following assumptions,

1. steady flow: time derivative is zero
2. neglect kinetic energy compared to enthalpy
3. neglect potential energy compared to enthalpy
4. no heat transfer since the maximum pressure is sought: $\dot{Q} = 0$
5. no work transfer: $\mathcal{P} = 0$,

(2.96) reduces to

$$\dot{m}_1 h_1 + \dot{m}_2 h_2 - \dot{m}_3 h_3 = 0. \tag{2.97}$$

Second law of thermodynamics:

$$\int_{\tau^*} \frac{\partial}{\partial t}(\rho s)d\tau + \int_{\sigma^*} \rho s(\vec{V} \cdot \vec{n})d\sigma \geq \frac{\dot{Q}}{T} \tag{2.98}$$

where $\dot{Q}$ is the heat flux per unit time. Assuming of steady flow and zero heat transfer, (2.98) reduces to

$$\dot{m}_3 s_3 - \dot{m}_1 s_1 - \dot{m}_2 s_2 \geq 0 \tag{2.99}$$

Let us define

$$\alpha = \frac{\dot{m}_2}{\dot{m}_1} = 0.1. \tag{2.100}$$

Use the variation of entropy (2.93):

$$s = s_1 + c_p \ln \frac{T}{T_1} - R \ln \frac{p}{p_1}. \tag{2.101}$$

From (2.95) and (2.100) :

$$\frac{\dot{m}_3}{\dot{m}_1} = 1 + \alpha. \tag{2.102}$$

From (2.97), (2.100), (2.102) and $h = c_p T$:

$$T_1 + \alpha T_2 - (1 + \alpha) T_3 = 0 \rightarrow T_3 = \frac{T_1 + \alpha T_2}{1 + \alpha} = 398.2 \text{ K}. \tag{2.103}$$

From (2.99) and (2.101):

$$(1 + \alpha) \left(s_1 + c_p \ln \frac{T_3}{T_1} - R \ln \frac{p_3}{p_1} \right) - s_1 - \alpha \left(s_1 + c_p \ln \frac{T_2}{T_1} - R \ln \frac{p_2}{p_1} \right) \geq 0.$$

The entropy s_1 cancels out, and after substituting $c_p = \gamma R/(\gamma - 1)$ yields

$$(1 + \alpha) \left(R\frac{\gamma}{\gamma - 1} \ln \frac{T_3}{T_1} - R \ln \frac{p_3}{p_1} \right) - \alpha \left(R\frac{\gamma}{\gamma - 1} \ln \frac{T_2}{T_1} - R \ln \frac{p_2}{p_1} \right) \geq 0.$$

Eliminating R and rearranging the terms yields

$$\rightarrow \frac{p_3}{p_1} \leq \left(\frac{T_3}{T_1} \right)^{\frac{\gamma}{\gamma - 1}} \cdot \left(\frac{p_2}{p_1} \right)^{\frac{\alpha}{\alpha + 1}} \cdot \left(\frac{T_1}{T_2} \right)^{\frac{\gamma}{\gamma - 1} \cdot \frac{\alpha}{\alpha + 1}}.$$

After substituting the numerical values $p_{3\text{max}} = 0.265$ MPa.

2.4.8.3 Thermodynamic Properties of Air

The thermodynamic properties of air vary with temperature and pressure. The table of Appendix B shows the variation of enthalpy, internal energy, and entropy with temperature at a pressure of 1 bar. The table values can be used to obtain the thermodynamic properties at pressure values different from 1 bar by using (2.93) as illustrated in the following example.

Example 2.4.6 Using the tables of air properties, calculate the entropy s of air knowing the temperature $T_1 = 288.16$ K and pressure $p_1 = 1.5$ bar. Note that the air gas constant is $R = 0.28716$ kJ/(kg K).

Solution

Let us start by reading from the tables of air properties the entropy value corresponding to temperature $T_1 = 288.16$ K:

$s_1 = 6.6608$ kJ/(kg K).

Recall that the properties in these tables are given for air at a pressure of 1 bar.

Figure 2.13 Temperature–entropy diagram.

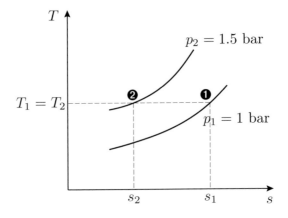

On the temperature–entropy diagram shown in Fig. 2.13, the temperature $T = 288.16$ K and entropy $s = 6.6608$ kJ/(kg K) define state 1. To find the entropy of state 2, defined by the same temperature and a pressure of 1.5 bar, we will use the entropy variation equation

$$s_2 - s_1 = c_p \ln \frac{T_2}{T_1} - R \ln \frac{p_2}{p_1} \tag{2.93}$$

where $T_2 = T_1 = T = 288.16$ K, $p_2 = 1.5$ bar and $p_1 = 1$ bar. Therefore,

$$s_2 = s_1 - R \ln \frac{p_2}{p_1} = 6.6608 - 0.28716 \ln 1.5 = 6.5444 \text{ kJ/(kg K)}.$$

Consequently, the entropy of air at temperature 288.16 K and pressure 1.5 bar is 6.5444 kJ/(kg K).

2.5   Summary of Fundamental Equations

Reynolds transport theorem:

$$\frac{d}{dt} \int_{\tau^*} \aleph \, d\tau = \int_{\tau^*} \frac{\partial \aleph}{\partial t} \, d\tau + \int_{\sigma^* \equiv \partial \tau^*} \aleph \vec{V}^* \cdot \hat{n} \, d\sigma.$$

Mass conservation:

$$\int_{\tau^*} \frac{\partial \rho}{\partial t} \, d\tau + \int_{\sigma^* \equiv \partial \tau^*} \rho \vec{V} \cdot \hat{n} \, d\sigma = 0 \quad \text{for} \ \vec{V}^* = 0.$$

Momentum conservation:

$$\int_{\tau^*} \frac{\partial}{\partial t} (\rho \vec{V}) \, d\tau + \int_{\sigma^* \equiv \partial \tau^*} \rho \vec{V} (\vec{V} \cdot \hat{n}) \, d\sigma = \vec{F}_V + \vec{F}_S \quad \text{for} \ \vec{V}^* = 0.$$

Moment of momentum conservation:

$$\int_{\tau^*} \frac{\partial}{\partial t} \left[\rho (\vec{r} \times \vec{V}) \right] d\tau + \int_{\sigma^* \equiv \partial \tau^*} \rho (\vec{r} \times \vec{V}) (\vec{V} \cdot \hat{n}) \, d\sigma = \vec{M}_F + \vec{M}_P \quad \text{for} \ \vec{V}^* = 0.$$

First law of thermodynamics:

$$\int_{\tau^*} \frac{\partial}{\partial t}(\rho e)\, d\tau + \int_{\sigma^*\equiv\partial\tau^*} \rho\left(h + \frac{V^2}{2} + gz\right)(\vec{V}\cdot\hat{n})\, d\sigma = \dot{Q} - \mathcal{P}_{\text{shaft}} - \mathcal{P}_{\text{shear}} \quad \text{for} \ \ \vec{V}^* = 0.$$

(2.104)

Second law of thermodynamics:

$$\int_{\tau^*} \frac{\partial}{\partial t}(\rho s)\, d\tau + \int_{\sigma^*\equiv\partial\tau^*} \rho s \cdot \vec{V}\cdot\hat{n}\, d\sigma \geq \frac{\dot{Q}}{T} \quad \text{for} \ \ \vec{V}^* = 0.$$

Problems

1. Calculate the drag on the symmetric airfoil shown in Fig. 2.14 assuming the angle of attack to be zero and using the control volume specified in the figure. The velocity upstream of the airfoil is $U = $ constant. The variation of the velocity downstream of the airfoil, $u(y)$, is defined as $u(y) = u_1(|y|)$, where $u_1(y)$ is a cubic polynomial function that satisfies the following conditions:
   1. $u_1(0) = U/2$
   2. $du_1/dy(0) = 0$
   3. $u_1(b/2) = U$
   4. $du_1/dy(b/2) = 0$.
   Assume the flow is steady and incompressible.
   Hint: calculate first the drag as a function of a general velocity $u(y)$ and then substitute $u(y)$ by its polynomial.

Figure 2.14 Alternate control volume for airfoil in wind tunnel.

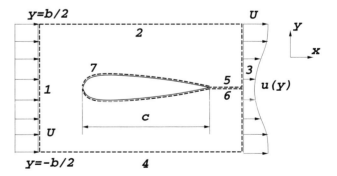

2. The jet of an airplane engine impinges on a deflecting wall as shown in Fig. 2.15. The velocity of the jet is V, the density of the jet gases is ρ, and the jet engine exhaust area (*i.e.*, the area of the jet) is σ. One can assume that the density is constant.
   1. What is the value of the normal force produced by the jet on the deflecting wall?
   2. What is the mass flow rate split between the upper and the lower ends of the deflecting wall?

(Hint: Consider using Bernoulli's equation in addition to mass and momentum conservation.)

Figure 2.15 Jet impinging on plate.

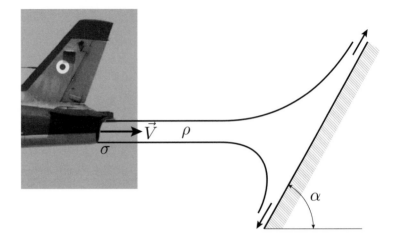

3. The velocity of the exhaust gases of the rocket engine shown in Fig. 2.16 is $V_e = 2700$ m/s, and the mass flow rate of the exhaust gases is $\dot{m}_e = 60$ kg/s. The area of the exhaust nozzle is $A = 0.2$ m^2. The pressure of the exhaust gases is $p_e = 0.16$ MPa, and the atmospheric pressure is $p_a = 0.101$ MPa. Derive the expression of the rocket engine thrust using the momentum conservation equation. What is the thrust of the rocket engine?

Figure 2.16 Rocket.

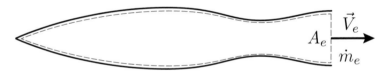

4. Determine the force with which a fluid with velocity V, section A, and density ρ impinges on the Pelton turbine blade shown in Fig. 2.17.

Figure 2.17 Jet impinging on a Pelton turbine.

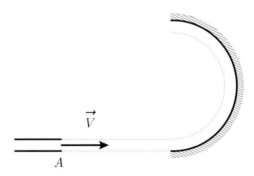

One can assume density is constant.

5. Gas with density ρ flows through an elbow nozzle of an F135 engine used on an F-35 airplane, as shown in Fig. 2.18. The following parameters are specified:

at inlet: velocity, v_1
       pressure, p_1
       area of nozzle, σ_1
       exterior normal, $\vec{n}_1$;

at exit : pressure, p_2
       area of nozzle, σ_2
       exterior normal, $\vec{n}_2$.

Figure 2.18 Elbow nozzle.

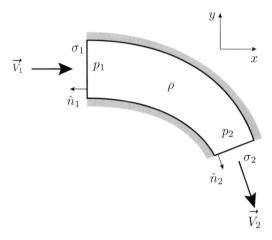

1. Derive a formula for the force of the gas acting on the nozzle, assuming that the density ρ is constant and that the atmospheric pressure is $p_a = 101,350$ Pa.
2. Find the value of the force using the following data:

$v_1 = 200$ m/s $\rho = 0.35$ kg/m^3
$p_1 = 0.16$ MPa $p_2 = 0.12$ MPa
$\sigma_1 = 0.12$ m^2 $\sigma_2 = 0.09$ m^2
$\vec{n}_1 = -\vec{i}$ $\vec{n}_2 = -\vec{j}$.

6. Two jets are merging as shown in Fig. 2.19. Consider incompressible flow and neglect gravitational forces. What are the velocity V and the angle β of the jet as a function of V_1, V_2, A_1, A_2, and α?

Figure 2.19 Merging jets.

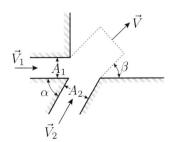

7. The jet coming out of the fan of a jet engine has a velocity $V_1 = 150$ m/s. The area of this jet is $A_1 = 0.1$ m². As shown in Fig. 2.20, this jet must be deflected by $\beta = 20$ deg using a second jet that has an area $A_2 = 0.02$ m². What should be the velocity of the actuating jet, V_2, and what would be the velocity of the deflected jet, V?

Figure 2.20 Deflected jet.

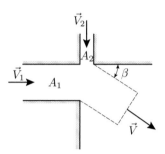

Consider incompressible flow and neglect gravitational forces.

8. The products of combustion in a jet engine combustor have a temperature of 1300 K and pressure of 15 bar. In the turbine, the products of combustion then expand to atmospheric pressure. During this expansion, the entropy increase is $\Delta s = 40$ J/(kg K).

1. Draw the Mollier diagram (or the enthalpy–entropy diagram) of this process.
2. What is the temperature of the combustion products once expanded to atmospheric pressure?
3. What is the lowest temperature the combustion products could reach at the end of the expansion?

One assumes the gas constant $R = 287$ J/(kg K) and the ratio of specific heat capacities $\gamma = 1.36$.

9. What is the power of the fan shown in Fig. 2.21?
Assume:

1. fluid density $\rho = 1.23$ kg/m³
2. exit velocity $u = 5$ m/s
3. fan diameter $D = 0.4$ m.

Explain and justify any other assumptions needed to solve the problem.

10. The exit nozzle of a small jet engine is made out of steel and has a mass of 1 kg. The mass flow rate through the engine is 2 kg/s. At start up, the temperature of the nozzle increases by 400 K in 15 s. The temperature of the gases at inlet in the exit nozzle is 800 K, and the speed of the gases is 200 m/s. The velocity at the exit of the nozzle is 220 m/s. The gases have a ratio of specific heat capacities $\gamma = 1.34$ and a gas constant $R = 293$ J/(kg K). The specific heat capacity of steel is $c_{steel} = 460$ J/(kg K).

Figure 2.21 Fan.

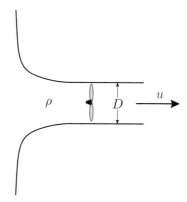

1. What is the rate of heat transfer in the nozzle?
2. What is the temperature of the gases at the exit from the nozzle?

11. Two jets are mixing in an energy-conversion machine, shown in Fig. 2.22. There is a single air-outlet stream discharging to the atmosphere. Power is being extracted from the machine through a shaft. The casing around the machine is well insulated, and therefore the heat transfer between the machine and the surrounding air can be neglected. The inlet and exit kinetic energy can be neglected compared to enthalpy. The shear stresses inside the energy-conversion machine can be neglected. The potential energy can also be neglected.

At inlet port 1, which has an area of 0.1 m^2, the pressure is 2 MPa, the velocity is 50 m/s, and the temperature is 1000 K. At inlet port 2, which has an area of 0.2 m^2, the pressure is 0.1 MPa, the velocity is 200 m/s, and the temperature is 300 K. At exit port 3, the temperature is 500 K.

The air gas constant is $R = 287.16$ J/(kg K), and the heat capacity at constant pressure is $c_p = 1004.5$ J/(kg K).

1. What is the shaft power extracted from the energy-conversion machine?
2. What should the temperature at the exit port be such that the shaft power is zero?

Figure 2.22 Energy-conversion machine.

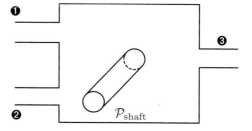

Bibliography

G. K. Batchelor. *Fluid Dynamics*. Cambridge University Press, 1967. 19, 20

A. Bejan. *Advanced Engineering Thermodynamics*. Wiley-Interscience, 1988. 29, 32

R. B. Bird, W. E. Steward, and E. N. Lightfoot. *Transport Phenomena*. John Wiley & Sons Inc., New York, 1960. Second printing, 1962, with corrections. 14

S. Carnot. *Reflection on the Motive Power of Fire*. Dover, 1960. 33

P. Hill and C. Peterson. *Mechanics and Thermodynamics of Propulsion*. Addison Wesley, second edition, 1992. 14, 44

O. Reynolds. *Papers on Mechanical and Physical Subjects*, volume III. Cambridge University Press, 1903. 13

A. H. Shapiro. *The Dynamics and Thermodynamics of Compressible Fluid Flow*, volume I. Ronald Press Company, 1953. 29

J. C. Slattery. *Advanced Transport Phenomena*. Cambridge University Press, 1999. 15

3   Steady One-Dimensional Gas Dynamics

Compressible Flows and Shock Waves

3.1   Introduction

The mass, momentum, and energy balance equations and the second law of thermodynamics equation are used to model the transport phenomena in propulsion systems. A simplified version of these equations that assumes the flow is steady and one-dimensional is frequently used for the pre-design and analysis of the propulsion systems.

The one-dimensional approximation implies that all fluid properties are uniform over any cross section of a duct or stream line. In other words, the rate of change of fluid properties normal to the stream line direction is much smaller than the rate of change along the stream line. It is known, however, that the properties vary over each cross section in duct flows. Consequently, we deal with averaged properties for each cross section. The errors introduced by the one-dimensional assumption are small if: (1) the relative change of area with respect to axial distance is small, $dA/A\, dx \ll 1$, (2) the radius of curvature of the duct or stream tube is large compared to the diameter of the passage, and (3) the velocity and temperature profiles are approximately constant along the axis.

The steady one-dimensional governing equations are derived in the first section of this chapter. Subsequently, the speed of sound, the critical speed and the Mach and λ numbers are introduced. Isentropic and nonisentropic flows are then presented, including flow through nozzles. The chapter ends by describing the features of normal and oblique shock waves.

3.2   Governing Equations

This section derives the steady one-dimensional mass, momentum, and energy balance equations. These governing equations are derived starting from the balance equations introduced in Chapter 2.

Figure 3.1 Control volume for one-dimensional flow.

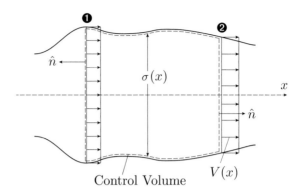

Control Volume

3.2.1 Mass Conservation Equation

The general form of the mass conservation equation[1] for a fixed control volume, that is, for $\vec{V}^* = 0$, was derived in Section 2.3.1:

$$\int_{\tau^*} \frac{\partial}{\partial t} \rho \, d\tau + \int_{\sigma^* \equiv \partial \tau^*} \rho \vec{V} \cdot \vec{n} \, d\sigma = 0. \tag{2.7}$$

For steady flow, the time derivative is zero, and (2.7) becomes

$$\int_{\sigma^* \equiv \partial \tau^*} \rho \vec{V} \cdot \vec{n} \, d\sigma = 0. \tag{3.1}$$

Let us calculate this integral by considering the control volume shown in Fig. 3.1. The integral along the upper and lower boundaries is zero since the flux through the boundaries is zero. Because the flow is one-dimensional, the density and velocity are functions of x only and have constant values at the inlet (station 1) and exit (station 2). Consequently, (3.1) becomes

$$-\rho(x_1) \, V(x_1) \, \sigma(x_1) + \rho(x_2) \, V(x_2) \, \sigma(x_2) = 0, \tag{3.2}$$

where the negative sign in front of the first term is due to the fact that the velocity and the normal at station 1 have opposite directions. The equation simply states that the mass flow rate at inlet equals the mass flow rate at exit of the control volume. In other words, the mass flow rate is constant, such that (3.2) can be written as

$$\rho(x) \, V(x) \, \sigma(x) = \text{const}, \tag{3.3}$$

or the variation of the mass flow rate $\rho V \sigma$ in the x direction is zero

$$\frac{d}{dx}(\rho V \sigma) = 0. \tag{3.4}$$

This derivative yields

$$V\sigma \frac{d\rho}{dx} + \rho\sigma \frac{dV}{dx} + \rho V \frac{d\sigma}{dx} = 0$$

---

[1] An alternative name for the mass conservation or mass balance equation is the continuity equation.

Figure 3.2 Control volume and flow parameters variation in a one-dimensional flow.

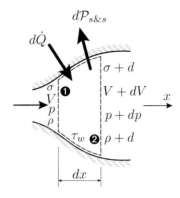

or after dividing by $\rho V \sigma$ and multiplying by dx,

$$\frac{d\rho}{\rho} + \frac{dV}{V} + \frac{d\sigma}{\sigma} = 0. \tag{3.5}$$

A more general way to derive (3.5) is the following. Let's apply continuity equation (2.7) to the control volume of thickness dx shown in Fig. 3.2, where the values at exit of the control volume are equal to the values at inlet plus the variation across the length dx:

$$-\rho V \sigma + (\rho + d\rho)(V + dV)(\sigma + d\sigma) = 0. \tag{3.6}$$

Note that the magnitude of the term ρ is much larger than that of the term $d\rho$. Likewise, the magnitudes of V and σ are much larger than those of dV and $d\sigma$, respectively. ρ, V, and σ are called zeroth-order terms, while $d\rho$, dV, and $d\sigma$ are called first-order terms. The product of two first-order terms, such as $d\rho\, dV$, is called a second-order term. The second-order terms are much smaller than the zero-order terms and will be neglected compared to them. Higher-order terms are even smaller and will also be neglected when compared with zeroth-order terms.

Neglecting the second order and higher terms in (3.6) yields

$$\rho \sigma \, dV + V\sigma \, d\rho + \rho V d\sigma = 0 \tag{3.7}$$

and after dividing by $\rho V \sigma$ yields

$$\frac{d\rho}{\rho} + \frac{dV}{V} + \frac{d\sigma}{\sigma} = 0,$$

which is identical to (3.5). Recall that the mass conservation equation is a scalar equation.

3.2.2   Momentum Conservation Equation

For steady flow and fixed control volume, $\vec{V}^* = 0$, the momentum conservation equation (2.29) becomes

$$\int_{\sigma^* \equiv \partial \tau^*} \rho \vec{V}(\vec{V} \cdot \vec{n}) \, d\sigma = \vec{F}_V + \vec{F}_S. \tag{3.8}$$

The momentum equation for two- and three-dimensional flows is a vectorial equation with two and three scalar components, respectively. For one-dimensional flow, the momentum equation has only one scalar component. The unit vectors normal to the left and right sides of the control volume are $-\hat{\imath}$ and $\hat{\imath}$, respectively. Assuming the flow is oriented in the x direction as shown in Fig. 3.2, the velocity vector on the left side of the control volume is $\vec{V} = \hat{\imath}V$. Consequently, the left-hand-side term of (3.8) is

$$\int_{\sigma^* \equiv \partial \tau^*} \rho \vec{V}(\vec{V} \cdot \vec{n})\, d\sigma = \hat{\imath}\left[-\rho V^2 \sigma + (\rho + d\rho)(V + dV)(V + dV)(\sigma + d\sigma)\right].$$

Neglecting the second- and third-order terms, the left-hand-side term of (3.8) becomes

$$\hat{\imath}(2\rho V\sigma dV + \sigma V^2 d\rho + \rho V^2 d\sigma).$$

Using the mass conservation equation (3.7) reduces the left-hand-side term to $\hat{\imath}V\sigma\rho dV$.

The volume (or body) force acting on the control volume is

$$\vec{F}_V = \hat{\imath}\rho f\sigma dx,$$

where f is the body force per unit mass. The surface forces acting on the surface of the control volume include the viscous force due to the shear stress at wall τ_w and the pressure force. The viscous force is acting on the surface $c\, dx$, where c is the length of the circumference of the control volume. This viscous force due to the shear stress is $-\hat{\imath}\tau_w c\, dx$, where the negative sign indicates that the force is acting in the direction opposite to the fluid flow movement. The expression of pressure force (2.12) becomes $\hat{\imath}[p\sigma - (p + dp)(\sigma + d\sigma)]$ and neglecting the second-order term, the pressure force reduces to $-\hat{\imath}(pd\sigma + \sigma dp)$. Consequently, the surface force is

$$\vec{F}_S = -\hat{\imath}\tau_w c dx - \hat{\imath}(pd\sigma + \sigma dp).$$

The linear momentum equation for steady one-dimensional flows (3.8) becomes

$$\rho V\sigma \frac{dV}{dx} = \rho f\sigma - \left(\tau_w c + \sigma \frac{dp}{dx} + p\frac{d\sigma}{dx}\right). \tag{3.9}$$

3.2.3 Energy Conservation Equation

For steady flows and fixed control volume, $\vec{V}^* = 0$, the energy conservation equation, aka the first law of thermodynamics (2.63), becomes

$$\dot{Q} - \mathcal{P}_{s\&s} = \int_{\sigma^* \equiv \partial \tau^*} \rho \left(h + \frac{V^2}{2} + gz\right)(\vec{V} \cdot \vec{n})\, d\sigma$$

where $\mathcal{P}_{s\&s} = \mathcal{P}_{shear} + \mathcal{P}_{shaft}$. Let's apply the energy conservation equation to the control volume shown in Fig. 3.2. Note that $\vec{V} = \hat{\imath}V$. Since for the control volume of Fig. 3.2 the heat transfer rate is $d\dot{Q}$ and the power is $d\mathcal{P}_{s\&s}$, and neglecting the gravitational potential energy, yields

$$d\dot{Q} - d\mathcal{P}_{s\&s} = -\rho V \sigma \left(h + \frac{V^2}{2} \right) + (\rho + d\rho)(V + dV)(\sigma + d\sigma) \left[h + dh + \frac{(V + dV)^2}{2} \right].$$

Neglecting the second- and third-order terms in the right-hand-side term yields

$$d\dot{Q} - d\mathcal{P}_{s\&s} = -\rho V \sigma \left(h + \frac{V^2}{2} \right) + \underbrace{(\rho V \sigma + \sigma V d\rho + \sigma \rho dV + \rho V d\sigma)}_{=0, \text{ from } (3.7)} \left(h + \frac{V^2}{2} + dh + V dV \right).$$

Using the mass conservation equation (3.7) further simplifies the right-hand-side term:

$$d\dot{Q} - d\mathcal{P}_{s\&s} = \rho V \sigma (dh + V dV),$$

which can be written as

$$\boxed{dq - dw_{s\&s} = dh + V dV} \tag{3.10}$$

where $dq := d\dot{Q}/\dot{m}$, $dw_{s\&s} := d\mathcal{P}_{s\&s}/\dot{m}$ and $\dot{m} = \rho V \sigma$.

3.2.3.1 Stagnation State

The expression (3.10) of the energy conservation equation allows us to introduce the notion of *stagnation state*. The stagnation state is the state that would be reached by a fluid if it were brought to rest reversibly, adiabatically, and without work.

Consequently, if the process is adiabatic, that is $dq = 0$, and without work transfer, that is $dw_{s\&s} = 0$, (3.10) becomes

$$dh + V dV = 0. \tag{3.11}$$

Integrating this equation yields

$$h + \frac{V^2}{2} = C,$$

where C is the integration constant. This constant is defined as the stagnation enthalpy. There are at least three notations used to denote stagnation enthalpy: h^*, h_T, and h_0. The last notation will be used herein, such that

$$h_0 = h + \frac{V^2}{2}. \tag{3.12}$$

With the assumption of a perfect gas, (3.12) can be rewritten as

$$c_p T_0 = c_p T + \frac{V^2}{2}, \tag{3.13}$$

where T_0 denotes the stagnation temperature. Note that stagnation values can also be defined for pressure, density, and speed of sound.

Example 3.2.1 Calculate the stagnation and static temperature at the inlet in a jet engine compressor, shown in Fig. 3.3, knowing that the airplane flies with a speed of 100 m/s and the air temperature at the altitude where the airplane flies is 281 K. The ratio between the area at inlet in the intake nozzle, A_I, and the area at inlet in the compressor, A_C, is 1.5. The air specific heat capacity at constant pressure is $c_p = 1004$ J/(kg K). One can neglect the density variation in the inlet.

Figure 3.3 Subsonic inlet.

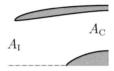

Solution
From the mass conservation equation applied between the inlet in the intake and the inlet in the compressor, neglecting the density variation, one obtains the velocity at inlet in the compressor, V_C:

$$V_C = V_I \frac{A_I}{A_C} = 100 \times 1.5 = 150 \text{ m/s}.$$

The stagnation enthalpy at inlet in the intake is

$$h_{0I} = h_I + 0.5 \, V_I^2 = c_p \, T_I + 0.5 \, V_I^2 = 1004 \times 281 + 0.5 \times 100^2 = 287,124 \text{ J/kg}.$$

Since there is no work done in the intake, and assuming an adiabatic process, the stagnation enthalpy $h_{0I} = h_{0C}$. Consequently, the stagnation temperatures at inlet in the compressor is

$$T_{0C} = h_{0C}/c_p = 287,124/1004 = 286 \text{ K}.$$

The static enthalpy is

$$h_C = h_{0C} - 0.5 \, V_C^2 = 287,124 - 0.5 \times 150^2 = 275,874 \text{ J/kg}$$

and the temperature is

$$T_C = h_C/c_p = 275,874/1004 = 274.8 \text{ K}.$$

The enthalpy–entropy diagram of the process that takes place in the intake is shown in Fig. 3.4. Although the problem did not provide enough information to estimate the entropy increase across the inlet diffuser, the entropy at inlet in the compressor is larger than the entropy at inlet in the diffuser because of the losses inside the diffuser. Note also that although the stagnation enthalpies h_{0I} and h_{0C} are equal, the stagnation pressure $p_{0C} < p_{0I}$.

Figure 3.4 Enthalpy–entropy diagram of engine intake static and stagnation states.

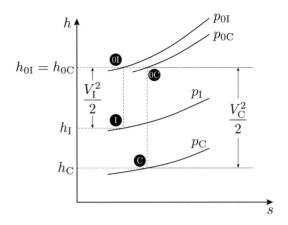

The process described for reaching the stagnation state is reversible and adiabatic, that is, isentropic. Consequently, the stagnation and static values of pressure, temperature, and densities can be related using equations (2.87)–(2.89):

$$\frac{p_0}{p} = \left(\frac{T_0}{T}\right)^{\frac{\gamma}{\gamma-1}} \tag{3.14}$$

$$\frac{\rho_0}{\rho} = \left(\frac{T_0}{T}\right)^{\frac{1}{\gamma-1}} \tag{3.15}$$

$$\frac{p_0}{p} = \left(\frac{\rho_0}{\rho}\right)^{\gamma}. \tag{3.16}$$

3.3 Dimensional and Nondimensional Reference Speeds

This section begins by deriving the expression of the speed of sound. Subsequently, the Mach number is defined. Three other reference speeds that are relevant to internal flows are then introduced: the maximum velocity, the speed of sound corresponding to stagnation temperature, and the critical speed. A dimensionless number of flow speed, called the λ number, is then introduced to address some of the limitations associated with the definition of the Mach number.

3.3.1 Speed of Sound and Mach Number

Let's derive the expression of the speed of sound. For this, let's assume that the sound wave is standing, that is, the wave propagation speed a is equal to the local flow speed, V. Since the wave is infinitesimally thin, that is, dx is of the order of 2000 Ångströms[2], the area variation of the control volume can be approximated as $d\sigma = 0$. Consequently, the mass conservation equation (3.5) becomes

[2] 1 Ångström $= 10^{-10}$m

$$\frac{d\rho}{\rho} + \frac{dV}{a} = 0. \tag{3.17}$$

With the observation that $V = a$ and $d\sigma = 0$, the momentum conservation equation (3.9) can be written as

$$\rho a \frac{dV}{dx} = - \left(\frac{dp}{dx} + \frac{\tau_w c}{\sigma} \right) + \rho f$$

or better

$$\rho a dV = - \left(dp + \frac{\tau_w c}{\sigma} dx \right) + \rho f \, dx.$$

The external forces f are bounded and small compared to the other forces. In addition the wall shear stresses τ_w are bounded. Therefore, since the wave is infinitesimally thin, the equation reduces to

$$\rho a dV = -dp. \tag{3.18}$$

Eliminating dV between (3.17) and (3.18) yields

$$\frac{d\rho}{\rho} + \frac{-dp}{\rho a} \frac{1}{a} = 0,$$

such that the speed of sound is

$$a = \sqrt{\frac{dp}{d\rho}}. \tag{3.19}$$

If the process is isentropic, then $p/\rho^\gamma \overset{(2.88)}{=} \text{const}$, such that

$$dp = \text{const} \; \gamma \, \rho^{\gamma-1} \, d\rho = \frac{p}{\rho^\gamma} \gamma \, \rho^{\gamma-1} \, d\rho,$$

which yields

$$\frac{dp}{d\rho} = \gamma \frac{p}{\rho}.$$

Consequently, if the flow is isentropic, the speed of sound (3.19) becomes

$$a = \sqrt{\gamma \frac{p}{\rho}} = \sqrt{\gamma R T}. \tag{3.20}$$

By definition, the ratio between the flow velocity and the speed of sound is the Mach number

$$\boxed{M := V/a}$$

in honor of Ernst Waldfried Josef Wenzel Mach (1838–1916), an Austrian physicist and philosopher. If M is less than 1, the flow is subsonic. Otherwise, the flow is supersonic.

Using (3.13) to write temperature T as a function of the stagnation temperature T_0 and velocity V, the Mach number becomes

$$M = \frac{V}{\sqrt{\gamma R(T_0 - V^2/(2c_p))}}. \tag{3.21}$$

Consequently, for a given stagnation temperature T_0, M is not proportional to velocity alone because the denominator of (3.21) also varies with V.

3.3.2 Maximum Velocity

Let's answer the following question: What is the maximum velocity a fluid with a given stagnation temperature, T_0, can reach? To answer this question, let us use the stagnation temperature equation (3.13) to obtain the expression of velocity:

$$V = \sqrt{2c_p(T_0 - T)} = \sqrt{\frac{2\gamma}{\gamma - 1}R(T_0 - T)}. \tag{3.22}$$

Therefore the maximum value of velocity corresponds to temperature $T = 0$:

$$\boxed{V_{\max} = \sqrt{\frac{2\gamma}{\gamma - 1}RT_0}}.$$

Note that for a given stagnation temperature, T_0, all states with the same static temperature, T, have the same velocity V.

3.3.3 Speed of Sound Corresponding to Stagnation Temperature

The speed of sound corresponding to the stagnation temperature is defined as

$$a_0 := \sqrt{\gamma R T_0}. \tag{3.23}$$

3.3.4 Critical Speed

The critical speed is the velocity at $M = 1$, that is, when $V = a$. Note that both velocity V and the speed of sound a depend on the static temperature, which is unknown. Using Eqs. (3.20) and (3.22), $V = a$ yields

$$\sqrt{\frac{2\gamma}{\gamma - 1}R(T_0 - T_{M=1})} = \sqrt{\gamma R T_{M=1}}$$

where $T_{M=1}$ is the static temperature at Mach $M = 1$. The static temperature at Mach $M = 1$ is also called the *critical temperature*, T_{cr}.

From the above equation, the critical temperature (or the static temperature at $M = 1$) is

$$T_{cr} \equiv T_{M=1} = \frac{2T_0}{\gamma + 1}, \tag{3.24}$$

such that the critical speed is

$$a_{cr} := \sqrt{\gamma R T_{cr}} = \sqrt{2\frac{\gamma}{\gamma + 1}RT_0}. \tag{3.25}$$

Example 3.3.1 Calculate the relationships between the maximum velocity, speed of sound corresponding to stagnation temperature, and critical speed. Rank these three velocities. Assume the ratio of specific heats to be $\gamma = 1.4$.

Solution

$$\frac{a_{\text{cr}}}{a_0} = \sqrt{\frac{2}{\gamma + 1}} = 0.913; \quad \frac{V_{\text{max}}}{a_0} = \sqrt{\frac{2}{\gamma - 1}} = 2.24; \quad \frac{V_{\text{max}}}{a_{\text{cr}}} = \sqrt{\frac{\gamma + 1}{\gamma - 1}} = 2.45, \tag{3.26}$$

$$\boxed{V_{\text{max}} > a_0 > a_{\text{cr}}}$$

Note that this relationship is valid as long as $\gamma \in (1, 3)$.

Having defined these velocities, one can revisit the stagnation temperature equation (3.13) which, after using the definition of c_{p} (2.74), becomes

$$V^2 + \frac{2}{\gamma - 1}\gamma RT = \frac{2}{\gamma - 1}\gamma RT_0.$$

Using the definition of the speed of sound based on stagnation temperature and maximum velocity yields

$$V^2 + \frac{2}{\gamma - 1}a^2 = \frac{2}{\gamma - 1}a_0{}^2 = V_{\text{max}}^2. \tag{3.27}$$

In coordinates (V, a), (3.27) represents an ellipse, shown in Fig. 3.5. This is called the *steady-flow adiabatic ellipse* and provides a representation of the different flow regimes ranging from incompressible to hypersonic flows.

Several flow regimes can be identified on the ellipse: incompressible, subsonic, transonic, supersonic, and hypersonic. In the incompressible flow region, velocity V is much smaller than the speed of sound a. The speed of sound changes much less than the velocity V. In the subsonic flow region, the velocity V and the speed of sound are of comparable

Figure 3.5 Steady-flow adiabatic ellipse.

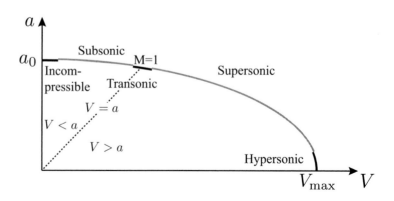

magnitude, although the former is smaller than the latter. In this region, Mach number varies primarily due to changes in velocity V. In the transonic flow region, the differences between V and a are smaller than either value. In the supersonic flow region, the velocity V and the speed of sound are of comparable magnitudes, although the former is larger than the latter. In this region, the Mach number varies due to variations in both V and a. In the hypersonic flow region, velocity V is much larger than the speed of sound. Changes of velocity are small while changes of the speed of sound are large. The variation of the Mach number is due almost exclusively to changes in the speed of sound.

3.3.5   λ Number

The Mach number has two limitations: (1) it is not proportional to the velocity alone because, for a given stagnation temperature, the speed of sound varies with velocity, as shown in (3.21), and (2) at high speeds, the Mach number tends to infinity, as shown in Fig. 3.5. To avoid these limitations, a new dimensionless velocity, called the λ number (or M* number), is defined as

$$\lambda := \frac{V}{a_{cr}}.$$

The λ number is proportional to velocity alone because the critical speed does not vary with velocity. To find out the bounds of the λ number, let us derive the relation between M and λ. λ can be rewritten as

$$\lambda^2 = \frac{V^2}{a_{cr}^2} = \frac{V^2}{a^2} \cdot \frac{a^2}{a_{cr}^2} = M^2 \frac{a^2}{a_{cr}^2}. \tag{3.28}$$

To determine the expression of the a^2/a_{cr}^2 term, let us use (3.26) in (3.27) such that

$$V^2 + \frac{2}{\gamma - 1}a^2 = \frac{2}{\gamma - 1}a_0{}^2 = V_{max}^2 = \frac{\gamma + 1}{\gamma - 1}a_{cr}^2.$$

Dividing this equation by a_{cr}^2 yields

$$\frac{V^2}{a_{cr}^2} + \frac{2}{\gamma - 1}\frac{a^2}{a_{cr}^2} = \frac{\gamma + 1}{\gamma - 1}, \tag{3.29}$$

Eliminating the a^2/a_{cr}^2 term between (3.28) and (3.29) yields

$$\lambda^2 = \frac{\dfrac{\gamma + 1}{2}M^2}{1 + \dfrac{\gamma - 1}{2}M^2} \tag{3.30}$$

Table 3.1 Comparison between the values of Mach and λ numbers.

Mach number	λ number
0	0
<1	<1
1	1
>1	>1
∞	$\sqrt{\frac{\gamma+1}{\gamma-1}}$

and

$$M^2 = \frac{\dfrac{2}{\gamma+1}\lambda^2}{1-\dfrac{\gamma-1}{\gamma+1}\lambda^2}.$$

It is apparent from (3.30) that when M goes to infinity, λ goes to $\sqrt{(\gamma+1)/(\gamma-1)}$.

A comparison between the values of Mach and λ numbers is shown in Table 3.1. It is remarkable that when M=1, λ is also equal to 1. Furthermore, the flow is subsonic when both M and λ are less than 1; the flow is supersonic when both M and λ are larger than 1.

3.4 Isentropic and Nonisentropic Flows

This section discusses both isentropic and nonisentropic flows. While the focus is on the former, and therefore a significant part is devoted to flow through nozzles, the study of nonisentropic flows is important since it helps us assess the losses that occur in the components of propulsion systems.

3.4.1 Isentropic Flow

The study of isentropic processes is important because they provide the limit for the ideal case process and therefore allow us to estimate how good the engine components are and how much room for improvement there is. There are components in a propulsion system where the processes are close to being isentropic. Examples of these components include diffusers, nozzles, and compressor and turbine vanes. In addition, since there are no moving parts in nozzles and diffusers, the work transfer is zero. If the flow is fast, then the fluid particles do not have time for heat transfer interaction, such that the flow can be approximated as adiabatic. In reality, a small heat transfer occurs, and even if we neglect it for a particle, the cumulative effect over a period of time is significant. Even if we assume the flow in the exit nozzle is adiabatic, the exit nozzle will be too hot to touch after the engine has run for a few seconds.

Assuming the heat and work transfers are zero, the energy equation for a perfect gas is reduced to

$$c_p \, T_0 = c_p \, T + \frac{V^2}{2}. \tag{3.13}$$

This relationship is exact only for perfect gases. For semi-perfect gases, (3.13) is prone to errors for large values of velocity, unless appropriate averaged values of the heat capacity are used. The equation can be rearranged as

$$\frac{T_0}{T} = 1 + \frac{V^2}{2c_p T}$$

or

$$\frac{T_0}{T} = 1 + \frac{(\gamma - 1) \, V^2}{2\gamma \, R \, T},$$

and using the definition of the speed of sound and Mach number yields

$$\boxed{\frac{T_0}{T} = 1 + \frac{\gamma - 1}{2} M^2}. \tag{3.31}$$

Using (3.20), (3.23), and (3.31) yields

$$\frac{a_0}{a} = \left(1 + \frac{\gamma - 1}{2} M^2\right)^{\frac{1}{2}}. \tag{3.32}$$

For isentropic flow, using (3.14) and (3.15), (3.31) yields

$$\boxed{\frac{p_0}{p} = \left(1 + \frac{\gamma - 1}{2} M^2\right)^{\frac{\gamma}{\gamma - 1}}} \tag{3.33}$$

$$\boxed{\frac{\rho_0}{\rho} = \left(1 + \frac{\gamma - 1}{2} M^2\right)^{\frac{1}{\gamma - 1}}}. \tag{3.34}$$

Let us find approximations for (3.33) and (3.34) in the case when M is small. For this, let us use the formula for binomial expansion

$$(1 + z)^\alpha = \sum_{k=0}^{\infty} C_\alpha^k \, z^k = C_\alpha^0 + C_\alpha^1 z + C_\alpha^2 z^2 + \cdots \tag{3.35}$$

where α is any real number, z is any complex number with modulus smaller than 1, and the binomial coefficients C_α^k are the combinations $C_\alpha^k = \alpha!/(k! \, (\alpha - k)!)$. Note that by definition $C_\alpha^0 = 1$.

Applying the binomial expansion to (3.33) and neglecting the high-order terms M^4 and above for low Mach numbers $M \ll 1$ yields

$$\frac{p_0}{p} = 1 + \frac{\gamma}{\gamma - 1} \cdot \frac{\gamma - 1}{2} M^2 + \mathcal{O}(M^4) = 1 + \frac{\gamma}{2} M^2 + \mathcal{O}(M^4),$$

and for low Mach numbers $M \ll 1$ neglecting the high-order terms M^4 and above yields

$$p_0 \simeq p + \frac{\gamma}{2} p \, M^2 = p + \frac{\gamma}{2} p \frac{V^2}{\gamma RT} = p + \rho \frac{V^2}{2},$$

which is a particular case of the Bernoulli equation (2.21).

Assuming low Mach number flows and using the binomial expansion (3.35) in (3.34) yields

$$\frac{\rho_0}{\rho} = 1 + \frac{1}{\gamma - 1} \left(\frac{\gamma - 1}{2} M^2 \right) + \mathcal{O}(M^4)$$

such that $\rho_0 \simeq \rho$, which implies the density is constant, although the flow was not assumed incompressible.

3.4.1.1 Mass Flow Rate per Unit Area

In internal flows, one often faces the following related questions: (1) given a mass flow rate, what is the smallest nozzle area that allows the flow to go through, and (2) given a nozzle area, what is the largest mass flow rate that can flow through. To answer these questions, let us calculate the mass flow rate per unit area

$$\frac{\dot{m}}{A} = \rho V \tag{3.36}$$

through the nozzle shown in Fig. 3.6.

Let us write the density, ρ, as a function of stagnation density using (3.34), and the velocity as a function of the Mach number and stagnation temperature

$$V = M\, a = M\sqrt{\gamma RT} = M\sqrt{\gamma R \frac{T_0}{1 + \frac{\gamma - 1}{2} M^2}}. \tag{3.37}$$

Equation (3.36) becomes

$$\frac{\dot{m}}{A} = \frac{\rho_0}{\left(1 + \frac{\gamma - 1}{2} M^2\right)^{\frac{1}{\gamma - 1}}} M \frac{\sqrt{\gamma RT_0}}{\left(1 + \frac{\gamma - 1}{2} M^2\right)^{\frac{1}{2}}} = \frac{\rho_0 \sqrt{\gamma}}{\sqrt{RT_0}} M \left(1 + \frac{\gamma - 1}{2} M^2\right)^{-\frac{\gamma + 1}{2(\gamma - 1)}}. \tag{3.38}$$

Figure 3.6 Nozzle.

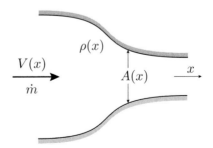

Figure 3.7 Mass flow rate parameter $\frac{\dot{m}\sqrt{RT_0}}{p_0 A}$ as a function of Mach number, for air and combustion products.

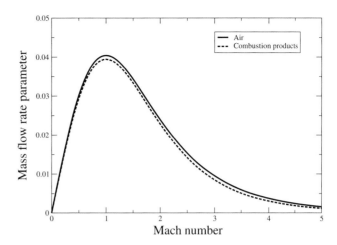

Equation (3.38) shows that for a given fluid defined by γ and R, and inlet state defined by p_0 and T_0, the mass flow rate per unit area is a function of the Mach number only. A dimensionless mass flow rate parameter can be obtained by grouping the terms in (3.38) as

$$
\frac{\dot{m}\sqrt{RT_0}}{p_0 A} = \sqrt{\gamma}\,\mathrm{M}\left(1 + \frac{\gamma-1}{2}\mathrm{M}^2\right)^{-\frac{\gamma+1}{2(\gamma-1)}}.
\tag{3.39}
$$

This dimensionless mass flow rate parameter $\dot{m}\sqrt{RT_0}/(p_0 A)$ or dimensional mass flow rate parameters such as $\dot{m}\sqrt{T_0}/(p_0 A)$ or $\dot{m}\sqrt{T_0}/p_0$ will be used in Chapter 7 to describe the performance of compressors and turbines.

The variation of the mass flow rate parameter $\dot{m}\sqrt{T_0}/(p_0 A)$ as a function of the Mach number is shown in Fig. 3.7 for air and combustion products. The combustion products have a ratio of specific heats of approximately 1.35 and a gas constant of 294 J/(kg K). Note that the maximum mass flow rate is higher for air than for combustion products. This means that for air and combustion products with the same stagnation temperature and pressure, a higher mass flow rate of air than combustion products can flow through a given area. Note also that for both air and combustion products, it appears that the maximum mass flow rate corresponds to Mach number $\mathrm{M} = 1$.

Example 3.4.1 Show that the maximum mass flow rate per unit area occurs when Mach number is $\mathrm{M} = 1$, that is, when the flow is critical.

Solution
The mass flow rate has an extreme value with respect to Mach number if the derivative $\partial(\dot{m}/A)/\partial\mathrm{M} = 0$. Let us calculate the value of Mach number for which $\partial(\dot{m}/A)/\partial\mathrm{M} = 0$.

$$\frac{\partial(\dot{m}/A)}{\partial M} = \frac{p_0\sqrt{\gamma}}{\sqrt{R\,T_0}}\left\{\left(1 + \frac{\gamma-1}{2}M^2\right)^{-\frac{\gamma+1}{2(\gamma-1)}} + \right.$$

$$\left. +M\left(-\frac{\gamma+1}{2(\gamma-1)}\right)\left[1 + \frac{\gamma-1}{2}M^2\right]^{-\frac{\gamma+1}{2(\gamma-1)}-1}\frac{\gamma-1}{2}2M\right\}$$

or

$$\frac{\partial(\dot{m}/A)}{\partial M} = \frac{p_0\sqrt{\gamma}}{\sqrt{R\,T_0}}\left(1 + \frac{\gamma-1}{2}M^2\right)^{-\frac{\gamma+1}{2(\gamma-1)}-1}\left(1 - M^2\right) = 0$$

such that indeed the maximum mass flow rate occurs at $M = 1$.

___

Since the maximum mass flow rate occurs when the flow is critical, that is, $M = 1$, this mass flow rate is called the *critical mass flow rate*, $\dot{m}_{cr}$. The area where the mass flow rate reaches its critical value is called the *critical area*, A_{cr}. From (3.39), the maximum mass flow rate per unit area is:

$$\left(\frac{\dot{m}}{A}\right)_{max} = \frac{\dot{m}_{cr}}{A_{cr}} = \frac{p_0\sqrt{\gamma}}{\sqrt{RT_0}}\left(1 + \frac{\gamma-1}{2}\right)^{-\frac{\gamma+1}{2(\gamma-1)}} = \frac{p_0\sqrt{\gamma}}{\sqrt{RT_0}}\left(\frac{2}{\gamma+1}\right)^{\frac{\gamma+1}{2(\gamma-1)}}, \qquad (3.40)$$

which also yields

$$\rho_{cr}a_{cr} = \frac{p_0\sqrt{\gamma}}{\sqrt{RT_0}}\left(\frac{2}{\gamma+1}\right)^{\frac{\gamma+1}{2(\gamma-1)}}. \qquad (3.41)$$

We can now answer the question posed at the beginning of section 3.4.1.1: for a given area A, the maximum mass flow rate that can flow through that area is the critical mass flow rate. For a given mass flow rate, $\dot{m}$, the smallest area that allows the flow of the mass flow rate $\dot{m}$ is A_{cr}.

___

Example 3.4.2 The gas at exit from the jet engine nozzle has a total temperature $T_0 = 900$ K and total pressure $p_0 = 1.86$ bar. The minimum diameter of the jet engine exit nozzle is $d = 0.17$ m. The ratio of specific heats is $\gamma = 1.35$ and the gas constant is $R = 294$ J/(kg K). What is the maximum mass flow rate of the engine for the given parameters?

Solution
The critical area is $A_{cr} = \pi d^2/4 = 2.27 \times 10^{-2}$ m^2. Using (3.40), the maximum mass flow rate is

$$\dot{m} = A_{cr}\frac{p_0\sqrt{\gamma}}{\sqrt{RT_0}}\left(\frac{2}{\gamma+1}\right)^{\frac{\gamma+1}{2(\gamma-1)}} = 2.27 \times 10^{-2}\frac{1.86 \times 10^5\sqrt{1.35}}{\sqrt{294 \times 900}}\left(\frac{2}{2.35}\right)^{3.357} = 5.55 \text{ kg/s}.$$

___

Equation (3.40) can be rewritten as

$$\frac{\dot{m}}{A_{\text{cr}}}\frac{\sqrt{T_0}}{p_0} = \frac{\sqrt{\gamma}}{\sqrt{R}}\left(\frac{\gamma+1}{2}\right)^{-\frac{\gamma+1}{2(\gamma-1)}}. \tag{3.42}$$

For air, which has $\gamma = 1.4$ and $R = 287.16$ J/(kg K), (3.42) becomes

$$\frac{\dot{m}}{A_{\text{cr}}}\frac{\sqrt{T_0}}{p_0} = 0.040407. \tag{3.43}$$

This is Fliegner's formula, which was experimentally determined in the nineteen century, before the theoretical arguments presented above were widely understood. Fliegner conducted his experiments on a converging nozzle, and his error in determining the constant of (3.43) was approximately 1%.

3.4.1.2 Area Ratio

The ratio between the area A and the critical area A_{cr} can be obtained from (3.38) and (3.40):

$$\frac{A}{A_{\text{cr}}} = \frac{1}{M}\left[\frac{2}{\gamma+1}\left(1+\frac{\gamma-1}{2}M^2\right)\right]^{\frac{\gamma+1}{2(\gamma-1)}}. \tag{3.44}$$

Equation (3.44) shows that the area ratio is a function of Mach number and gas properties. Note also that always $A/A_{\text{cr}} \geq 1$, as shown in Fig. 3.8.

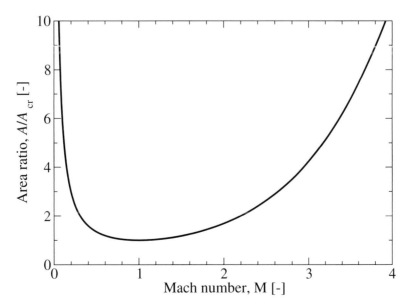

Figure 3.8 Area ratio A/A_{cr} as a function of Mach number for $\gamma = 1.4$.

Because of mass conservation, (3.44) yields

$$\frac{A}{A_{cr}} = \frac{1}{M}\left[\frac{2}{\gamma+1}\left(1 + \frac{\gamma-1}{2}M^2\right)\right]^{\frac{\gamma+1}{2(\gamma-1)}} = \frac{\rho_{cr}a_{cr}}{\rho V}. \tag{3.45}$$

Example 3.4.3 Derive the expression of the ratio between the area A and the critical area A_{cr} as a function of the pressure ratio p/p_0.

Solution
Substitute M by p_0/p using (3.33) so that

$$\frac{A}{A_{cr}} = \sqrt{\frac{\gamma-1}{2}} \cdot \frac{\left(\frac{2}{\gamma+1}\right)^{\frac{\gamma+1}{2(\gamma-1)}}}{\left(\frac{p}{p_0}\right)^{\frac{1}{\gamma}}\sqrt{1 - \left(\frac{p}{p_0}\right)^{\frac{\gamma-1}{\gamma}}}}.$$

It is easy to use (3.44) to find the ratio A/A_{cr} for a given Mach number. Finding the Mach number at a given A/A_{cr}, however, requires the solution of the nonlinear equation (3.44) in M. For a given A/A_{cr}, there are always two values of M.

Example 3.4.4 The critical area of a jet engine exit nozzle is 2.27×10^{-2} m^2. Assuming the flow in the exit nozzle has reached its maximum mass flow rate value, what is the Mach number at a section with area 3×10^{-2} m^2, if the ratio of specific heats is $\gamma = 1.35$?

Solution
After substituting the values of the areas and ratio of specific heats in (3.44) squared and using the notation $x = M^2$, one obtains

$$x - 0.1939(1 + 0.175x)^{6.714} = 0.$$

The solution of this nonlinear equation can be obtained using the Newton–Raphson method [Chapra and Canale, 1998, p. 148]. The iterative process for computing the solution is

$$x_{i+1} = x_i - \frac{x_i - 0.1939(1 + 0.175\, x_i)^{6.714}}{1 - 0.2278(1 + 0.175\, x_i)^{5.714}}$$

where x_i is the approximation of the solution at step i, and x_{i+1} is the approximation of the solution at step $i + 1$. For an error $|x_i - x_{i+1}| \leq 10^{-5}$, the solution $x = 0.2619$ is obtained in less than five iterations. As a result, the Mach number is $M = \sqrt{x} = 0.5118$.

Example 3.4.5 The axial Mach number at the end of the engine inlet diffuser, just in front of the compressor, is $M_2 = 0.5$. The airplane Mach number M_0 varies between zero, at takeoff,

Table 3.2 Ratio between areas A_0 and A_2 as a function of airplane Mach number M_0 for an inlet Mach number $M_2 = 0.5$.

M_0	0	0.1	0.2	0.3	0.4	0.6	0.8
A_0/A_2	∞	4.345	2.212	1.519	1.187	0.887	0.775

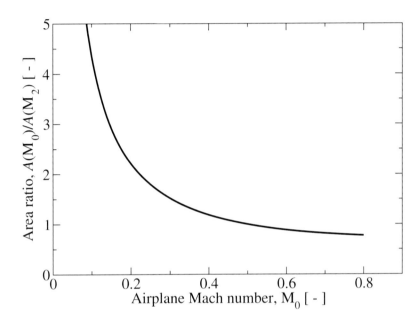

Figure 3.9 Area ratio A_0/A_2 as a function of airplane Mach number M_0 for an inlet Mach number $M_2 = 0.5$.

and 0.8, in cruise. How should the area A_0 in front of the inlet diffuser vary so that it captures all the streamlines getting into the engine?

Solution
Let us use the notation $A(M_0) = A_0$ and $A(M_2) = A_2$. Using (3.44) for M_0 and M_2 yields

$$A_0 = A_{cr}\frac{1}{M_0}\left[\frac{2}{\gamma+1}\left(1+\frac{\gamma-1}{2}M_0^2\right)\right]^{\frac{\gamma+1}{2(\gamma-1)}}$$

and

$$A_2 = A_{cr}\frac{1}{0.5}\left[\frac{2}{\gamma+1}\left(1+\frac{\gamma-1}{2}0.5^2\right)\right]^{\frac{\gamma+1}{2(\gamma-1)}} = 1.3398A_{cr}$$

for $\gamma = 1.4$. The variation of A_0/A_2 is shown in Table 3.2 and Fig. 3.9.

Figure 3.10 Schematic of a control volume.

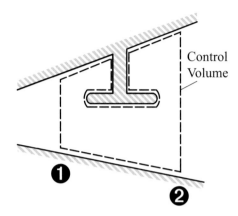

3.4.1.3 The Impulse Function

The impulse function is a quantity that is convenient to use for some jet propulsion problems. To introduce the definition of the impulse function, let us determine the net thrust produced by the flow between sections 1 and 2, shown in Fig. 3.10.

Using the momentum conservation equation for steady flow (3.8) and neglecting volume forces yields

$$-\rho_1 A_1 V_1^2 + \rho_2 A_2 V_2^2 = p_1 A_1 - p_2 A_2 + \mathcal{T} \tag{3.46}$$

where $\mathcal{T}$ is the net thrust produced by the flow. The thrust $\mathcal{T}$ is the net force due to the flow acting on the internal solid surfaces wetted by the fluid. The thrust includes the pressure forces and the viscous forces on the duct walls, together with the total drag of any obstacles in the stream. Rearranging (3.46) as

$$\mathcal{T} = p_2 A_2 + \rho_2 A_2 V_2^2 - p_1 A_1 - \rho_1 A_1 V_1^2$$

suggests that it is convenient to define the impulse function as

$$F := pA + \rho A V^2 \tag{3.47}$$

such that

$$\mathcal{T} = F_2 - F_1. \tag{3.48}$$

The thrust formula is valid for adiabatic and non-adiabatic processes as well as for reversible or irreversible processes.

For a perfect gas, the impulse function becomes

$$F = pA + \frac{p}{RT} A V^2 = pA(1 + \gamma M^2).$$

The impulse function can be nondimensionalized as either

$$\frac{F}{p_0 A_{cr}} = \frac{p}{p_0} \cdot \frac{A}{A_{cr}} (1 + \gamma M^2)$$

or

$$\frac{F}{F_{cr}} = \frac{p}{p_{cr}} \cdot \frac{A}{A_{cr}} \cdot \frac{1 + \gamma M^2}{1 + \gamma}. \qquad (3.49)$$

Since

$$\frac{p}{p_{cr}} = \frac{p}{p_0} \frac{p_0}{p_{cr}}$$

and for M = 1, (3.33) yields

$$\frac{p_0}{p_{cr}} = \left(\frac{\gamma + 1}{2}\right)^{\frac{\gamma}{\gamma - 1}}, \qquad (3.50)$$

using (3.33), (3.44), and (3.50), the dimensionless impulse function (3.49) becomes

$$\frac{F}{F_{cr}} = \frac{1 + \gamma M^2}{M\sqrt{2(\gamma + 1)(1 + \frac{\gamma - 1}{2} M^2)}}.$$

3.4.2   Nonisentropic Flow

Variation of stagnation pressure in a propulsion system provides important information on the efficiency and losses of a system. In this section, we want to determine the relationship between entropy increase and the variation of the stagnation pressure p_0 for nonisentropic steady flows. Let us start with the ratio between stagnation and static temperature

$$\frac{T_0}{T} = 1 + \frac{\gamma - 1}{2} M^2, \qquad (3.31)$$

and let us define z as

$$z := 1 + \frac{\gamma - 1}{2} M^2$$

such that (3.31) becomes

$$T_0 = T z. \qquad (3.51)$$

By taking the logarithm and then the derivative, (3.51) becomes

$$\frac{dT}{T} = \frac{dT_0}{T_0} - \frac{dz}{z}. \qquad (3.52)$$

The ratio of the stagnation and static pressure can also be written as a function of z,

$$\frac{p_0}{p} = z^{\frac{\gamma}{\gamma - 1}},$$

such that

$$p_0 = p \, z^{\frac{\gamma}{\gamma - 1}}.$$

Taking the logarithm and then the derivative yields

$$\frac{dp_0}{p_0} = \frac{dp}{p} + \frac{\gamma}{\gamma - 1}\frac{dz}{z}. \tag{3.53}$$

Using the first law of thermodynamics in the particular case when the energy e is equal to the internal energy u,

$$\delta q = du + \delta w,$$

and assuming all the work is due to normal forces, yields

$$T ds = du + p dv.$$

Substituting internal energy by the expression of enthalpy, $u = h - pv$, yields

$$T ds = dh - v dp.$$

Let us divide the equation by RT such that we get

$$\frac{ds}{R} = \frac{c_p \, dT}{RT} - \frac{v dp}{RT} = \frac{\gamma}{\gamma - 1} \cdot \frac{dT}{T} - \frac{dp}{p} \overset{(3.52,3.53)}{=}$$
$$= \frac{\gamma}{\gamma - 1}\left(\frac{dT_0}{T_0} - \frac{dz}{z}\right) - \left(\frac{dp_0}{p_0} - \frac{\gamma}{\gamma - 1} \cdot \frac{dz}{z}\right)$$

which simplifies to

$$\frac{ds}{R} = \frac{\gamma}{\gamma - 1} \cdot \frac{dT_0}{T_0} - \frac{dp_0}{p_0}. \tag{3.54}$$

If the process is adiabatic and the work transfer is zero, then from (3.10) it follows that $dh_0 = 0$ and therefore $dT_0 = 0$. Consequently, (3.54) yields

$$\frac{dp_0}{p_0} = -\frac{ds}{R},$$

and since the entropy variation is nonnegative, the stagnation pressure will decrease or, if the process is isentropic, be constant.

If the process is not adiabatic and/or the work transfer is nonzero, the variation of the stagnation pressure depends on the sign of the term $\frac{\gamma}{\gamma-1} \cdot \frac{dT_0}{T_0} - \frac{ds}{R}$. If the variation of the stagnation temperature is large enough, this term can become positive and therefore lead to an increase of stagnation pressure.

3.4.3 Flow through Nozzles

Many components of a propulsion system are similar to nozzles. Therefore, understanding the flow through nozzles is beneficial to comprehending propulsion systems. The flow through nozzles can be modeled as a steady, isentropic, one-dimensional flow. To determine the relationship between the variation of the nozzle area and the velocity, let us use the mass conservation equation (3.5), the energy conservation equation (3.11), and the relation between the pressure and density for isentropic processes (2.88). The work done by the normal forces on the nozzle is zero since the nozzle walls are assumed to be fixed. In addition, to a first approximation one can neglect the work done by the viscous forces. Consequently, by substituting the enthalpy by $\gamma p/[(\gamma - 1)\rho]$, the energy equation (3.11) becomes

$$\frac{\gamma}{\gamma - 1} d\left(\frac{p}{\rho}\right) + V\, dV = 0.$$

Substituting the pressure from (2.88) yields

$$\gamma \frac{p}{\rho^2}\, d\rho + V\, dV = 0,$$

which can be rewritten as

$$a^2 \frac{d\rho}{\rho} + V dV = 0. \tag{3.55}$$

Combining (3.5) and (3.55) yields

$$-\frac{V}{a^2} dV + \frac{dV}{V} + \frac{d\sigma}{\sigma} = 0,$$

which can be rewritten as

$$\boxed{(1 - M^2)\frac{dV}{V} + \frac{d\sigma}{\sigma} = 0}. \tag{3.56}$$

The above equation relates velocity variation to nozzle area variation. The sign of the term $1 - M^2$ changes depending whether the flow is subsonic or supersonic. The sign of the $1 - M^2$ term is crucial for how the variation of the nozzle area affects the velocity variation. If the flow is subsonic, the area increase leads to a decrease of velocity.

If in (3.56) the area term $d\sigma/\sigma$ is substituted using (3.5), this yields

$$M^2 \frac{dV}{V} + \frac{d\rho}{\rho} = 0, \tag{3.57}$$

which shows that the variation of density is opposite to that of velocity irrespective of the Mach number. If in (3.56) the velocity term dV/V is substituted using (3.5), this yields

$$M^2 \frac{d\sigma}{\sigma} = (1 - M^2)\frac{d\rho}{\rho}. \tag{3.58}$$

Similar relations linking the variation of area to the variation of pressure and temperature can be obtained by using the isentropic equations (2.90) and (2.91).

Problem 3.4.1   Using (3.58), (2.90), and (2.91), show that the pressure and temperature variations depend on the area variation as follows:

$$M^2 \frac{d\sigma}{\sigma} = (1 - M^2)\frac{1}{\gamma}\frac{dp}{p}$$

$$M^2 \frac{d\sigma}{\sigma} = (1 - M^2)\frac{1}{\gamma - 1}\frac{dT}{T}.$$

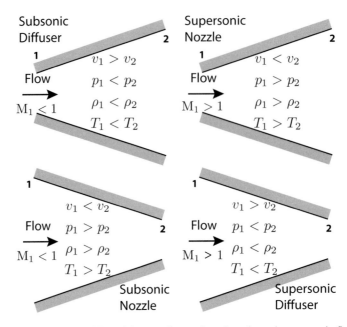

Figure 3.11 Variation of flow variables with area change in subsonic and supersonic flow.

Consequently, if the flow is subsonic, the area increase leads to an increase of pressure and temperature, as shown in Fig. 3.11. This figure also includes the other three possible cases: converging nozzle, subsonic and supersonic flow; and diverging nozzle, supersonic flow.

Recall that σ was introduced in (2.2) as the area of a closed surface, the boundary of volume τ. Since the area in (3.56) is not necessarily the area of a closed surface, we will use the notation A instead of σ. As a result, (3.56) becomes

$$(1 - M^2)\frac{dV}{V} + \frac{dA}{A} = 0. \tag{3.59}$$

Example 3.4.6 Find the expression of the mass flow rate in the nozzle shown in Fig. 3.12 as a function of the pressure ratio p/p_0, nozzle area A, stagnation pressure and stagnation density, and ratio of specific heat capacities.

Hint: Start with mass flow rate written as $\dot{m} = \rho V A$ and replace V and ρ using Equations (3.22) and (2.88). The solution is the Saint-Venant equation:

$$\dot{m} = A\sqrt{\frac{2\gamma}{\gamma - 1}p_0\rho_0\left(\frac{p}{p_0}\right)^{\frac{2}{\gamma}}\left[1 - \left(\frac{p}{p_0}\right)^{\frac{\gamma-1}{\gamma}}\right]}. \tag{3.60}$$

This equation will be used in Chapter 10 to derive the maximum thrust of a rocket engine.

Figure 3.12 Convergent nozzle.

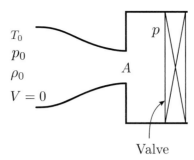

3.4.3.1 Converging Nozzle

Let us consider that a large tank is connected to the nozzle shown in Fig. 3.12. The flow out of the tank and through the nozzle is controlled by a valve. The gas in the tank has the pressure p_0, density ρ_0, and temperature T_0. The pressure outside of the tank is lower than the pressure p_0. If the valve is closed, $p = p_0$ and the mass flow rate is zero, $\dot{m} = 0$. If the valve opens, the pressure p decreases and the mass flow rate $\dot{m}$ increases, as specified by (3.60). The maximum mass flow rate is reached when $\partial\dot{m}/\partial(p/p_0) = 0$, and the value of p/p_0 corresponding to the maximum mass flow rate is

$$\left(\frac{p}{p_0}\right)_{\max \dot{m}} = \left(\frac{2}{\gamma+1}\right)^{\frac{\gamma}{\gamma-1}}. \tag{3.61}$$

There should be no surprise to notice that the pressure ratio (3.61) is the same as the value given by (3.33) for $M = 1$. Consequently, the maximum mass flow rate occurs at critical conditions, that is, when $M = 1$, and for this reason it is called the critical mass flow rate. Therefore,

$$\left(\frac{p}{p_0}\right)_{\max \dot{m}} = \frac{p_{\mathrm{cr}}}{p_0} = \left(\frac{2}{\gamma+1}\right)^{\frac{\gamma}{\gamma-1}}.$$

The isentropic relationship between pressure and temperature ratios yields the critical temperature, T_{cr}, which occurs at maximum mass flow rate

$$\frac{T_{\mathrm{cr}}}{T_0} = \frac{2}{\gamma+1}.$$

The variation of the mass flow rate as a function of p/p_0, obtained by experiment, is shown in Fig. 3.13. The mass flow rate predicted by (3.60) is in agreement with the experimental results for pressure values larger than the critical pressure. For pressure values smaller than the critical pressure, the experiment shows that the mass flow rate is constant and equal to the critical value, while (3.60) predicts that the mass flow rate will decrease, as indicated by the dashed line in Fig. 3.13. Certainly, the experimental results are correct. Once the flow reaches the critical state, the velocity of the fluid equals the speed of sound. As a result, the opening of the valve past the critical section is no longer known upstream, at the nozzle, since perturbations that travel with the speed of sound cannot reach the nozzle. For this reason, the mass flow rate remains constant while the pressure decreases below the critical value.

Figure 3.13 Variation of mass
flow rate vs. pressure in a nozzle.

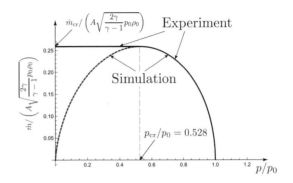

3.4.3.2 Converging-Diverging Nozzle

The converging-diverging nozzle, also called the de Laval nozzle after Karl Gustaf Patrik de Laval (1845–1913), a Swedish engineer, is used to accelerate the flow beyond the critical velocity. The convergent-divergent nozzle is used on most modern rocket engines, as well as on some types of steam turbines and supersonic jet engines.

Four different flow regimes are possible in the converging-diverging nozzle, depending on the ratio between the exit static pressure and the inlet stagnation pressure. These four regimes, A through D, and the flow conditions, 0 through 7, are described using the nozzle shown in Fig. 3.14 where the stagnation pressure, p_0, and stagnation temperature, T_0, are kept constant while the static back pressure, p_b, is varied.

The mass flow rate through the nozzle is zero as long as the static back pressure is equal to the stagnation pressure at inlet, as shown by condition 0 in Fig. 3.14. As the static back pressure is reduced, the velocity and mass flow rate increase, as shown by condition 1. Lowering the static back pressure further, the Mach number becomes unity at the nozzle throat, as shown at condition 2. Until reaching critical conditions at the throat, the flow in regime A, defined between conditions 0 and 2, is entirely subsonic. In this regime, the mass flow rate keeps increasing as the static back pressure is decreased.

As the static back pressure is reduced below that of condition 2, a discontinuity (a normal shock, which will be introduced in the next section) appears downstream of the throat. The flow downstream of this shock is subsonic, as shown at condition 3. As the static back pressure is lowered, the shock moves down the nozzle, until it reaches the exit of the nozzle, at condition 4. During regime B, defined between conditions 2 and 4, the mass flow rate is constant and is not affected by the static back pressure variation, as shown in Section 3.4.3.1. In regime B, as in regime A, the static back pressure, p_b is the same as the static pressure at the exit plane, p_e.

The flow in the entire nozzle is supersonic for condition 5, but the pressure at the exit plane, p_e, is lower than the back pressure, p_b. As a result of the pressure difference, compression through oblique shocks occurs outside of the nozzle. This is valid for the entire regime C, defined between conditions 4 and 6. Condition 6 is called the design condition for the nozzle under supersonic conditions because the back pressure is equal to the exit-plane pressure.

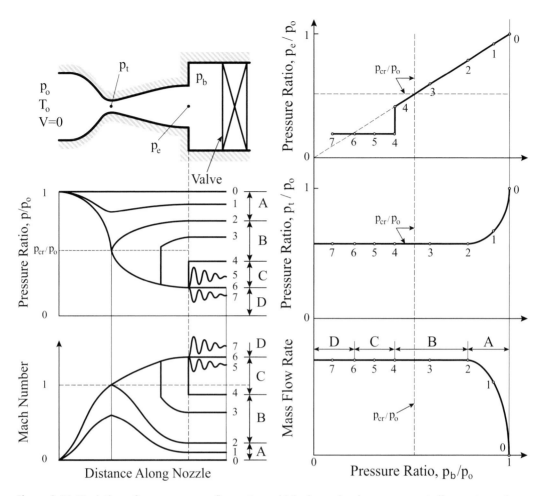

Figure 3.14 Variation of pressure, mass flow rate, and Mach number in a convergent-divergent nozzle.

If the back pressure is reduced further, as shown for condition 7 of regime D, the flow pattern within the nozzle is not affected. The pressure difference between p_e and p_b results in oblique expansion waves that occur outside of the nozzle.

The mass flow rate depends on the ratio between the static back pressure and the stagnation inlet pressure only during regime A when the flow in the nozzle is subsonic. Furthermore, for regime A, there is an infinite number of pressure variation curves. For regimes B, C, and D, the mass flow rate is independent of the back pressure, and the pressure variation vs. axial location curve is unique.

3.5 Shock Waves

Discontinuities in real fluids are smoothed out by viscous, heat conduction, and mass diffusion effects and therefore cannot exist for finite periods of time. Experimental investigations,

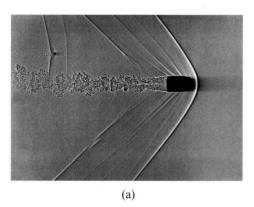
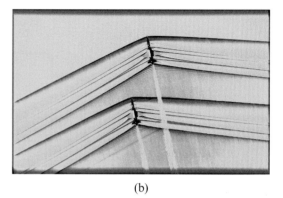

(a)   (b)

Figure 3.15 Schlieren photographs (courtesy to NASA). (a) Bow shock waves around a bullet flying at M = 1.5 and (b) shock waves interacting between two aircraft.

such as Schlieren photographs in a supersonic wind tunnel or of a supersonic projectile shown in Fig. 3.15, reveal that normal shock waves and oblique shock waves are extremely thin. Consequently, changes of pressure, velocity, temperature, and density occur across a distance that is approximately 2,000 Ångströms.[3] Since for engineering applications the primary concern is the net change of properties across the shock, the viscous and heat conduction phenomena inside the shock wave can be disregarded. As a result, by ignoring the details of the interior of the shock wave, a simplified model of a pure discontinuity can be employed.

The shock wave is a discontinuity in a partly supersonic flow. When the fluid goes through a shock wave, the pressure, density, and temperature increase, the velocity decreases, and the direction is modified, except for the case of normal shocks. The process is irreversible.

Many components of propulsion systems operate in transonic or supersonic flow where shock waves occur. For this reason, it is important to understand how shock waves form and affect the flow, as well as which types of shock waves are preferable. The following sections will discuss normal and oblique shock waves.

3.5.1   Normal Shocks

Shock waves are often oblique to the flow direction but to start with, let us consider only one-dimensional motion, that is, only discontinuities normal to the direction of the flow. This section will derive the governing equations for normal shock waves and will apply them to predict the flow for several components of propulsion systems.

3.5.1.1 Governing Equations

It is typical when deriving the governing equations to rely on the mass, momentum, and energy conservation equations. For shock waves, however, the second law of thermodynamics is also needed to determine the direction of the process. To derive the mass, momentum, and

---

[3] 1 Ångström = 10^{-10}m

Figure 3.16 Control volume for deriving mass, momentum, and energy conservation equations for a normal shock (note: dimension of dx is 2000 Ångströms).

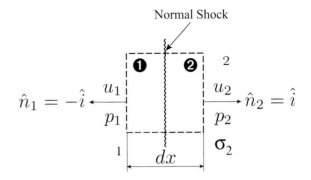

energy conservation equations, let us consider the control volume shown in Fig. 3.16. The shock wave is indicated by the wavy line. Note that dx, the width of the control volume, is of the order of 2,000 Ångströms.

Mass Conservation Equation

The starting point for deriving the mass conservation equation is (2.7), where the unsteady term is zero. Let us consider the control volume shown in Fig. 3.16, where the thickness dx is slightly larger than that of the shock wave. Since the thickness of the shock wave discontinuity is 2,000 Ångströms, and therefore the dimension dx is extremely small, the difference between the left and right surfaces σ_1 and σ_2 can be neglected, such that $\sigma_1 = \sigma_2$. The fluxes through the upper and lower surfaces of the control volume are zero because the velocities and the normals to these surfaces are orthogonal. In addition, since the flow is assumed to be one-dimensional, (2.7) reduces to

$$\rho_1 u_1 = \rho_2 u_2. \tag{3.62}$$

Momentum Conservation Equation

To derive the expression of the linear momentum conservation equation for normal shocks, let us start with (2.31) applied to the control volume shown in Fig. 3.16. Since the flow is one-dimensional, (2.31) reduces to

$$\rho_1 u_1 \hat{i}(-u_1)\sigma_1 + \rho_2 u_2 \hat{i}(u_2)\sigma_2 = -[p_1(-\hat{i})\sigma_1 + p_2 \hat{i}\sigma_2]$$

where $\hat{i}$ is the unit vector in the x-direction. Using $\sigma_1 = \sigma_2$ yields

$$\rho_2 u_2^2 - \rho_1 u_1^2 = p_1 - p_2,$$

which, using mass conservation (3.62) can be rearranged as

$$p_2 - p_1 = \rho_1 u_1 (u_1 - u_2). \tag{3.63}$$

The momentum conservation equation shows that if $u_1 > u_2$ then $p_2 > p_1$.

Energy Conservation Equation

The starting point for deriving the energy conservation equation is (2.63), which reduces to

$$
\int_{\sigma \equiv \partial \tau} \rho \left(h + \frac{V^2}{2} \right) \vec{V} \cdot \hat{n} \, d\sigma = 0 \tag{3.64}
$$

by assuming the heat transfer rate, the shear and shaft power, and the unsteady term are all zero. In addition, the potential energy is neglected compared to the kinetic energy. For the control volume shown in Fig. 3.16, (3.64) becomes

$$
-\rho_1 (h_1 + u_1^2/2) u_1 \sigma_1 + \rho_2 (h_2 + u_2^2/2) u_2 \sigma_2 = 0.
$$

Using the mass conservation equation (3.62) yields the energy conservation for normal shocks,

$$
h_1 + u_1^2/2 = h_2 + u_2^2/2. \tag{3.65}
$$

Using the definition of stagnation enthalpy (3.12), an alternate form of the energy conservation equation for normal shocks is

$$
h_{01} = h_{02}, \tag{3.66}
$$

which for a perfect gas yields

$$
T_{01} = T_{02}. \tag{3.67}
$$

Using the definition of enthalpy variation for a perfect gas (2.76) in (3.65) yields

$$
c_p T_1 + \frac{u_1^2}{2} = c_p T_2 + \frac{u_2^2}{2} = c_p T_0. \tag{3.68}
$$

Using the perfect gas relations

$$
c_p T = \frac{\gamma}{\gamma - 1} RT = \frac{\gamma}{\gamma - 1} \frac{p}{\rho} \tag{3.69}
$$

in (3.68) yields

$$
\frac{\gamma}{\gamma - 1} \left(\frac{p_2}{\rho_2} - \frac{p_1}{\rho_1} \right) = (u_1^2 - u_2^2)/2. \tag{3.70}
$$

Using the definition of critical speed (3.25) and the perfect gas relations (3.69) in (3.68) yields

$$
\frac{\gamma}{\gamma - 1} \frac{p_1}{\rho_1} + \frac{u_1^2}{2} = \frac{\gamma}{\gamma - 1} \frac{p_2}{\rho_2} + \frac{u_2^2}{2} = \frac{\gamma + 1}{2(\gamma - 1)} a_{cr}^2. \tag{3.71}
$$

Second Law of Thermodynamics

The mass, momentum, and energy conservation equations presented above place no restriction on the direction of the process, that is, whether it evolves from state 1 to 2 or 2 to 1. The second law of thermodynamics 2.4.3 describes the direction of natural processes. To determine the direction of the process across the normal shock, let us introduce two new concepts: the Fanno line and the Rayleigh line.

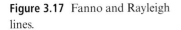

Figure 3.17 Fanno and Rayleigh lines.

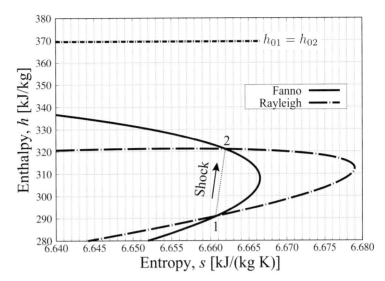

The Fanno Line – Let us consider fixed the conditions at state 1 ahead of the normal shock and determine the locus of states that evolve from 1 and satisfy the mass and energy conservation equations and the equation of state. The equation of state introduced in 2.4.8.1 would be written implicitly in the form

$$h = h(s, \rho)$$

$$s = s(p, \rho).$$

The Fanno line, shown in Fig. 3.17, is defined as the locus of states on the $h - s$ diagram that pass through state 1 and satisfy the equation of state and the mass and energy conservation equations. The Fanno line represents the states with the same flow per unit area and the same stagnation enthalpy, but not necessarily with the same impulse function, because the momentum conservation equation is not enforced.

The Fanno line that goes through a specified state 1 is determined by an iterative process. To find the location of another point on the Fanno line, let us choose a particular value of the velocity u_2, calculate the density ρ_2 from (3.62), calculate the enthalpy h_2 from (3.65) and then the temperature T_2, calculate the pressure p_2 from the equation of state, and then calculate the entropy s_2 from the entropy variation equation (2.93). To determine other points on the Fanno line, this process is repeated by choosing different velocities u_2.

The Rayleigh Line – The Rayleigh line, shown in Fig. 3.17, is the locus of states that pass through a given point 1 and are defined by the mass and momentum conservation equations and the equation of state. To compute this locus, let us choose a particular value of the velocity u_2, calculate the density ρ_2 from (3.62), calculate the pressure p_2 from (3.63), and then calculate the entropy s_2 from the entropy variation equation (2.93). The Rayleigh line of states reachable from state 1 is obtained by repeating these calculations for other u_2 velocity values. Since the states on the Rayleigh line have different stagnation enthalpies, they can be reached by continuous changes through heat transfer.

The Fanno and Rayleigh lines were introduced not because the shock process occurs along either line, but because the two lines help determine the direction of the process across the shock. Since the normal shock must satisfy mass, momentum, and energy conservation equations and the equation of state, the initial and final states across the shock must be located at the intersection between the Fanno and Rayleigh lines. Before we can determine the direction of the shock wave, the variation of entropy as a function of the initial Mach number must be derived. This will be done in the following section.

3.5.1.2 Working Formulas for Normal Shocks

A first useful relationship would be to predict the Mach number after the shock, knowing the Mach number ahead of the shock. To obtain this relationship, let us first introduce three more or less obvious equations by using $M = V/\sqrt{\gamma RT}$, and $p/\rho = RT$, in (3.62), (3.63), and (3.67):

$$\frac{p_2}{p_1} = \frac{M_1}{M_2}\sqrt{\frac{T_2}{T_1}} \tag{3.72}$$

$$\frac{p_2}{p_1} = \frac{1 + \gamma M_1^2}{1 + \gamma M_2^2} \tag{3.73}$$

$$\sqrt{\frac{T_2}{T_1}} = \frac{\sqrt{1 + \frac{\gamma - 1}{2}M_1^2}}{\sqrt{1 + \frac{\gamma - 1}{2}M_2^2}}. \tag{3.74}$$

Eliminating p_2/p_1 and T_2/T_1 from equations (3.72–3.74) yields

$$\frac{M_2\sqrt{1 + \frac{\gamma - 1}{2}M_2^2}}{1 + \gamma M_2^2} = \frac{M_1\sqrt{1 + \frac{\gamma - 1}{2}M_1^2}}{1 + \gamma M_1^2}$$

or

$$\boxed{\left(\frac{\gamma - 1}{2} + \frac{1}{M_1^2}\right)\left(\frac{\gamma - 1}{2} + \frac{1}{M_2^2}\right) = \left(\frac{\gamma + 1}{2}\right)^2} \tag{3.75}$$

so that

$$M_2^2 = \frac{M_1^2 + \frac{2}{\gamma - 1}}{\frac{2\gamma}{\gamma - 1}M_1^2 - 1}. \tag{3.76}$$

Substituting M_2^2 of (3.76) into (3.73) yields

$$\boxed{\frac{p_2}{p_1} = \frac{2\gamma}{\gamma+1} M_1^2 - \frac{\gamma-1}{\gamma+1}}. \tag{3.77}$$

Using (3.77) and (3.33) yields

$$\boxed{\frac{p_{02}}{p_{01}} = \frac{\left[\dfrac{\frac{\gamma+1}{2}M_1^2}{\left(1+\frac{\gamma-1}{2}M_1^2\right)}\right]^{\frac{\gamma}{\gamma-1}}}{\left(\dfrac{2\gamma}{\gamma+1}M_1^2 - \dfrac{\gamma-1}{\gamma+1}\right)^{\frac{1}{\gamma-1}}}}. \tag{3.78}$$

Substituting M_2^2 of (3.76) into (3.74) yields

$$\boxed{\frac{T_2}{T_1} = \frac{\left(1+\dfrac{\gamma-1}{2}M_1^2\right)\left(\dfrac{2\gamma}{\gamma-1}M_1^2 - 1\right)}{\dfrac{(\gamma+1)^2}{2(\gamma-1)}M_1^2}}. \tag{3.79}$$

3.5.1.3 Rankine–Hugoniot Equations

Rankine–Hugoniot equations relate the pressure ratio and the density ratio across the shock. To derive the Rankine–Hugoniot equations, the velocity terms from the momentum and energy conservation equations will be eliminated. Substituting the mass conservation equation (3.62) into the momentum conservation equation (3.63) yields

$$p_2 - p_1 \stackrel{(3.63)}{=} \rho_1 u_1^2 \left(1 - \frac{\rho_2}{\rho_1}\frac{u_2^2}{u_1^2}\right) \stackrel{(3.62)}{=} \rho_1 u_1^2 \left(1 - \frac{\rho_1}{\rho_2}\right)$$

such that

$$u_1^2 = \frac{p_2 - p_1}{\rho_2 - \rho_1}\frac{\rho_2}{\rho_1} = \frac{p_1}{\rho_1}\left(\frac{p_2/p_1 - 1}{\rho_2/\rho_1 - 1}\right)\frac{\rho_2}{\rho_1}. \tag{3.80}$$

In a similar way, factoring out the $\rho_2 u_2^2$ term in the momentum conservation equation (3.63) yields

$$u_2^2 = \frac{p_2 - p_1}{\rho_2 - \rho_1}\frac{\rho_1}{\rho_2} = \frac{p_1}{\rho_1}\left(\frac{p_2/p_1 - 1}{\rho_2/\rho_1 - 1}\right)\frac{1}{\rho_2/\rho_1}. \tag{3.81}$$

With the energy conservation equation (3.70) written as

$$\frac{\gamma}{\gamma-1}\frac{p_1}{\rho_1}\left(\frac{p_2}{p_1}\frac{\rho_1}{\rho_2} - 1\right) = (u_1^2 - u_2^2)/2 \tag{3.70}$$

Figure 3.18 Rankine–Hugoniot equation (3.82) vs. isentropic relation (2.88).

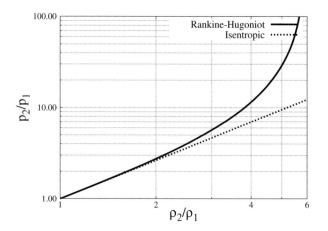

and the squared velocities substituted by (3.80) and (3.81), the pressure ratio is obtained as a function of the density ratio only, and vice versa:

$$\boxed{\frac{p_2}{p_1} = \frac{\dfrac{\gamma+1}{\gamma-1}\cdot\dfrac{\rho_2}{\rho_1}-1}{\dfrac{\gamma+1}{\gamma-1}-\dfrac{\rho_2}{\rho_1}}} \qquad \boxed{\frac{\rho_2}{\rho_1} = \frac{\dfrac{\gamma+1}{\gamma-1}\cdot\dfrac{p_2}{p_1}+1}{\dfrac{\gamma+1}{\gamma-1}+\dfrac{p_2}{p_1}}}. \tag{3.82}$$

The variation of pressure ratio across the shock vs. the density ratio is shown in Fig. 3.18. The pressure ratio across the shock is compared to the isentropic pressure ratio (2.88). It is apparent that for density ratios larger than 2, the process across the shock starts deviating rapidly from an isentropic one.

The density ratio after and before the shock can be written as a function of the Mach number before the shock by substituting (3.77) into (3.82):

$$\frac{\rho_2}{\rho_1} = \frac{(\gamma+1)M_1^2}{(\gamma-1)M_1^2+2} \tag{3.83}$$

or

$$\frac{\rho_1}{\rho_2} = \frac{2}{(\gamma+1)M_1^2} + \frac{\gamma-1}{\gamma+1}. \tag{3.84}$$

The variation of M_2, p_{02}/p_{01}, T_2/T_1, p_2/p_1, and ρ_2/ρ_1 as a function of initial Mach number is shown in Fig. 3.19. It is apparent that p_{02}/p_{01} decreases abruptly with the increase of M_1 and goes to 0 as M_1 goes to infinity. As M_1 increases, M_2 asymptotically reaches $\sqrt{(\gamma-1)/(2\gamma)}$, and ρ_2/ρ_1 reaches $(\gamma+1)/(\gamma-1)$. Both T_2/T_1 and p_2/p_1 tend to infinity as M_1 tends to infinity.

Figure 3.19 Parameter variation across normal shock as a function of initial Mach number.

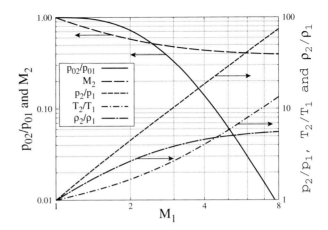

Example 3.5.1 What is the maximum possible increase of air density through a normal shock?

Solution

For a normal shock:

$$\frac{\rho_2}{\rho_1} = \frac{(\gamma + 1)M_1^2}{(\gamma - 1)M_1^2 + 2}.$$

Let us use the following notation:

$$M_1^2 \equiv x$$

$$\frac{\rho_2}{\rho_1} \equiv f(x).$$

Therefore, the equivalent problem is to find the max of $f(x)$ with $x \in [0, \infty)$. Start with

$$f(x) = \frac{(\gamma + 1)x}{(\gamma - 1)x + 2};$$

take the derivative:

$$f'(x) = \frac{(\gamma + 1)[(\gamma - 1)x + 2] - (\gamma + 1)x(\gamma - 1)}{[(\gamma - 1)x + 2]^2} = \frac{2(\gamma + 1)}{[(\gamma - 1)x + 2]^2} > 0$$

because $\gamma + 1$ is larger than 0.

Since $f(x)$ is monotonically increasing,

$$\max f(x) = \lim_{x \to \infty} f(x) = \left(\frac{\rho_2}{\rho_1}\right)_{\max} = \frac{\gamma + 1}{\gamma - 1}.$$

For air $\gamma = 1.4$, such that $\left(\frac{\rho_2}{\rho_1}\right)_{\max} = 6$.

3.5.1.4 Prandtl Equation

Using mass and momentum equations (3.62) and (3.63) yields

$$u_1 u_2 = \frac{p_2 - p_1}{\rho_2 - \rho_1}.$$
(3.85)

Using energy conservation equation (3.71) yields

$$p_1 = \rho_1 \left(\frac{\gamma + 1}{2\gamma} a_{cr}^2 - \frac{\gamma - 1}{2\gamma} u_1^2 \right)$$
(3.86)

$$p_2 = \rho_2 \left(\frac{\gamma + 1}{2\gamma} a_{cr}^2 - \frac{\gamma - 1}{2\gamma} u_2^2 \right).$$
(3.87)

Substituting (3.86) and (3.87) in (3.85) and using mass conservation equation (3.62) yields

$$\boxed{u_1 u_2 = a_{cr}^2},$$
(3.88)

which is known as the Prandtl equation.

Using the λ number, the Prandtl equation can be written as

$$\lambda_1 \lambda_2 = 1.$$
(3.89)

It is evident from (3.89) that if the incoming flow is supersonic, the flow downstream of the normal shock must be subsonic.

Example 3.5.2   Ahead of a normal shock, the air velocity is 400 m/s and the stagnation temperature is 260 K. Assuming the ratio of specific heat capacities is $\gamma = 1.4$ and the gas constant is $R = 287.16$ J/(kg K), what is the λ number after the shock?

Solution
Let us find the critical temperature:

$$T_{cr} = \frac{2}{\gamma + 1} T_0 = \frac{2}{2.4} 260 = 216.7 \text{ K}.$$

The critical speed is then

$$a_{cr} = \sqrt{\gamma R T_{cr}} = \sqrt{1.4 \times 287.16 \times 216.7} = 295.1 \text{ m/s}$$

such that the lambda number ahead of the shock wave is

$$\lambda_1 = \frac{u_1}{a_{cr}} = \frac{400}{295.1} = 1.355.$$

Using (3.89) yields

$$\lambda_2 = \frac{1}{\lambda_1} = 0.738.$$

3.5.1.5 Entropy Variation and Impossibility of a Rarefaction Shock

To estimate the losses occurring in a normal shock, let us calculate the entropy variation by rearranging (2.93) as

$$s_2 - s_1 = c_p \ln \left[\frac{T_2}{T_1} \left(\frac{p_1}{p_2} \right)^{\frac{\gamma-1}{\gamma}} \right].$$

Using (3.31) and (3.33) yields

$$s_2 - s_1 = c_p \ln \left[\frac{T_{02}}{T_{01}} \left(\frac{p_{01}}{p_{02}} \right)^{\frac{\gamma-1}{\gamma}} \right],$$

which can be simplified further using (3.67) and (2.74):

$$s_2 - s_1 \overset{(3.67)}{=} c_p \ln \left[\left(\frac{p_{01}}{p_{02}} \right)^{\frac{\gamma-1}{\gamma}} \right] \overset{(2.74)}{=} R \ln \frac{p_{01}}{p_{02}}. \tag{3.90}$$

Therefore, the entropy variation, which is a measure of the losses across the normal shock, increases as the stagnation pressure p_{02} decreases with respect to p_{01}. As shown in Fig. 3.19, the ratio p_{02}/p_{01} decreases abruptly as the initial Mach number increases. Consequently, the entropy variation increases rapidly as M_1 increases.

Let us substitute (3.78) into (3.90):

$$\frac{s_2 - s_1}{R} = \frac{1}{\gamma - 1} \ln \left(\frac{2\gamma}{\gamma + 1} M_1^2 - \frac{\gamma - 1}{\gamma + 1} \right) - \frac{\gamma}{\gamma - 1} \ln \frac{\frac{\gamma + 1}{2} M_1^2}{\left(1 + \frac{\gamma - 1}{2} M_1^2 \right)}.$$

The study of the $(s_2 - s_1)/R$ function, which is defined for $M_1 \in (\sqrt{(\gamma - 1)/(2\gamma)}, \infty)$, reveals that it is zero for $M_1 = 1$, irrespective of the γ value, and that

$$\lim_{M_1 \to \sqrt{(\gamma-1)/(2\gamma)}} (s_2 - s_1)/R = -\infty, \qquad \lim_{M_1 \to \infty} (s_2 - s_1)/R = \infty.$$

The variation of the $(s_2 - s_1)/R$ function is shown in Fig. 3.20. Since according to the second law of thermodynamics entropy always increases,[4] only the shock from supersonic to subsonic speed is possible. Furthermore, it is impossible for a rarefaction shock to occur because that would result in a decrease of entropy.

To determine how fast the losses increase in a normal shock, let us calculate the entropy variation by substituting (3.77) and (3.84) in (2.85):

$$s_2 - s_1 = c_v \ln \left[\frac{p_2}{p_1} \left(\frac{\rho_1}{\rho_2} \right)^{\gamma} \right] = c_v \ln \left[\left(\frac{2\gamma}{\gamma + 1} M_1^2 - \frac{\gamma - 1}{\gamma + 1} \right) \left(\frac{(\gamma - 1)M_1^2 + 2}{(\gamma + 1)M_1^2} \right)^{\gamma} \right].$$

---

[4] Unless the process is isentropic

Figure 3.20 Dimensionless entropy variation vs. initial Mach.

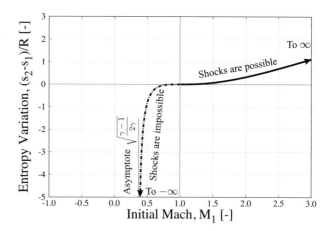

Assuming the initial flow is slightly supersonic, that is, for $M_1^2 = 1 + \varepsilon$, where ε is a small number compared to 1, the entropy variation can be written using series expansion of the logarithmic function as

$$s_2 - s_1 = c_v \frac{2\gamma(\gamma - 1)}{(\gamma + 1)^2} \cdot \frac{(M_1^2 - 1)^3}{3} + O(\varepsilon^2).$$

It is apparent that the entropy increases with the cube of $M_1^2 - 1$. Because the entropy increases rapidly with Mach number, normal shock waves should be avoided and replaced, if possible, by oblique shocks.

3.5.2   Oblique Shocks

The normal shocks discussed in the previous section are only particular forms of pressure discontinuities within a fluid. More often in practice, the discontinuities are inclined to the direction of the incoming flow. These shocks are called oblique shocks. Oblique shocks can be either attached to the surface of a body or detached. Most of the time, oblique shocks are detached. They are attached when flow is forced to suddenly change direction, as in the case of a sharp corner.

3.5.2.1 Governing Equations

Let us start by deriving the mass, momentum, and energy conservation equations by using the control volume shown in Fig. 3.21. The shock wave is indicated by the wavy ss' line. Two coordinate systems could be used for the derivation: the $x - y$ coordinate system or the $n - t$ coordinate system aligned to the shock. The latter system is preferred, since it simplifies the derivation.

Figure 3.21 Control volume for deriving mass, momentum, and energy conservation equations for an oblique shock.

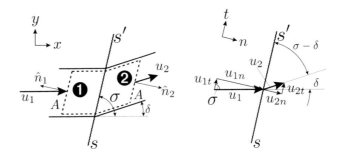

Mass Conservation Equation

To derive the mass conservation equation for an oblique shock, let us start with the general form of mass conservation (2.7), where the unsteady term is zero. Using the same arguments presented for normal shocks, the surfaces σ_1 and σ_2 are equal. The dot product between the normal to the control volume surface and the velocity vector yields the velocity component normal to the shock surface, such that

$$\rho_1 u_{1_n} = \rho_2 u_{2_n}. \tag{3.91}$$

Momentum Conservation Equation

The starting point for deriving the momentum conservation equation for an oblique shock is (2.31):

$$\int_{\sigma \equiv \partial \tau} \rho \vec{V}(\vec{V} \cdot \vec{n}) \, d\sigma = \vec{F}_V + \vec{F}_S \tag{2.31}$$

where the surface forces are $\vec{F}_S = \vec{F}_p + \vec{F}_{vis}$. If we neglect the volume forces $\vec{F}_V$ and the viscous forces $\vec{F}_{vis}$ compared to the pressure forces $\vec{F}_p$, the momentum conservation equation projected on the direction normal to the shock is

$$\rho_1 u_{1_n}(-u_{1_n})A + \rho_2 u_{2_n} \cdot u_{2_n} A = -(-p_1 + p_2)A,$$

where A is the area of the inlet and outlet of the control volume shown in Fig. 3.21. Consequently, the normal component of the momentum conservation equation is

$$\rho_2 u_{2_n}^2 - \rho_1 u_{1_n}^2 = p_1 - p_2. \tag{3.92}$$

The momentum conservation equation projected on the direction tangential to the shock is

$$-\rho_1 u_{1_t} u_{1_n} A + \rho_2 u_{2_t} u_{2_n} A = 0$$

such that using the mass conservation equation (3.91) yields

$$u_{1_t} = u_{2_t} = u_t. \tag{3.93}$$

The momentum conservation equation (2.31) is a vectorial equation. Since the oblique shock leads to a two-dimensional flow, the momentum conservation equation has two components, a normal and a tangent component. Recall that in the case of normal shocks, the flow does not change direction as it goes through the shock. Consequently the flow is one dimensional and the momentum conservation equation (3.63) has only one component.

Energy Conservation Equation

Let us start with the general form of energy conservation (2.63), which reduces to

$$
\int_{\sigma \equiv \partial \tau} \rho \left(h + \frac{V^2}{2} \right) \vec{V} \cdot \hat{n} \, d\sigma = 0,
\tag{3.64}
$$

since the heat transfer rate, the shear and shaft power, and the unsteady term are all zero. Given the control volume in Fig. 3.21, (3.64) yields

$$
-\rho_1 u_{1_n} h_{01} A + \rho_2 u_{2_n} h_{02} A = 0,
$$

which using the mass conservation equation (3.91) gives

$$
h_{01} = h_{02},
\tag{3.94}
$$

or

$$
T_{01} = T_{02}.
$$

Using the definition of stagnation enthalpy (3.12) and the perfect gas relations (3.69), Eq. (3.94) becomes

$$
\frac{u_1^2}{2} + \frac{\gamma}{\gamma - 1} \cdot \frac{p_1}{\rho_1} = \frac{u_2^2}{2} + \frac{\gamma}{\gamma - 1} \cdot \frac{p_2}{\rho_2} = \frac{\gamma}{\gamma - 1} \cdot \frac{p_0}{\rho_0},
$$

and using (3.93) yields

$$
\frac{u_{1_n}^2}{2} + \frac{\gamma}{\gamma - 1} \cdot \frac{p_1}{\rho_1} = \frac{u_{2_n}^2}{2} + \frac{\gamma}{\gamma - 1} \cdot \frac{p_2}{\rho_2} = \frac{\gamma}{\gamma - 1} \cdot \frac{p_0}{\rho_0} - \frac{u_t^2}{2}.
$$

Using the definition of critical speed (3.25) yields

$$
\frac{u_{1_n}^2}{2} + \frac{\gamma}{\gamma - 1} \cdot \frac{p_1}{\rho_1} = \frac{u_{2_n}^2}{2} + \frac{\gamma}{\gamma - 1} \cdot \frac{p_2}{\rho_2} = \frac{\gamma + 1}{2(\gamma - 1)} a_{cr}^2 - \frac{u_t^2}{2}.
\tag{3.95}
$$

3.5.2.2 Working Formulas for Oblique Shocks

An oblique shock acts as a normal shock to the component of the flow perpendicular to it, while the tangential component is unchanged. Consequently, the normal shock relations, except those for the ratios of static and stagnation pressures, apply to oblique shocks if M_1 and M_2 are replaced by their normal components $M_1 \sin \sigma$ and $M_2 \sin(\sigma - \delta)$:

$$
\boxed{M_{1_n} = M_1 \sin \sigma}
$$

$$
\boxed{M_{2_n} = M_2 \sin(\sigma - \delta)}.
$$

Consequently, (3.75) becomes

$$\left(\frac{\gamma-1}{2}+\frac{1}{M_1^2\sin^2\sigma}\right)\left(\frac{\gamma-1}{2}+\frac{1}{M_2^2\sin^2(\sigma-\delta)}\right)=\left(\frac{\gamma+1}{2}\right)^2,\tag{3.96}$$

(3.77) becomes

$$\frac{p_2}{p_1}=\frac{2\gamma}{\gamma+1}M_1^2\sin^2\sigma-\frac{\gamma-1}{\gamma+1},\tag{3.97}$$

(3.78) becomes

$$\frac{p_{02}}{p_{01}}=\frac{\left[\dfrac{\dfrac{\gamma+1}{2}M_1^2\sin^2\sigma}{\left(1+\dfrac{\gamma-1}{2}M_1^2\sin^2\sigma\right)}\right]^{\frac{\gamma}{\gamma-1}}}{\left(\dfrac{2\gamma}{\gamma+1}M_1^2\sin^2\sigma-\dfrac{\gamma-1}{\gamma+1}\right)^{\frac{1}{\gamma-1}}},\tag{3.98}$$

(3.79) becomes

$$\frac{T_2}{T_1}=\frac{\left(1+\dfrac{\gamma-1}{2}M_1^2\sin^2\sigma\right)\left(\dfrac{2\gamma}{\gamma-1}M_1^2\sin^2\sigma-1\right)}{\dfrac{(\gamma+1)^2}{2(\gamma-1)}M_1^2\sin^2\sigma},\tag{3.79}$$

and (3.84) becomes

$$\frac{\rho_1}{\rho_2}=\frac{2}{(\gamma+1)M_1^2\sin^2\sigma}+\frac{\gamma-1}{\gamma+1}.\tag{3.99}$$

Rankine–Hugoniot Equations

As in the case of normal shocks, the Rankine–Hugoniot equations relate the pressure ratio p_2/p_1 to the density ratio ρ_2/ρ_1, and vice versa. To derive the Rankine–Hugoniot equations, let us eliminate the velocity terms from the momentum and energy conservation equations. Substituting the mass conservation equation (3.91) into the momentum conservation equation (3.92) yields

$$p_2-p_1\overset{(3.92)}{=}\rho_1u_{1_n}^2\left(1-\frac{\rho_2}{\rho_1}\frac{u_{2_n}^2}{u_{1_n}^2}\right)\overset{(3.91)}{=}\rho_1u_{1_n}^2\left(1-\frac{\rho_1}{\rho_2}\right)$$

such that

$$u_{1_n}^2=\frac{p_2-p_1}{\rho_2-\rho_1}\frac{\rho_2}{\rho_1}=\frac{p_1}{\rho_1}\left(\frac{p_2/p_1-1}{\rho_2/\rho_1-1}\right)\frac{\rho_2}{\rho_1}.\tag{3.100}$$

In a similar way, factoring out the $\rho_2 u_{2n}^2$ term in the momentum conservation equation (3.92) yields

$$u_{2n}^2 = \frac{p_2 - p_1}{\rho_2 - \rho_1} \frac{\rho_1}{\rho_2} = \frac{p_1}{\rho_1} \left(\frac{p_2/p_1 - 1}{\rho_2/\rho_1 - 1} \right) \frac{1}{\rho_2/\rho_1}. \tag{3.101}$$

Once the energy conservation equation (3.95) is rewritten as

$$\frac{\gamma}{\gamma - 1} \frac{p_1}{\rho_1} \left(\frac{p_2}{p_1} \frac{\rho_1}{\rho_2} - 1 \right) = (u_{1n}^2 - u_{2n}^2)/2$$

and the velocities are substituted by (3.100) and (3.101), the pressure ratio is obtained as a function of the density ratio only, and vice versa:

$$\frac{p_2}{p_1} = \frac{\dfrac{\gamma + 1}{\gamma - 1} \cdot \dfrac{\rho_2}{\rho_1} - 1}{\dfrac{\gamma + 1}{\gamma - 1} - \dfrac{\rho_2}{\rho_1}} \qquad \frac{p_2}{p_1} = \frac{\dfrac{\gamma + 1}{\gamma - 1} \cdot \dfrac{p_2}{p_1} + 1}{\dfrac{\gamma + 1}{\gamma - 1} + \dfrac{p_2}{p_1}}. \tag{3.82}$$

Note that the Rankine–Hugoniot equations derived for oblique shocks are identical to those derived for normal shocks. This comes as no surprise, since a shock with a given pressure ratio has associated with it a density ratio that does not depend on the obliquity of the shock.

Meyer Equation

To derive Meyer's equation, we need to eliminate pressures and densities from the governing equations. Using mass and momentum equations (3.91) and (3.92) yields

$$u_{1n} u_{2n} = \frac{p_2 - p_1}{\rho_2 - \rho_1}. \tag{3.102}$$

Using energy conservation equation (3.95) yields

$$p_1 = \rho_1 \left(\frac{\gamma + 1}{2\gamma} a_{cr}^2 - \frac{\gamma - 1}{2\gamma} (u_{1n}^2 + u_t^2) \right)$$

$$p_2 = \rho_2 \left(\frac{\gamma + 1}{2\gamma} a_{cr}^2 - \frac{\gamma - 1}{2\gamma} (u_{2n}^2 + u_t^2) \right)$$

which substituted in (3.102) and using (3.91) yields the Meyer equation

$$\boxed{u_{1n} u_{2n} = a_{cr}^2 - \frac{\gamma - 1}{\gamma + 1} u_t^2.}$$

It is apparent that the Prandtl equation (3.88) is a particular form of the Meyer equation for the case when $u_t = 0$.

Other Useful Equations

The link between the wedge angle, the shock angle, and the Mach number in front of the shock wave is given by

$$\boxed{\frac{1}{\tan \delta} = \left(\frac{\gamma + 1}{2} \cdot \frac{M_1^2}{M_1^2 \sin^2 \sigma - 1} - 1 \right) \tan \sigma.} \tag{3.103}$$

Given M_1 and σ, the wedge angle δ can be easily obtained from (3.103). Often, however, one needs to calculate the shock angle σ as a function of M_1 and δ. In this case, since (3.103) is nonlinear in σ, the solution can be obtained using a Newton–Raphson method, as shown in the following example.

Example 3.5.3 The incoming flow at the inlet of a ramjet must be deflected by $30°$, as shown in Fig. 3.22. Calculate the stagnation pressure ratio, the density ratio, the final Mach number, and the total entropy variation assuming that the initial Mach number is $M = 3.5$, the ratio of specific heats is $\gamma = 1.35$, and the gas constant is $R = 287.16$ J/(kg K).

Figure 3.22 Inlet configuration.

Solution
The shock angle σ is found from (3.103), by using a Newton–Raphson method. This iterative method is built by using (3.103) to define the function

$$f(\sigma) = \left(\frac{\gamma + 1}{2} \frac{M_1^2}{M_1^2 \sin^2 \sigma - 1} - 1 \right) \tan \sigma - \frac{1}{\tan \delta}$$

and its derivative

$$f'(\sigma) = \left(\frac{\gamma + 1}{2} \frac{M_1^2}{M_1^2 \sin^2 \sigma - 1} - 1 \right) \frac{1}{\cos^2 \sigma} - \frac{\gamma + 1}{2} \frac{M_1^4 \sin 2\sigma}{\left(M_1^2 \sin^2 \sigma - 1 \right)^2} \tan \sigma$$

such that

$$\sigma_{i+1} = \sigma_i - \frac{f(\sigma_i)}{f'(\sigma_i)}, \quad i = 1, 2, 3, \ldots$$

The initial guess for σ was $\sigma_1 = 44°$. The iterative process ended in four iterations and the shock angle was $\sigma = 46.49°$.

To calculate the stagnation pressure ratio, let us apply (3.98), where

$$M_1 \sin \sigma = 3.5 \sin 46.49° = 2.538.$$

The stagnation pressure ratio is then

$$\frac{p_{02}}{p_{01}} = \frac{\left[\dfrac{\dfrac{\gamma + 1}{2} M_1^2 \sin^2 \sigma}{\left(1 + \dfrac{\gamma - 1}{2} M_1^2 \sin^2 \sigma \right)} \right]^{\frac{\gamma}{\gamma - 1}}}{\left(\dfrac{2\gamma}{\gamma + 1} M_1^2 \sin^2 \sigma - \dfrac{\gamma - 1}{\gamma + 1} \right)^{\frac{1}{\gamma - 1}}} = 0.465.$$

The density ratio is calculated using (3.99)

$$\frac{\rho_1}{\rho_2} = \frac{2}{(\gamma + 1)M_1^2 \sin^2 \sigma} + \frac{\gamma - 1}{\gamma + 1} = 0.2811$$

so that $\rho_2/\rho_1 = 3.558$.
The ratio of static pressures (3.97) is needed to calculate the entropy variation:

$$\frac{p_2}{p_1} = \frac{2\gamma}{\gamma + 1}M_1^2 \sin^2 \sigma - \frac{\gamma - 1}{\gamma + 1} = 7.254.$$

The entropy variation is calculated using (2.85):

$$s_2 - s_1 = c_v \ln\left[\frac{p_2}{p_1}\left(\frac{\rho_1}{\rho_2}\right)^\gamma\right] = 219.8 \text{ kJ/(kg K)}.$$

The initial guess for the shock angle was obtained by using the graph $\sigma(M_1)$ vs. δ of the NACA Report 1135 [AMES Research Staff, 1953, p. 654]. Although this graph was generated for a ratio of specific heat capacities, $\gamma = 1.4$, the shock angle value was close enough to the value corresponding to $\gamma = 1.35$ such that the iterative method converged.

When the working fluid is air, an expeditious method for solving shock wave problems is to use the tables and graphs provided in either the NACA 1135 report or Figs. 3.23–3.26.

Figures 3.23–3.26 show the variation of dependent variables σ, M_2, p_2/p_1, and p_{02}/p_{01} as a function of independent variables M_1 and δ. It is apparent from Figs. 3.23–3.26 that for a given incoming Mach number M_1 and turning angle δ, there could be two solutions to the oblique shock relations, one solution or no solution. For a given turning angle δ, there is a minimum M_1 for which a solution exists. Conversely, for a given M_1, there is maximum δ for which a solution exists. When two solutions are possible, one is a *strong shock* and the other is a *weak shock*, as shown in Fig. 3.27, which illustrates the two shocks for $M_1 = 3$ and $\delta = 20°$. As shown in Fig. 3.23, for these M_1 and δ values, the wave angle is 82.2° for the strong shock and 37.7° for the weak shock if $\gamma = 1.4$. The pressure downstream of the shock is a main factor in determining whether the shock is weak or strong. If the back pressure is sufficiently high, a strong shock occurs; otherwise the shock is weak. Although there is no definitive and clear-cut answer to what type of shock develops, typically weak shocks happen in external flows while strong shocks happen in internal flows.

Figure 3.23 Shock wave angle σ vs. initial Mach number M_1 and turning angle δ for air, $\gamma = 1.4$.

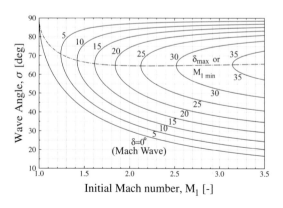

Figure 3.24 Final Mach number M_2 vs. initial Mach number M_1 and turning angle δ for air, $\gamma = 1.4$.

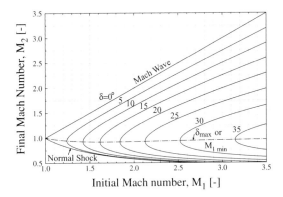

Figure 3.25 Static pressure ratio p_2/p_1 vs. initial Mach number M_1 and turning angle δ for air, $\gamma = 1.4$.

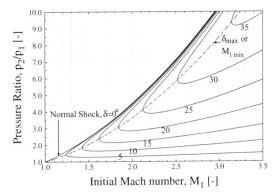

Figure 3.26 Stagnation pressure ratio p_{02}/p_{01} vs. initial Mach number M_1 and turning angle δ for air, $\gamma = 1.4$.

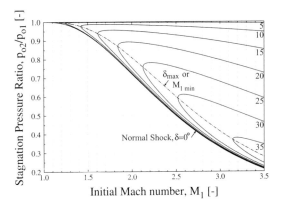

Detached Shocks

Figures 3.23–3.26 show that for a given initial Mach number, there is maximum value of the turning angle, δ_{max}, for which an oblique shock solution exists. Conversely, for a given turning angle, there is a minimum initial Mach number $M_{1\ min}$ for which an oblique shock solution exists. When this limiting condition occurs, the weak and strong shock become identical. The results of the limiting condition are summarized in Fig. 3.28 where it is impossible for an attached oblique shock to exist above the curve. One can also note that for $\gamma = 1.4$, the turning angle reaches an asymptotic value of $45.22°$ as the Mach number goes to infinity.

Figure 3.27 Weak and strong shocks.

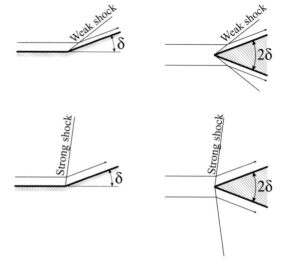

Figure 3.28 Minimum initial Mach number for a given turning angle or maximum turning angle for a given initial Mach number for $\gamma = 1.4$.

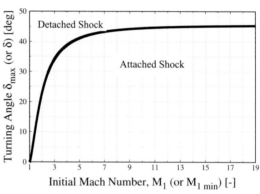

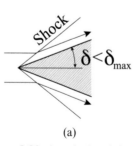

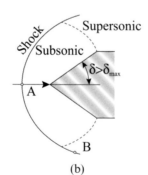

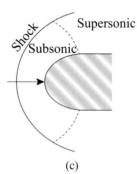

(a)   (b)   (c)

Figure 3.29 Attached and detached shocks: (a) attached, (b) detached on wedge with large angle, and (c) detached on blunt body.

Consider the case of a wedge of half-angle δ with an incoming uniform supersonic flow at Mach number M_1. If the half-angle δ is smaller than δ_{max} corresponding to M_1, then the shock wave is attached, as shown in Fig. 3.29a. Since this is external flow, most likely the shock is weak, and considering Fig. 3.24, most likely the flow downstream of the shock is supersonic. If the wedge half-angle δ is larger than δ_{max} corresponding to M_1, then the shock wave is detached, as shown in Fig. 3.29b. The flow behind the detached shock is in part subsonic and in part supersonic. In addition, the shock is always curved as it weakens from a normal shock at A to a Mach wave of vanishing strength at B, far away from the body. In the case of a blunt body, the shock is always detached, as shown in Fig. 3.29c.

A similar approach is used for the flow past a concave corner in a duct in order to determine whether the shock wave is attached or detached.

Example 3.5.4 The air inside a duct with a rectangular cross section has an initial Mach number $M = 1.8$. A concave corner inside the duct has an angle of $20°$, as shown in Fig. 3.30. Assuming the flow can be approximated as two-dimensional, determine whether the shock is attached or detached.

Figure 3.30 Cross section through duct with concave corner.

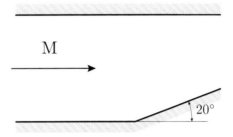

M

$20°$

Solution
Using the graph from Fig. 3.23, one can determine that for a deflection of $20°$, the initial Mach number must be larger than 1.85 for an attached shock to develop. Consequently, for $M = 1.8$ the shock is detached.

Problems

1. Show that for a normal shock,

   $$\frac{p_2}{p_1} = \frac{1 + \gamma M_1^2}{1 + \gamma M_2^2}$$

   where 1 is the state before the shock and 2 is the state after the shock.
2. Using mass and momentum conservation equations, show that for a normal shock,

   $$u_1 u_2 = \frac{p_2 - p_1}{\rho_2 - \rho_1}$$

   where 1 is the state before the shock and 2 is the state after the shock.
3. Show that Mach number is unity at the point of maximum entropy on the Fanno and Rayleigh lines.

4. Ahead of a normal shock, the air velocity is 400 m/s and the stagnation temperature is 260 K. Assuming the ratio of specific heat capacities is $\gamma = 1.4$ and the gas constant is $R = 287.16$ J/(kg K), what is the λ number after the shock?

5. The two-dimensional diffuser of a jet engine supersonic inlet is shown in Fig. 3.31. The inlet Mach number is $M_1 = 3$ and the specific-heat ratio is $\gamma = 1.33$. What is the static pressure ratio across the two-shock system?

Figure 3.31 Supersonic wedge inlet.

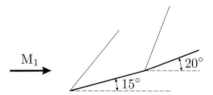

6. Calculate the static pressure and density ratios, final Mach number, and total entropy variation for the three inlet wedges in a ramjet shown in Figure 3.32, knowing that the upstream Mach number is $M = 3.5$, the ratio of specific heats is $\gamma = 1.35$, and the gas constant is $R = 287.16$ J/(kg K). Comment on the results.

Figure 3.32 Inlet configurations.

7. Consider a symmetrical, double-wedged airfoil, shown in Fig. 3.33, which has a relative thickness of 5%. The airfoil flies at an angle of attack of 1 degree and a Mach number $M = 4$. What is the angle of the shock wave on the suction side of the leading edge, if the ratio of heat specific capacities γ is 1.37?

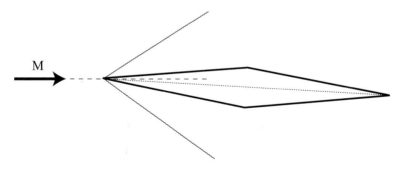

Figure 3.33 Symmetrical, double-wedged airfoil.

8. At the inlet in an engine, shown in Fig. 3.34, the incoming flow is at Mach 2.4. A two-shock system develops in the inlet: a first oblique shock that originates at the tip of the

inlet wedge and a second oblique shock that reflects from the engine cowling. One can assume that the gas constant is $R = 287.16$ J/(kg K), the ratio of specific heats is $\gamma = 1.35$, and the boundary layer effects are negligible.

1. What is the Mach number downstream of the second shock wave?
2. What is the stagnation pressure ratio across the two-shock system?

Figure 3.34 Engine inlet configuration.

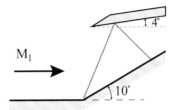

9. A fighter jet flies at Mach 2.2. The atmospheric pressure is 50,000 Pa. The inlet in the engine has a 10° cone (that we will approximate as a 10° wedge). The air entering the jet engine inlet has a gas constant $R = 287.16$ J/(kg K), and the ratio of specific heats is $\gamma = 1.4$.
    1. What is the pressure fluctuation at the compressor face, that is, downstream of the shock wave, when the fighter jet fires its ammunition, knowing that the mixture of the air and gases ingested in the engine has a ratio of specific heats $\gamma = 1.36$? (Hint: calculate the pressure after the shock, before and after firing.)
    2. How would the pressure fluctuation vary if the fighter jet would be flying closer to the ground? (Justify your answer qualitatively.)
10. The incoming flow at the inlet of a ramjet must be deflected by 30°. Five inlet cones are being considered that turn the flow using between one and 10 oblique shock waves: (1) a single 30° turn, (2) two 15° turns, (3) three 10° turns, shown in Fig. 3.35, (4) five 6° turns, and (5) ten 3° turns. Compare the stagnation pressure and density ratios, final Mach number and total entropy variation for the five inlet configurations assuming that the upstream Mach number is M = 3.5, the ratio of specific heats is $\gamma = 1.35$, and the gas constant is $R = 287.16$ J/(kg K). Comment on the results.

Figure 3.35 Inlet configurations.

The results of the last problem are extremely relevant as they underlie the importance of turning a flow through multiple shocks. Figure 3.36 shows that the entropy decreases from approximately 220 J/(kg K) to 3 J/(kg K), while the stagnation pressure ratio increases from 0.465 to 0.99 when the flow is turned using 10 shock waves as opposed to a single shock wave. The implication of these results will be further explored while discussing supersonic inlets in Chapter 7.

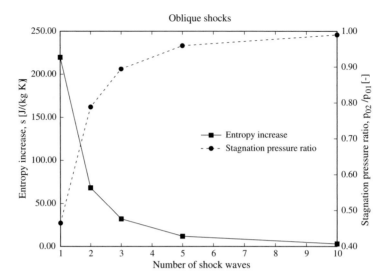

Figure 3.36 Variation of stagnation pressure ratio and entropy vs. number of shock waves.

Bibliography

AMES Research Staff. Equations, tables, and charts for compressible flow. Technical report, NACA Ames Aeronautical Laboratory, Moffett Field, CA, 1953. 100

S. C. Chapra and R. P. Canale. *Numerical Methods for Engineers*. McGraw-Hill, third edition, 1998. 74

4 Viscous Boundary Layer and Thermal Boundary Layer

4.1 Introduction

An important challenge of compressor design is flow separation. A significant challenge of turbine design is heat transfer from the hot gases to the metal blades. To understand the physics of these two challenges, this chapter will introduce the viscous boundary layer and thermal boundary layer concepts.

The boundary layer concept was introduced by Ludwig Prandtl in 1904 [Prandtl, 1904]. The boundary layer theory had huge impact on the development of aerodynamics, since it reconciled the analytical solutions of inviscid flows with the experimental results. Among other things, the boundary layer theory helped explain d'Alembert's paradox [d'Alembert, 1752], which, at the end of the nineteenth century, generated a lot of controversy. d'Alembert's paradox stems from a proof generated in 1752 by the French mathematician Jean le Rond d'Alembert that showed that for incompressible, inviscid potential flows, the drag force is zero on a body moving with a constant velocity relative to the fluid. This result contradicts experimental investigations, which measure a significant drag on bodies. The result of d'Alembert's paradox was a split between the fields of hydraulics and fluid mechanics, "when fluid dynamicists were divided into hydraulic engineers who observed things that could not be explained and mathematicians who explained things that could not be observed," as summarized by Sir Cyril Hinshelwood [Lighthill, 1956].

Consider the flow over a flat plate at zero angle of attack, shown in Fig. 4.1. With the exception of the immediate region near the wall, the velocities are of the order of the free-stream velocity, U_∞. At the wall, however, the velocity is zero because the fluid adheres to the surface. Therefore, the transition from zero velocity at the wall to the free-stream velocity occurs in a very thin layer, called the viscous boundary layer. It is apparent that the velocity gradient $\partial u/\partial y$ is largest at the wall, where the velocity variation is largest. The velocity gradient decreases as the distance from the wall increases. At some distance, δ, indicated by the dashed line, the velocity stops varying and the gradient tends to zero. The value of boundary layer thickness, δ, is typically defined as the distance measured in the y-direction from the wall to the point where the velocity u is 0.99 of U_∞. Additional information on the definition of the boundary layer thickness is provided in Section 4.2.1.1.

Figure 4.1 Velocity variation over a flat plate.

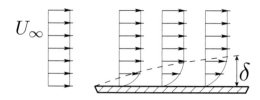

Prandtl introduced the viscous boundary layer by arguing that the shear stress at the contact between a fluid and a solid boundary, $\tau = \mu \partial u / \partial y$, cannot be neglected in spite of the fact the viscosity μ is very small (approximately $15 \cdot 10^{-6}$ Pa·s for air at standard conditions). Prandtl noticed that although the viscosity is small, the velocity gradient $\partial u / \partial y$ is large enough such that the shear stress becomes significant. Consequently, two regions can be identified: (1) a very thin layer near the wall where the velocity gradient normal to the wall, $\partial u / \partial y$, is very large, and (2) the remaining region where the velocity gradient is smaller and the effect of viscosity is unimportant.

This chapter presents aspects of boundary layer theory relevant to propulsion systems, and is divided into two parts: viscous boundary layer and thermal boundary layer. The first section of the viscous boundary layer defines the salient parameters. The governing equations of the viscous boundary layer are subsequently derived and the laminar and turbulent viscous boundary layers are presented. The second part of the chapter presents the thermal boundary layers, both laminar and turbulent.

4.2   Viscous Boundary Layer

This section presents the viscous boundary layer theory. The first part describes the parameters of the viscous boundary layer. The second part illustrates the derivation of the boundary layer equations for steady, two-dimensional, incompressible flows. The third part of the section presents the laminar boundary layer theory. The last part of this section briefly describes the turbulent boundary layer theory.

4.2.1   Boundary Layer Parameters

There are several parameters that can be used to describe a boundary layer. The most used parameters are the boundary layer thickness, the displacement thickness, the momentum thickness, and the shape factor. Before defining these parameters, let us present the variation of the velocity in the boundary layer.

The velocity variation in the boundary layer is shown in Fig. 4.2, where u is the x-component of the velocity. In the boundary layer, u varies from zero at a fixed wall to U_∞ at the upper limit. The velocity U_∞ is the free-stream velocity, that is, the velocity at the edge of the boundary layer.

Figure 4.2 Velocity variation and displacement thickness in a boundary layer.

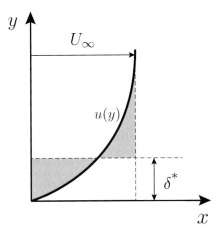

4.2.1.1 Boundary Layer Thickness

The value of boundary layer thickness, δ, is defined as the distance measured in the y-direction from the wall to the point where the velocity u is 0.99 of U_∞. The 0.99 value is somewhat arbitrary and sometimes is replaced by either 0.995 or 0.999.

4.2.1.2 Displacement Thickness

The displacement thickness, δ^* (or δ_1), is defined as the distance a streamline at the edge of the boundary layer is displaced away from the wall. In other words, the displacement thickness is the distance the wall would have to be displaced to produce the same solution for the flow outside the boundary layer as the boundary layer equations produce. For incompressible flows, an equivalent way to define the displacement thickness is the distance that, when multiplied by the free-stream velocity, equals the integral of velocity defect, $U_\infty - u$, across the boundary layer:

$$U_\infty \delta^* = \int_0^\infty (U_\infty - u)\, dy \tag{4.1}$$

which yields

$$\boxed{\delta^* = \int_0^\infty \left(1 - \frac{u}{U_\infty}\right) dy}\,. \tag{4.2}$$

According to (4.1), the two gray areas in Fig. 4.2 must be equal.

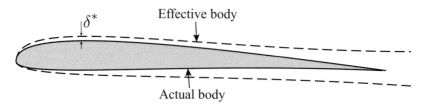

Figure 4.3 Actual body vs. effective body.

The upper limit of the integral can be substituted by a finite value as long as this value is larger than the boundary layer thickness, that is,

$$\delta^* = \int_0^h \left(1 - \frac{u}{U_\infty}\right) dy, \quad h > \delta,$$ (4.3)

because for $h > \delta$ the velocity u equals U_∞.

The displacement thickness is a measure of the reduction of the mass flow rate due to the presence of the boundary layer. This lost mass flow rate is indicated in Fig. 4.2 by the gray area below the $u(y)$ curve.

The displacement thickness has an important practical application. Let's suppose one needs to predict the lift and drag on the airfoil shown in Fig. 4.3. By adding the displacement thickness to the dimensions of the actual airfoil, an equivalent airfoil (the effective body in Fig. 4.3) is being generated. Then a new inviscid solution is computed using the equivalent airfoil. This generates a slightly different free-stream condition compared to the initial calculation. The boundary layer solution is then recalculated with the new free-stream solutions for the actual (real) airfoil. The process is repeated until the variation of the displacement thickness is smaller than an imposed error. The inviscid prediction of the lift and drag on the equivalent airfoil is a good approximation of the viscous lift and drag on the actual airfoil.

For compressible flows, the displacement thickness is defined as

$$\delta^* = \int_0^\infty \left(1 - \frac{\rho u}{\rho_\infty U_\infty}\right) dy,$$ (4.4)

where ρ_∞ is the fluid density at the edge of the boundary layer.

4.2.1.3 Momentum Thickness

The momentum thickness, θ (or δ_2), is the distance that, when multiplied by the density and the free-stream velocity squared, equals the integral of the momentum defect across the boundary layer. For incompressible flows, this yields

$$\theta = \int_0^\infty \frac{u}{U_\infty}\left(1 - \frac{u}{U_\infty}\right) dy.$$ (4.5)

The momentum thickness can be visualized similarly to the displacement thickness, except that in Fig. 4.2 the velocity distribution $u(y)$ should be replaced by the specific momentum flux distribution $u^2(y)$.

As will be shown in the following sections, the momentum thickness is useful for determining the skin friction drag on a surface. We will show in Example 4.2.1 that if the flow is incompressible and the free-stream velocity U_∞ is constant, the airfoil total skin friction drag per unit depth is equal to $\rho U_\infty^2 \theta$ at the trailing edge.

4.2.1.4 Shape Factor

The shape factor, H (or H_{12}), is defined as the ratio between the displacement thickness and the momentum thickness

$$H = \frac{\delta^*}{\theta}. \tag{4.6}$$

The shape factor provides information about the status of the boundary layer, that is, whether the boundary layer is attached, is on the verge of separation, or is already separated. The values of the shape factor vary depending on whether the boundary layer is laminar or turbulent. For example, on a flat plate, the laminar boundary layer has a shape factor $H = 2.59$, while for the turbulent boundary layer, $H = 1.30$.

4.2.2 Viscous Boundary Layer Equations for Steady, Incompressible Flow

There are two approaches for modeling boundary layers: using differential equations or integral equations. The differential approach can predict the variation of all dependent variables inside the boundary layer, but it is computationally expensive. The integral approach can only predict the variation of certain integral parameters, such as displacement thickness, momentum thickness, or skin friction, but it is less computationally expensive compared to the differential approach. The next two sections present both the differential and integral forms of the boundary layer equations.

4.2.2.1 Differential Form: Prandtl Boundary Layer Equations

To derive the boundary layer equations, several assumptions are necessary. First, it is assumed that the boundary layer is a two-dimensional shear layer aligned in the free-stream flow direction. This assumption is only valid in the absence of strong cross-flow pressure gradients. Second, it is assumed that Cartesian coordinates can be used in the boundary layer such that x and u are the tangential distance and velocity and y and v are the normal distance and velocity. This assumption is valid if the local boundary layer thickness is small compared to the surface radius of curvature. Third, it is assumed that the boundary layer is thin, that is, the thickness of the boundary layer is much smaller than a characteristic length, such as the chord of an airfoil. This assumption is valid at relatively high Reynolds numbers, $\mathrm{Re} = UL/v$, where U and L are the reference velocity and length, and v is the kinematic viscosity.

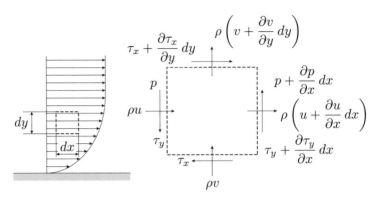

Figure 4.4 Control volume used for derivation of viscous boundary layer equations.

For the derivation of the viscous boundary layer equations, let us use the mass and linear momentum conservation equations applied to a control volume shown in Figure 4.4.

The mass conservation equation for steady, two-dimensional, incompressible flows has been derived in Chapter 2 and is

$$\nabla \vec{V} = 0 \tag{2.10}$$

or in differential form

$$\frac{\partial u}{\partial x} + \frac{\partial v}{\partial y} = 0. \tag{4.7}$$

Next, let us project the momentum conservation equation

$$\int_{\sigma^* \equiv \partial \tau^*} \rho \vec{V}(\vec{V} \cdot \vec{n}) \, d\sigma = \vec{F}_V + \vec{F}_S \tag{2.31}$$

in the x-direction. The integral over the surface σ^* of the control volume will be approximated by assuming that flow variables are averaged at the middle of each face. The x-component of the velocity in the middle of the west side of the control volume is u and on the east side it is $u + \frac{\partial u}{\partial x} dx$. To calculate the x-components of the velocity on the north and south sides, one uses the x-component in the middle of the control volume, $u_c = u + \frac{1}{2}\frac{\partial u}{\partial x} dx$. The x-component of the velocity on the south side is

$$u_s = u_c - \frac{1}{2}\frac{\partial u_c}{\partial y} dy = u + \frac{1}{2}\frac{\partial u}{\partial x} dx - \frac{1}{2}\frac{\partial (u + \frac{1}{2}\frac{\partial u}{\partial x} dx)}{\partial y} dy = u + \frac{1}{2}\frac{\partial u}{\partial x} dx - \frac{1}{2}\frac{\partial u}{\partial y} dy - \frac{1}{4}\frac{\partial^2 u}{\partial x \partial y} dx dy.$$

Neglecting the second-order term $\frac{1}{4}\frac{\partial^2 u}{\partial x \partial y} dx dy$ yields

$$u_s = u + \frac{1}{2}\frac{\partial u}{\partial x} dx - \frac{1}{2}\frac{\partial u}{\partial y} dy.$$

Similarly, the velocity on the north face is

$$u_n = u + \frac{1}{2}\frac{\partial u}{\partial x} dx + \frac{1}{2}\frac{\partial u}{\partial y} dy.$$

Table 4.1 Fluxes of x-momentum.

west	$-\rho u^2 dy$
north	$\rho \left(u + \frac{1}{2}\frac{\partial u}{\partial x}dx + \frac{1}{2}\frac{\partial u}{\partial y}dy\right)\left(v + \frac{\partial v}{\partial y}dy\right)dx$
east	$\rho \left(u + \frac{\partial u}{\partial x}dx\right)^2 dy$
south	$-\rho \left(u + \frac{1}{2}\frac{\partial u}{\partial x}dx - \frac{1}{2}\frac{\partial u}{\partial y}dy\right)vdx$

Let us calculate the x-component of the integral $\int_{\sigma^*_{south}} \rho\vec{V}(\vec{V}\cdot\hat{n})d\sigma$ on the south side. This integral is called the momentum flux through the south side of the control volume. The projection of $\vec{V}$ on the x-direction on the south side is u_s. On the south side, the normal is $\hat{n} = -\hat{j}$. Consequently, $\vec{V}\cdot\hat{n} = (u_s\hat{i} + v\hat{j})\cdot(-\hat{j}) = -v$. As a result, the flux through the south side is $-\rho u_s v dx = -\rho(u + \frac{1}{2}\frac{\partial u}{\partial x}dx - \frac{1}{2}\frac{\partial u}{\partial y}dy)vdx$. The x-components of the fluxes through the other sides of the control volume are calculated similarly and the results are shown in Table 4.1.

If the volume forces $\vec{F}_V$ are due to gravitation, then they can be neglected because they are much smaller than the surface forces. The surface forces include the viscous forces $\vec{F}_{vis}$ and the pressure forces $\vec{F}_p$. Using the fluxes from Table 4.1, the projection of the momentum conservation equation (2.31) in the x-direction yields

$$\rho\frac{\partial u}{\partial y}v\,dxdy + \rho u\frac{\partial v}{\partial y}\,dxdy + 2\rho u\frac{\partial u}{\partial x}\,dxdy = F_{p_x} + F_{vis_x}.$$

Using the mass conservation equation (4.7) and substituting F_{p_x} and F_{vis_x} as a function of pressure and shear stress, the x-component of the momentum conservation equation becomes

$$\rho\left(u\frac{\partial u}{\partial x} + v\frac{\partial u}{\partial y}\right)dxdy = p\,dy - (p + \frac{\partial p}{\partial x}dx)\,dy + (\tau_x + \frac{\partial \tau_x}{\partial y}dy)\,dx - \tau_x\,dx$$

or

$$\rho\left(u\frac{\partial u}{\partial x} + v\frac{\partial u}{\partial y}\right) = -\frac{\partial p}{\partial x} + \frac{\partial \tau_x}{\partial y}, \tag{4.8}$$

where the shear stress in the x-direction is mainly due to the variation of the velocity u in the y-direction such that it can be approximated as

$$\tau_x \simeq \mu\frac{\partial u}{\partial y}. \tag{4.9}$$

Before simplifying (4.8) further, let us note that the y-momentum equation can be derived similarly, yielding

$$\rho\left(u\frac{\partial v}{\partial x} + v\frac{\partial v}{\partial y}\right) = -\frac{\partial p}{\partial y} + \frac{\partial \tau_y}{\partial x}. \tag{4.10}$$

where the shear stress in the y direction is mainly due to the variation of the velocity v in the x direction

$$\tau_y \simeq \mu \frac{\partial v}{\partial x}.$$

Assuming dynamic viscosity μ is not a function of space, (4.10) becomes

$$\rho \left(u \frac{\partial v}{\partial x} + v \frac{\partial v}{\partial y} \right) = -\frac{\partial p}{\partial y} + \mu \frac{\partial^2 v}{\partial x^2}. \tag{4.11}$$

The y-momentum equation (4.10) can be simplified by using an order-of-magnitude analysis to determine which terms are small enough to neglect. Let us start by observing that velocity u is proportional to U_∞:

$$u \propto U_\infty, \qquad v \propto u \frac{\delta}{\ell} \tag{4.12}$$

where ℓ is a characteristic length, such as the length of the airfoil or flat plate.

The next step is to define dimensionless dependent and independent variables. The dependent dimensionless variables $\bar{u}$ and $\bar{v}$ are defined using (4.12):

$$\bar{u} = \frac{u}{U_\infty}, \quad \bar{v} = \frac{v}{U_\infty} \frac{\ell}{\delta}.$$

In addition, the dimensionless pressure is defined as

$$\bar{p} = \frac{p}{\rho U_\infty^2}.$$

The dimensionless independent variables are defined as

$$\bar{x} = \frac{x}{\ell}, \quad \bar{y} = \frac{y}{\delta}.$$

Substituting the dimensionless variables in (4.11) yields

$$\mathrm{Re} \frac{\delta^3}{\ell^3} \left(\bar{u} \frac{\partial \bar{v}}{\partial \bar{x}} + \bar{v} \frac{\partial \bar{v}}{\partial \bar{y}} \right) = -\mathrm{Re} \frac{\delta}{\ell} \frac{\partial \bar{p}}{\partial \bar{y}} + \frac{\delta^3}{\ell^3} \frac{\partial^2 \bar{v}}{\partial \bar{x}^2}, \tag{4.13}$$

where the Reynolds number, Re, is defined as $\mathrm{Re} = \rho U_\infty \ell / \mu$. Since $\delta/\ell \ll 1$, (4.13) reduces to

$$\mathrm{Re} \frac{\delta}{\ell} \frac{\partial \bar{p}}{\partial \bar{y}} = 0$$

such that the projection of the momentum equation in the y-direction yields

$$\boxed{\frac{\partial p}{\partial y} = 0}, \tag{4.14}$$

which states that in the viscous boundary layer, the pressure does not vary in the direction normal to the wall.

Combining (4.8) and (4.9) and assuming again that dynamic viscosity μ is not a function of space yields

$$\rho \left(u\frac{\partial u}{\partial x} + v\frac{\partial u}{\partial y} \right) = -\frac{dp}{dx} + \mu \frac{\partial^2 u}{\partial y^2}. \tag{4.15}$$

Note that the partial derivative of pressure was replaced by a derivative since the pressure is a function of x only.

In summary, (4.7) and (4.15) are known as the *Prandtl boundary layer equations* for steady flow:

$$\frac{\partial u}{\partial x} + \frac{\partial v}{\partial y} = 0$$

$$\rho \left(u\frac{\partial u}{\partial x} + v\frac{\partial u}{\partial y} \right) = -\frac{dp}{dx} + \mu \frac{\partial^2 u}{\partial y^2}.$$

4.2.2.2 Integral Form

The momentum integral method can be applied to compressible and incompressible boundary layers. Let us integrate the x-component of the momentum equation (4.15). The integration will be done in the y direction, from zero (wall surface) to h, where $h > \delta$:

$$\int_0^h \left(u\frac{\partial u}{\partial x} + v\frac{\partial u}{\partial y} \right) dy = \int_0^h -\frac{1}{\rho}\cdot\frac{dp}{dx} dy + \int_0^h v\frac{\partial^2 u}{\partial y^2} dy. \tag{4.16}$$

By differentiating with respect to x the Bernoulli equation for a steady, incompressible flow, $p + \rho U_\infty^2/2 = \text{const}$, one obtains

$$-\frac{1}{\rho}\frac{dp}{dx} = U_\infty \frac{dU_\infty}{dx} \tag{4.17}$$

such that (4.16) becomes

$$\int_0^h \left(u\frac{\partial u}{\partial x} + v\frac{\partial u}{\partial y} \right) dy - \int_0^h U_\infty \frac{dU_\infty}{dx} dy = \int_0^h v\frac{\partial^2 u}{\partial y^2} dy. \tag{4.18}$$

Using the continuity equation (4.7), one obtains

$$v = \int_0^y \frac{\partial v}{\partial y} dy = -\int_0^y \frac{\partial u}{\partial x} dy \tag{4.19}$$

such that the second term of the LHS of (4.16) becomes

$$-\int_0^h \left(\int_0^y \frac{\partial u}{\partial x} dy \right) \frac{\partial u}{\partial y} dy = -\int_0^h \left(\int_0^y \frac{\partial u}{\partial x} dy \right) du. \tag{4.20}$$

Using integration by parts, (4.20) becomes

$$
\int_0^h u \frac{\partial u}{\partial x} dy - \left(\int_0^y \frac{\partial u}{\partial x} dy \right) u \Big|_0^h = \int_0^h u \frac{\partial u}{\partial x} dy - \left(\int_0^h \frac{\partial u}{\partial x} dy \right) U_\infty.
\tag{4.21}
$$

Substituting (4.21) in (4.18) yields

$$
\int_0^h \left(u \frac{\partial u}{\partial x} + u \frac{\partial u}{\partial x} - U_\infty \frac{\partial u}{\partial x} - U_\infty \frac{dU_\infty}{dx} \right) dy = \int_0^h v \frac{\partial^2 u}{\partial y^2} dy,
\tag{4.22}
$$

which can be written as

$$
\int_0^h \left(\frac{\partial u^2}{\partial x} - U_\infty \frac{\partial u}{\partial x} - U_\infty \frac{dU_\infty}{dx} \right) dy = \int_0^h v \frac{\partial^2 u}{\partial y^2} dy.
\tag{4.23}
$$

Adding and deducting $u \frac{dU_\infty}{dx}$ to the LHS of (4.23) yields

$$
\int_0^h \left((u - U_\infty) \frac{dU_\infty}{dx} \right) dy + \int_0^h \left(\frac{\partial u^2}{\partial x} - U_\infty \frac{\partial u}{\partial x} - u \frac{dU_\infty}{dx} \right) dy = \int_0^h v \frac{\partial^2 u}{\partial y^2} dy,
\tag{4.24}
$$

which can be rearranged as

$$
\frac{dU_\infty}{dx} \int_0^h (u - U_\infty) \, dy + \int_0^h \left(\frac{\partial u^2}{\partial x} - \frac{\partial (u U_\infty)}{\partial x} \right) dy = \int_0^h v \frac{\partial^2 u}{\partial y^2} dy,
\tag{4.25}
$$

or better yet

$$
U_\infty \frac{dU_\infty}{dx} \int_0^h \left(\frac{u}{U_\infty} - 1 \right) dy + \int_0^h \frac{\partial}{\partial x} \left(u^2 - u U_\infty \right) dy = \int_0^h v \frac{\partial^2 u}{\partial y^2} dy.
\tag{4.26}
$$

Changing the order of integral and differential operators in the second term of the LHS and multiplying and dividing by U_∞^2 yields

$$
U_\infty \frac{dU_\infty}{dx} \int_0^h \left(\frac{u}{U_\infty} - 1 \right) dy + \frac{\partial}{\partial x} \left[U_\infty^2 \int_0^h \frac{u}{U_\infty} \left(\frac{u}{U_\infty} - 1 \right) dy \right] = \int_0^h v \frac{\partial^2 u}{\partial y^2} dy.
\tag{4.27}
$$

Using the definitions of the displacement thickness (4.3) and momentum thickness (4.5) yields

$$
-U_\infty \frac{dU_\infty}{dx} \delta^* - \frac{\partial}{\partial x} \left(U_\infty^2 \theta \right) = \int_0^h v \frac{\partial^2 u}{\partial y^2} dy
\tag{4.28}
$$

or

$$U_\infty \frac{dU_\infty}{dx} \delta^* + 2 U_\infty \theta \frac{dU_\infty}{dx} + U_\infty^2 \frac{d\theta}{dx} = -\int_0^h \nu \frac{\partial^2 u}{\partial y^2} \, dy. \tag{4.29}$$

Assuming that the viscosity μ does not vary with y, the RHS of (4.29) can be written as

$$-\frac{1}{\rho} \int_0^h \frac{\partial}{\partial y} \left(\mu \frac{\partial u}{\partial y} \right) dy,$$

and approximating the shear stress in the boundary layer as

$$\tau = \mu \frac{\partial u}{\partial y},$$

the RHS of (4.29) becomes

$$-\frac{1}{\rho} \int_0^h \frac{\partial \tau}{\partial y} dy = -\frac{1}{\rho} \left(\tau(h) - \tau(0) \right). \tag{4.30}$$

Since the shear stress outside of the boundary layer, that is, at $y = h$, can be approximated as being zero, (4.29) becomes

$$U_\infty \frac{dU_\infty}{dx} \delta^* + 2 U_\infty \theta \frac{dU_\infty}{dx} + U_\infty^2 \frac{d\theta}{dx} = \frac{\tau_w}{\rho} \tag{4.31}$$

where the notation τ_w was used for the shear stresses at wall, $\tau(0)$. Dividing (4.31) by U_∞^2 yields the *momentum integral equation*

$$\boxed{\frac{d\theta}{dx} + \left(2 + \frac{\delta^*}{\theta} \right) \frac{\theta}{U_\infty} \frac{dU_\infty}{dx} = \frac{\tau_w}{\rho U_\infty^2}.} \tag{4.32}$$

Example 4.2.1   Calculate the airfoil total skin friction drag per unit depth as a function of momentum thickness θ if the free-stream velocity U_∞ on the airfoil is constant.

Solution

If U_∞ is constant, (4.32) simplifies to

$$\frac{d\theta}{dx} = \frac{\tau_w}{\rho U_\infty^2}$$

so that

$$\rho U_\infty^2 d\theta = \tau_w dx,$$

which after integration from the leading edge, $x = 0$, to the trailing edge, $x = x_{TE}$, yields

$$\rho U_\infty^2 \theta(x_{TE}) = \int_0^{x_{TE}} \tau_w dx.$$

4.2.3 Laminar Boundary Layer

This section will consider the case of very small viscosity or very large Reynolds number flows. The first part of the section will derive the solution for incompressible flow over a flat plate. This solution will be obtained using an analytical method. The second part of this section will introduce an integral method, valid for both incompressible and compressible flows. This integral method will be obtained as a combination between an analytical method and experimental results.

4.2.3.1 Incompressible Flow over a Flat Plate: Blasius Solution

Let us consider now the case of laminar, incompressible boundary layers over the simplest geometry, that is, over a flat plate. In this case, the governing equations can be further simplified since the free-stream velocity U_∞ is constant and the pressure variation $dp/dx = 0$. The solution of this boundary layer problem is called the Blasius solution.

The governing equations of the incompressible flow are the mass conservation equation

$$\frac{\partial u}{\partial x} + \frac{\partial v}{\partial y} = 0 \tag{4.7}$$

and the momentum conservation equation in the x direction

$$u\frac{\partial u}{\partial x} + v\frac{\partial u}{\partial y} = \nu\frac{\partial^2 u}{\partial y^2}. \tag{4.33}$$

In addition to the governing equations, boundary conditions must be specified in order to uniquely define the problem. At the wall surface, both velocity components must be zero. The boundary condition that requires that $u=0$ at $y=0$ is called the "no-slip" boundary condition. The boundary condition that requires that $v = 0$ at $y = 0$ is called the "no-penetration" boundary condition. In addition, the u velocity at the edge of the boundary layer must be equal to U_∞. In summary, the boundary conditions are

$$u = v = 0 \quad \text{at} \quad y = 0$$
$$u = U_\infty = \text{const} \quad \text{at} \quad y \to \infty. \tag{4.34}$$

It should be noted that the no-slip boundary condition must only be enforced for viscous flows, while the no-penetration boundary condition must be enforced whether the flow is viscous or not.

Rather than solving the two equations (4.7) and (4.33), let us use a stream function ψ and write the velocities u and v as a function of ψ:

$$u = \frac{\partial \psi}{\partial y} \qquad v = -\frac{\partial \psi}{\partial x}. \tag{4.35}$$

As a result, the stream function ψ satisfies the mass conservation equation (4.7). Consequently, one only has to solve (4.33). After substituting u and v by ψ, (4.33) becomes:

$$\frac{\partial \psi}{\partial y}\frac{\partial^2 \psi}{\partial x \partial y} - \frac{\partial \psi}{\partial x}\frac{\partial^2 \psi}{\partial y^2} = \nu \frac{\partial^3 \psi}{\partial y^3} \tag{4.36}$$

Blasius argued that for an infinite flat plate there is no true characteristic length [Blasius, 1908]. Consequently, the solution of the boundary layer should be scalable such that the solutions at different x locations along the plate are similar. Likewise, the velocity profiles at different x locations should be scalable. To satisfy this *flow similarity* requirement, Blasius suggested the similarity transformation

$$u(x, y) = U_\infty F(\eta) \tag{4.37}$$

where

$$\eta = \frac{y}{g(x)}.$$

$g(x)$ is called the stretch or scaling factor. Since y is much smaller than x, a natural choice for $g(x)$ is such that

$$\eta = \frac{y}{x}\sqrt{\text{Re}_x} = y\sqrt{\frac{U_\infty}{\nu x}} \tag{4.38}$$

where

$$\text{Re}_x = U_\infty x / \nu. \tag{4.39}$$

Since it is assumed that the y coordinate can be scaled at any x location to yield a similar velocity profile, it follows that the stream function should be a function of x and η, $\psi(x, \eta)$. Therefore

$$u = \frac{\partial \psi}{\partial y} = \frac{\partial \psi}{\partial \eta}\frac{\partial \eta}{\partial y} = \frac{\partial \psi}{\partial \eta}\sqrt{\frac{U_\infty}{\nu x}}. \tag{4.40}$$

Equations (4.40) and (4.37) yield

$$\frac{\partial \psi}{\partial \eta} = \sqrt{\nu x U_\infty} F(\eta). \tag{4.41}$$

Integrating (4.41) with respect to η yields

$$\psi = \sqrt{\nu x U_\infty} f(\eta) \tag{4.42}$$

where

$$F(\eta) = f'(\eta). \tag{4.43}$$

Using (4.42), the stream function ψ is replaced by the non-dimensional stream function f:

$$f(\eta) := \psi/\sqrt{\nu x U_\infty}. \tag{4.44}$$

Note that f is only a function of η.

Example 4.2.2 Derive the expression of the x-momentum conservation equation (4.36) as a function of the coordinates (x, η) and the dimensionless stream function f.

Solution

Let us express $\frac{\partial \psi}{\partial x}$ and $\frac{\partial \psi}{\partial y}$ using (4.38) and (4.44). Using the chain rule,

$$\frac{\partial \psi}{\partial x} = \frac{\partial}{\partial x}\left(f(\eta)\sqrt{\nu x U_\infty}\right) = f\sqrt{\nu U_\infty}\frac{1}{2\sqrt{x}} + \sqrt{\nu x U_\infty}f'\frac{\partial \eta}{\partial x} = \tag{4.45}$$

$$= \frac{1}{2}f\sqrt{\frac{\nu U_\infty}{x}} + \sqrt{\nu x U_\infty}f'\left(-\frac{1}{2}y\sqrt{\frac{U_\infty}{\nu x^3}}\right) = \frac{1}{2}\sqrt{\frac{\nu U_\infty}{x}}\left(f - \eta f'\right)$$

where prime $'$ denotes derivative with respect to η. The other terms of (4.36) are calculated similarly:

$$\frac{\partial^2 \psi}{\partial x \partial y} = -\frac{1}{2}\frac{U_\infty}{x}f''\eta, \tag{4.46}$$

$$\frac{\partial \psi}{\partial y} = \sqrt{\nu x U_\infty}f'\frac{\partial \eta}{\partial y} = \sqrt{\nu x U_\infty}f'\sqrt{\frac{U_\infty}{\nu x}} = U_\infty f', \tag{4.47}$$

$$\frac{\partial^2 \psi}{\partial y^2} = \frac{\partial}{\partial y}\left(U_\infty f'\right) = U_\infty f''\frac{\partial \eta}{\partial y} = U_\infty f''\sqrt{\frac{U_\infty}{\nu x}}, \tag{4.48}$$

and

$$\frac{\partial^3 \psi}{\partial y^3} = \frac{\partial}{\partial y}\left(U_\infty f''\sqrt{\frac{U_\infty}{\nu x}}\right) = \frac{U_\infty^2}{\nu x}f'''. \tag{4.49}$$

Substituting (4.45)–(4.49) in (4.36) yields

$$ff'' + 2f''' = 0 \tag{4.50}$$

which is a third-order ordinary differential equation.

It is certainly easier to solve the ordinary differential equation (4.50) than to solve either the system of partial differential equations (4.7) and (4.33), or the partial differential equation (4.36).

Having the x-momentum conservation equation written as (4.50) using the dimensionless stream function f, the next step is to determine the boundary conditions for the dimensionless stream function. For this, the boundary conditions (4.34) expressed by velocities u and v

must be written as a function of f. Velocity u is obtained by combining (4.37) and (4.43), such that

$$u(x,\eta) = U_\infty f'(\eta). \tag{4.51}$$

Velocity v is obtained from

$$v(x,\eta) \overset{(4.35)}{=} -\frac{\partial \psi}{\partial x} \overset{(4.45)}{=} -\frac{1}{2}\sqrt{\frac{\nu U_\infty}{x}}\,(f - \eta f'). \tag{4.52}$$

Using (4.51) and (4.52), the boundary conditions (4.34) become

$$\eta = 0 \quad f = f' = 0 \tag{4.53}$$
$$\eta \to \infty \quad f' = 1.$$

The ordinary differential equation (ODE) (4.50) does not have an analytical solution. However, a numerical solution of the ODE (4.50) can be obtained. This numerical solution was calculated for the first time by Paul Richard Heinrich Blasius in his Ph.D. dissertation (1908) and was published in [Blasius, 1908]. For this reason, the solution of this equation is called the Blasius solution. P. R. H. Blasius (1883–1970), a student of L. Prandtl, was a German fluid dynamicist.

Example 4.2.3 Find the Blasius solution and list the values of f, f' and f''.

Solution
Using the NDSolve function of Mathematica (or any similar software package that solves ordinary differential equations),

```
{s} = NDSolve[{f[e]f''[e]+2f'''[e]==0, f[0]==0, f'[0]==0,
f'[8]==1}, f,{e,0,8}],
```

the results are as shown in Table 4.2.

The variation of the $f'(\eta)$ function is shown in Fig. 4.5. In addition, Table 4.2 shows that at $\eta = 5, f'(\eta)$ is approximately 0.99, that is, $u/U_\infty = 99\%$. Then, according to the definition of the boundary layer thickness and the definition of η (4.38),

$$\delta = 5.0\sqrt{\frac{\nu x}{U_\infty}} \quad \text{for } u = 0.99\, U_\infty.$$

Using the values of f' from Table 4.2, the displacement thickness and the momentum thickness are calculated using their definitions (4.3) and (4.5):

$$\delta^* = 1.72\sqrt{\frac{\nu x}{U_\infty}}$$

Table 4.2 Blasius solution.

η	$f(\eta)$	$f'(\eta)$	$f''(\eta)$
0.00	0.00000	0.00000	0.33200
0.20	0.00664	0.06640	0.33200
0.40	0.02660	0.13300	0.33100
0.60	0.05970	0.19900	0.33000
0.80	0.10600	0.26500	0.32700
1.00	0.16600	0.33000	0.32300
1.20	0.23800	0.39400	0.31700
1.40	0.32300	0.45600	0.30800
1.60	0.42000	0.51700	0.29700
1.80	0.53000	0.57500	0.28300
2.00	0.65000	0.63000	0.26700
2.20	0.78100	0.68100	0.24800
2.40	0.92200	0.72900	0.22800
2.60	1.07000	0.77200	0.20600
2.80	1.23000	0.81200	0.18400
3.00	1.40000	0.84600	0.16100
3.20	1.57000	0.87600	0.13900
3.40	1.75000	0.90200	0.11800
3.60	1.93000	0.92300	0.09810
3.80	2.12000	0.94100	0.08010
4.00	2.31000	0.95600	0.06420
4.20	2.50000	0.96700	0.05050
4.40	2.69000	0.97600	0.03900
4.60	2.89000	0.98300	0.02950
4.80	3.09000	0.98800	0.02190
5.00	3.28000	0.99200	0.01590
5.20	3.48000	0.99400	0.01130
5.40	3.68000	0.99600	0.00793
5.60	3.88000	0.99700	0.00543
5.80	4.08000	0.99800	0.00365
6.00	4.28000	0.99900	0.00240

$$\theta = 0.664\sqrt{\frac{\nu x}{U_\infty}}.$$

The shape factor is

$$H = \frac{\delta^*}{\theta} = \frac{1.72}{0.664} = 2.59.$$

Taking into account that $U_\infty = \text{const}$, the non-dimensional wall shear stress is obtained from (4.32):

$$\frac{\tau_w}{\rho U_\infty^2} = 0.332\sqrt{\frac{\nu}{U_\infty x}}. \tag{4.54}$$

Figure 4.5 Blasius solution: velocity variation.

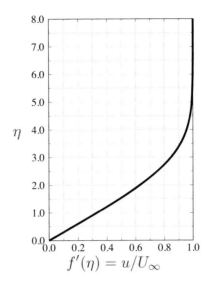

$$f'(\eta) = u/U_\infty$$

Using the wall shear stress, τ_w, one can define the local skin friction coefficient as

$$c_f := \frac{\tau_w}{0.5\rho U_\infty^2} \qquad (4.55)$$

and the skin friction drag coefficient, a global skin friction coefficient, as

$$C_f := \frac{1}{c}\int_0^c c_f\, dx$$

where c is the chord length.

Using the wall shear stress (4.54) obtained from the Blasius solution, the local skin friction (4.55) is

$$c_f = \frac{0.664}{\sqrt{\text{Re}_x}},$$

and the skin friction coefficient is

$$C_f = \frac{1.328}{\sqrt{\text{Re}_c}}.$$

Note that Re_x is the local Reynolds number, that is, the Reynolds number based on the x location. Re_c is the Reynolds number based on the chord length, c.

A more general case than the case of the boundary layer on a flat plate is when the free-stream velocity has a linear variation $U_\infty = U_1 - \alpha x$. This case was studied by L. Howarth, who used an expansion of stream function (4.44) as a function of $\alpha x/U_1$ [Howarth, 1938]. The variation of the velocity profile for different values of the α parameter is shown in Fig. 4.6.

Figure 4.6 Velocity variation in laminar boundary layer as a function of Howarth's parameter (adapted from [Howarth, 1938]).

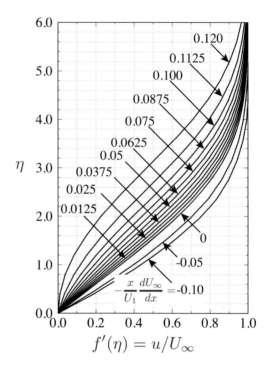

$$f'(\eta) = u/U_\infty$$

4.2.3.2 Thwaites Integral Momentum Method

The Thwaites integral momentum method is a fast way to determine the variation of the momentum thickness of the boundary layer. This method is valid for any variation of the velocity $U_\infty(x)$. The method was developed as a combination between an analytical method and experimental results [Thwaites, 1949].

Let us start by multiplying the momentum equation (4.32) by $2\theta/\nu$. One obtains

$$\frac{d}{dx}\left(\frac{\theta^2}{\nu}\right) + \frac{2}{U_\infty}\left[(2+H)\frac{\theta^2}{\nu}\frac{dU_\infty}{dx} - T\right] = 0, \tag{4.56}$$

where T, the Thwaites parameter, is defined as

$$\boxed{T := \frac{\tau_w \theta}{\mu U_\infty}}. \tag{4.57}$$

Using experimental results, T can be approximated as a function of the shape factor, H and the λ parameter as

$$T = (H-1)\lambda + 0.225, \tag{4.58}$$

where λ is defined as

$$\boxed{\lambda := \frac{\theta^2}{\nu}\frac{dU_\infty}{dx}}. \tag{4.59}$$

Let us replace λ from equation (4.59) into the expression of T given by (4.58), and then substitute T into the momentum equation (4.56). The result is

$$\frac{d}{dx}\left(\frac{\theta^2}{\nu}\right) + \frac{6}{U_\infty}\frac{\theta^2}{\nu}\frac{dU_\infty}{dx} = \frac{0.45}{U_\infty}. \tag{4.60}$$

Let us denote by U_1 the value of the edge velocity U_∞ at $x = x_1$. Then we multiply equation (4.60) by $(U_\infty/U_1)^6$ to obtain

$$\frac{d}{dx}\left[\left(\frac{\theta^2}{\nu}\right)\left(\frac{U_\infty}{U_1}\right)^6\right] = \frac{0.45}{U_\infty}\left(\frac{U_\infty}{U_1}\right)^6. \tag{4.61}$$

By integrating equation (4.61) from x_1 to x, one obtains

$$\boxed{\frac{\theta^2(x)}{\nu} = \frac{\theta_1^2}{\nu}\left(\frac{U_1}{U_\infty(x)}\right)^6 + \frac{0.45}{U_1}\left(\frac{U_1}{U_\infty(x)}\right)^6 \int_{x_1}^{x}\left[\frac{U_\infty(\xi)}{U_1}\right]^5 d\xi} \tag{4.62}$$

where θ_1 is the value of θ at $x = x_1$.

Example 4.2.4 Use Thwaites' method to calculate $\theta(x)$ on a flat plate.

Solution
Let us choose $x_1 = 0$ such that $\theta_1 = 0$. Since on a flat plate $\frac{U_1}{U_\infty} = 1$, equation (4.62) becomes

$$\frac{\theta^2}{\nu} = \frac{0.45}{U_\infty}x \rightarrow \theta = 0.67\sqrt{\frac{x\nu}{U_\infty}}.$$

Note that the coefficient 0.67 is close to 0.664, the coefficient obtained from Blasius analytical solution. This confirms that the experimentally determined Thwaites parameter, T, is plausible, at least for flat plates.

Example 4.2.5 The free stream velocity over a fan blade varies as $U(x) = 120 - 60x$ m/s, where x is the distance from the leading edge. Assuming the flow is laminar in the vicinity of the leading edge, and the kinematic viscosity is $\nu = 14.6 \cdot 10^{-6}$ m²/s, calculate:

1. the variation of momentum thickness as a function of the distance from the leading edge, x;
2. the momentum thickness, θ, at 10 cm downstream from the leading edge of the fan blade.

Solution

1. Since at $x_1 = 0$, $\theta_1 = 0$, and substituting $U(x) = 120 - 60x$ in

$$\frac{\theta^2(x)}{\nu} = \frac{\theta_1^2}{\nu} \left(\frac{U_1}{U_\infty(x)} \right)^6 + \frac{0.45}{U_1} \left(\frac{U_1}{U_\infty(x)} \right)^6 \int_{x_1}^{x} \left[\frac{U_\infty(\xi)}{U_1} \right]^5 d\xi \qquad (4.63)$$

yields

$$\frac{\theta^2(x)}{\nu} = \frac{0.45}{120} \left(\frac{120}{120 - 60x} \right)^6 \int_{0}^{x} \left[\frac{120 - 60\xi}{120} \right]^5 d\xi$$

or

$$\frac{\theta^2(x)}{\nu} = \frac{0.45}{120} \left(\frac{1}{1 - 0.5x} \right)^6 \int_{0}^{x} (1 - 0.5\xi)^5 \, d\xi. \qquad (4.64)$$

With the change of variables $1 - 0.5\xi = v$, (4.64) becomes

$$\frac{\theta^2(x)}{\nu} = \frac{0.45}{60} \left(\frac{1}{1 - 0.5x} \right)^6 \int_{1-0.5x}^{1} v^5 \, dv = \frac{0.45}{360} \left(\frac{1}{1 - 0.5x} \right)^6 \left[1 - (1 - 0.5x)^6 \right]$$

or

$$\theta(x) = \frac{1}{(1 - 0.5x)^3} \sqrt{v0.00125 \left[1 - (1 - 0.5x)^6 \right]}.$$

2. The momentum thickness at $x = 0.1$ is

$$\theta(x = 0.1) = \frac{1}{(1 - 0.5 \cdot 0.1)^3} \sqrt{14.6 \cdot 10^{-6} \times 0.00125 \left[1 - (1 - 0.5 \cdot 0.1)^6 \right]}$$

$$= 8.1 \cdot 10^{-5} \text{ m}.$$

A curve fit of the experimental results of the Thwaites T parameter yields [White, 2006, p. 269]

$$T = (\lambda + 0.09)^{0.62}, \qquad (4.65)$$

which is independent of the shape factor H. By eliminating the T parameter from (4.58) and (4.65), one obtains the variation of the displacement thickness δ^* as a function of the momentum thickness and the λ parameter:

$$\boxed{\delta^* = \frac{\theta}{\lambda} \left[\lambda + (\lambda + 0.09)^{0.62} - 0.225 \right],} \qquad (4.66)$$

or, using the shape factor definition (4.6),

$$H = \frac{1}{\lambda} \left[\lambda + (\lambda + 0.09)^{0.62} - 0.225 \right]. \qquad (4.67)$$

K. Pohlhausen [Pohlhausen, 1921] showed that $\left(\frac{\partial u}{\partial y}\right)_{y=0} = \frac{U_\infty}{6\delta}(12 + \lambda)$ and therefore separation occurs at $\lambda = -12$. Furthermore, if $\lambda < -12$, then $\left(\frac{\partial u}{\partial y}\right)_{y=0} < 0$. Consequently, only values of $\lambda > -12$ are of interest. In addition, λ has an upper value of 12 [Thwaites, 1949] such that λ varies between -12 and 12.

Example 4.2.6 Calculate the shape factor H for the flow over a flat plate using (4.66).

Solution
On a flat plate, $\lambda = 0$ because $\frac{dU_\infty}{dx} = 0$. Applying l'Hospital's rule yields

$$H = \lim_{\lambda \to 0} \frac{1 + 0.62(\lambda + 0.09)^{-0.38}}{1} = 2.55, \tag{4.68}$$

which is not far from the $H = 2.59$ obtained using the Blasius solution. This good approximation confirms that the approximations (4.58) and (4.65) proposed by Thwaites and White, respectively, are plausible.

4.2.4 Turbulent Boundary Layer

Laminar flows and the corresponding laminar boundary layers are rather limited in propulsion systems, as most flows in propulsion systems are turbulent. In turbulent flows, an irregular fluctuation is superimposed on the main stream. Consequently, the velocity components (u, v, w) can be written as a sum between a mean motion, which is no longer a function of time, and a fluctuating motion:

$$
\begin{aligned}
u(x, y, z, t) &= \bar{u}(x, y, z) + u'(x, y, z, t) \\
v(x, y, z, t) &= \bar{v}(x, y, z) + v'(x, y, z, t) \\
w(x, y, z, t) &= \bar{w}(x, y, z) + w'(x, y, z, t),
\end{aligned} \tag{4.69}
$$

where the mean value, for example, $\bar{u}$, is

$$\bar{u} = \frac{1}{t_1} \int_{t_0}^{t_0 + t_1} u\, dt$$

where t_1 should be large enough such that the average value does not depend on time. The pressure can also be split into a mean and fluctuating value,

$$p = \bar{p} + p'. \tag{4.70}$$

For compressible flows, density and temperature have also fluctuating components, that is,

$$\rho = \bar{\rho} + \rho', \quad T = \bar{T} + T'. \tag{4.71}$$

By definition, the time average of the fluctuations is zero, that is,

$$\overline{u'} = 0, \quad \overline{v'} = 0, \quad \overline{w'} = 0, \quad \overline{p'} = 0, \quad \overline{\rho'} = 0, \quad \overline{T'} = 0.$$

Assuming two turbulent quantities f and g, the following rules of averaging apply:

$$\overline{\overline{f}} = \overline{f} \qquad\qquad \overline{f'} = 0 \qquad\qquad \overline{f + g} = \overline{f} + \overline{g}$$

$$\overline{\overline{f}g} = \overline{f}\overline{g} \qquad\qquad \overline{f'\overline{g}} = 0 \qquad\qquad \overline{fg} = \overline{f}\overline{g} + \overline{f'g'}. \qquad (4.72)$$

$$\overline{\frac{\partial f}{\partial s}} = \frac{\partial \overline{f}}{\partial s} \qquad\qquad \overline{\int f ds} = \int \overline{f} ds$$

Of fundamental importance for turbulent flows is the fact that the fluctuations affect the mean motion such that an apparent increase in the resistance against deformation occurs. This apparent viscosity is the crucial concept for modeling turbulent flows.

After substituting (4.69)–(4.71) in the momentum conservation equations and using the time average rules (4.72), additional shear stress-like terms appear that involve turbulent fluctuations. These terms supplement the laminar shear stresses. The shear stress in the x direction due to the turbulent flow is [Schlichting, 1979, p. 559]

$$\tau_t = -\rho\overline{u'v'},$$

where the bar on top of the $u'v'$ denotes the mean of the product of turbulent fluctuations. As a result, the shear stress is

$$\tau = \tau_l + \tau_t = \mu\frac{\partial \overline{u}}{\partial y} - \rho\overline{u'v'}.$$

Note that upward fluctuations, $v' > 0$, on average carry flow with a local negative u fluctuation, $u' < 0$. Similarly, downward flow fluctuations, $v' < 0$, will have on average a local positive u fluctuation, $u' > 0$. Consequently, the time average of the product $\overline{u'v'}$ is negative, and therefore the turbulent stress is positive, $\tau_t > 0$. Note that away from the wall, $\tau_t \gg \tau_l$.

The turbulent flow can be modeled using different turbulence models. Without going into the details, the most widely used approaches can be grouped into four main categories: (1) the eddy viscosity model, (2) the stress-transport model, (3) large eddy simulation, and (4) direct numerical simulation.

The eddy viscosity model, also called the Boussinesq model, was introduced by Boussinesq [1896] and assumes that the turbulent shear stress looks similar to the laminar counterpart $\rho\nu\frac{\partial \overline{u}}{\partial y}$, that is

$$\tau_t = -\rho\overline{u'v'} = \rho\nu_T\frac{d\overline{u}}{dy},$$

where ν_T is the eddy viscosity. The Boussinesq assumption led to the development of several turbulence models: (i) the zero-equation model, (ii) the one-equation model, and (iii) the two-equation model. The one- and two-equation models require the solution of one and two equations, respectively, to predict the eddy viscosity.

The *zero-equation model* is based on the Prandtl mixing length concept [Prandtl, 1925], which represents the eddy viscosity as

$$\nu_T = \ell^2 \left| \frac{d\bar{u}}{dy} \right|,$$

where ℓ is the mean free path. Examples of zero-equation models are the Cebeci and Smith [1974], Baldwin and Lomax [1978], and Johnson and King [1985] models.

In *one-equation models*, the eddy viscosity, ν_T, is expressed as a product of a characteristic velocity and a length derived from a transport equation, usually the turbulence kinetic energy k, which is defined as $k = (\overline{u'^2} + \overline{v'^2} + \overline{w'^2})/2$. Modeling eddy viscosity that depends on the flow history yields a more physically realistic model than the zero-equation models. The lack of a turbulent scale, however, makes some of the one-equation models incomplete. Examples of one-equation models include Prandtl [1945], Baldwin and Barth [1991], and Spalart and Allmaras [1992].

In *two-equation models*, the eddy viscosity, ν_T, is expressed as a product of a characteristic velocity and a length derived from two transport equations. One transport equation is for k and the second equation is for either the turbulence dissipation per unit mass, $\epsilon = \nu \sum_{i=1}^{3} \sum_{k=1}^{3} \overline{\frac{\partial u_i'}{\partial x_k} \frac{\partial u_i'}{\partial x_k}}$, or the root mean square fluctuating vorticity, ω, of the turbulence scale, ℓ. The two-equation models are complete. The first two-equation model was introduced by Kolmogorov [1942] and modeled the k and ω parameters. Other two-equation models are the k-ϵ model by Launder and Spalding [1972] and the k-ω model by Saffman [1970], Wilcox [1988], and Menter [1994].

Stress-transport models were introduced by Chou [1945] and Rotta [1951], who obviated the use of the Boussinesq approximation. The stress-transport models describe the differential equation governing the evolution of the tensor that represents the turbulent stresses, that is, the Reynolds-stress tensor [Donaldson and Rosenbaum, 1968; Launder et al., 1975]. The stress-transport models introduce one equation for the turbulence scale and six equations for the components of the Reynolds-stress tensor. Consequently, the computational cost of using stress-transport models is much higher than using two-equation models, and for this reason their use in applications is rather limited compared to the zero-, one-, or two-equation models.

In Large Eddy Simulation (LES), it is assumed that the largest eddies are dependent on the boundary conditions and must be computed, while the small-scale turbulence is nearly isotropic, has nearly universal characteristics and can be modeled [Smagorinsky, 1963; Deardorff, 1974; Sagaut, 2009]. It is also assumed that the largest eddies carry most of the Reynolds stress, while the smallest eddies contribute less to the Reynolds stress. LES is currently used for the analysis of flows in propulsion systems; however, the computational cost of such simulations is as yet too high for design.

Direct Numerical Simulation (DNS) solves the time-dependent three-dimensional Navier–Stokes equations without a turbulence model. Consequently, the whole range of spatial and temporal scales of the turbulence must be resolved [Moin and Mahesh, 1998]. As a result, severe restrictions are imposed on the grid size as the number of grid points scales with Re$^{9/4}$ [Kim et al., 1987]. In addition, the time step must be of the same order as the Kolmogorov time scale, $(v/\epsilon)^{1/2}$. Because of these restrictions, with current computers, DNS can only be used for Reynolds numbers well below those relevant for engineering applications.

4.2.4.1 Turbulent Flow in a Pipe

A large number of experiments have been done to study turbulent flows in pipes and to derive various empirical correlations. This section presents a set of these results and correlations. These experimental results were instrumental in understanding the fundamentals of turbulent flow. Therefore, the methods developed for dealing with turbulent flows in pipes were expanded to other turbulent flows, such as the flows over flat plates and airfoils.

Let us apply the momentum conservation equation (3.8) to the flow in a pipe with constant radius R, shown in Fig. 4.7. Ignoring the volume forces yields

$$\pi R^2 \Delta p - 2\pi RL\tau_w = 0, \tag{4.73}$$

where R is the pipe radius, L is the pipe length, and Δp is pressure variation over the length L. The wall shear stress obtained from equation (4.73) is

$$\tau_w = \frac{R\Delta p}{2L}. \tag{4.74}$$

To obtain the shear stress at radius r, the momentum equation will be written for a tube of radius r:

$$\pi r^2 \Delta p - 2\pi rL\tau = 0. \tag{4.75}$$

The shear stress at radius r, where $0 < r < R$, becomes

$$\tau = \frac{r\Delta p}{2L}. \tag{4.76}$$

From equations (4.74) and (4.76) one obtains

$$\frac{\tau}{\tau_w} = \frac{r}{R}. \tag{4.77}$$

Figure 4.7 Pipe dimensions.

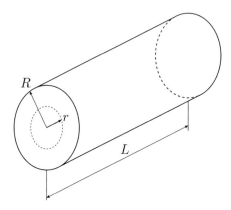

It is obvious from equations (4.74) and (4.76) that the value of the shear stress can be calculated by measuring the pressure variation Δp.

Going back to (4.74), one can write the pressure variation per length

$$\frac{\Delta p}{L} = \tau_w \frac{4}{D}$$

where D is the pipe diameter. It is beneficial to introduce a *dimensionless coefficient of resistance for pipe flow*, f, also called the friction factor

$$\frac{\Delta p}{L} = \frac{f}{D} \frac{\rho}{2} \overline{U}^2,$$

where $\overline{U}$ is the mean (or bulk) velocity

$$\overline{U} := \frac{\int\limits_0^R 2\pi r u(r)\, dr}{\int\limits_0^R 2\pi r\, dr} = \frac{2\int\limits_0^R r u(r)\, dr}{R^2} \tag{4.78}$$

and $u(r)$ is the velocity at radius r, such that

$$\tau_w = \frac{1}{8} f \rho \overline{U}^2. \tag{4.79}$$

The relationship between the friction factor, f, and the local skin friction coefficient, c_f of (4.55), is

$$f = 4c_f, \tag{4.80}$$

After surveying the experimental results, Blasius [1913] established an empirical equation for f as a function of Re only,

$$f = 0.3164 \frac{1}{\text{Re}^{0.25}}, \tag{4.81}$$

which is valid for Re $= \overline{U}D/\nu < 100,000$. The expression (4.81) is called the *Blasius resistance formula*.

Prandtl expanded the validity of f prediction up to a Re number of 3.2×10^6 by proposing that the dimensionless coefficient of resistance for pipe flow be obtained from

$$\frac{1}{\sqrt{f}} = 2.0 \log(\text{Re}\sqrt{f}) - 0.8, \tag{4.82}$$

which is called *Prandtl's universal law of friction for smooth pipes*.

Figure 4.8 shows the variation of the resistance coefficient of pipe flow as function of the Reynolds number based on the pipe diameter. For Reynolds numbers less than 2000, f drops rapidly with the increase of Reynolds number according to

$$f = 64/\text{Re}, \tag{4.83}$$

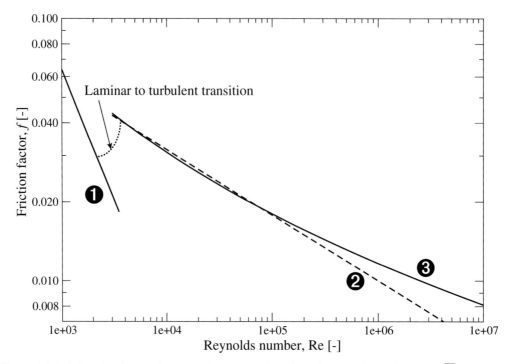

Figure 4.8 Frictional resistance in a smooth pipe as a function of Reynolds number, Re $= \overline{U}D/\nu$: (1) Hagen–Poiseuille laminar flow solution (4.83), (2) Blasius turbulent flow solution (4.81), and (3) Prandtl turbulent flow solution (4.82).

the Hagen–Poiseuille solution [Poiseuille, 1846]. At a Reynolds number of approximately 2000, the flow begins to transition from laminar to turbulent and the resistance coefficient suddenly increases. At a Reynolds number of approximately 4,000, the flow is fully turbulent, and as the Reynolds number continues to increase, f decreases but at a slope smaller than that during the laminar flow. The Blasius resistance formula (4.81) provides a good approximation of the experimental data up to Re $= 10^5$ while Prandtl's universal law of friction formula (4.82) works well up to Re $= 3.2 \times 10^6$.

Let us determine the velocity variation, that is, the velocity profile, in the turbulent boundary layer. Consider a simple dimensional analysis of the wall shear stress

$$\tau_w = \mu \frac{\partial u}{\partial y}\bigg|_w \sim \mu \frac{u}{y}$$

that allows us to write

$$\frac{u}{\sqrt{\frac{\tau_w}{\rho}}} \sim \frac{y\sqrt{\frac{\tau_w}{\rho}}}{\nu}.$$

Based on this equation, the following definitions are introduced:

$$u^* := \sqrt{\frac{\tau_w}{\rho}}$$ Friction velocity

$$y^+ := \frac{yu^*}{\nu}$$ The y^+ number

$$u^+ := \frac{u}{u^*}$$ The u^+ velocity.

Note that both y^+ and u^+ are dimensionless parameters. The y^+ number resembles the Reynolds number, where the velocity is replaced by the friction velocity, u^*. The y^+ and u^+ parameters are used to describe the structure of the turbulent boundary layer. The y^+ number is also important in computational fluid dynamics where it is used to verify that the computational grid near walls is fine enough to capture velocity variation in the boundary layer.

The turbulent boundary layer is typically divided into three regions, from the solid wall to the exterior boundary:

1. Purely laminar friction sublayer. This region is located next to the solid wall and is characterized by purely laminar friction, that is, the contribution from laminar friction is much larger than that of the turbulent friction. The latter can be completely neglected in this sublayer. In this region, which spans $0 < y^+ < 5$ the u^+, velocity varies as

$$u^+ = y^+. \tag{4.84}$$

2. Laminar-turbulent friction sublayer, which spans between $5 \leq y^+ < 70$. In this region both laminar and turbulent friction contributions are of the same order of magnitude. The velocity variation can be approximated as [von Karman, 1939]

$$u^+ = -3.05 + 5.00 \ln y^+. \tag{4.85}$$

3. Purely turbulent friction sublayer, which spans above $y^+ \geq 70$. In this region the laminar friction contribution can be neglected compared with turbulent friction. In this sublayer the velocity variation is given by

$$u^+ = 2.5 \ln y^+ + 5.5 \tag{4.86}$$

or, using logarithm in base 10,

$$u^+ = 5.75 \log y^+ + 5.5.$$

Either form is called the *law of the wall*.

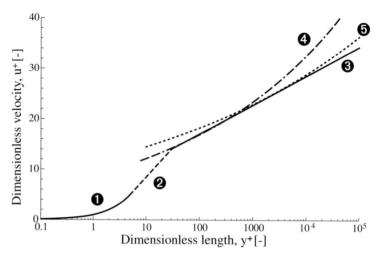

Figure 4.9 The universal velocity-distribution law for smooth pipes, u^+ vs. $\log(y^+)$: (1) laminar (4.84), (2) transition from laminar to turbulent, $u^+ = -3.05 + 5.00 \ln y^+$, (3) turbulent (4.86), (4) turbulent, $\text{Re} < 10^5$, $u^+ = 8.74(y^+)^{1/7}$, (5) $u^+ = 11.5(y^+)^{1/10}$.

The velocity variation near the center of the pipe is approximated by

$$\frac{U_c - u}{u^*} = -2.5 \ln \frac{y}{R} \tag{4.87}$$

where U_c is the velocity at the center of the pipe.

The variation of the dimensionless velocity u^+ vs. the y^+ number is shown in Fig. 4.9. The figure includes the analytical approximations for the three regions of the boundary layer. Two additional analytical approximations are provided: (i) $u^+ = 8.74(y^+)^{1/7}$ for $\text{Re} < 10^5$, and (ii) $u^+ = 11.5(y^+)^{1/10}$ for higher Re numbers. The former better fits the experimental data at lower Re numbers while the latter is closer to the experimental data at larger Re numbers.

Using the Blasius resistance formula (4.81), the velocity variation in the turbulent boundary layer can be approximated as

$$\frac{u}{U_c} = \left(\frac{y}{R}\right)^{\frac{1}{7}} \tag{4.88}$$

where u is the velocity at distance y from the wall. Equation (4.88) is called the $\frac{1}{7}$-*power velocity-distribution law*. Since the Blasius resistance formula was derived based on experimental data up to a Re number of 100,000, the validity of (4.88) is limited by this Re number. For Re numbers different from 100,000, the power velocity-distribution law is written as

$$\boxed{\frac{u}{U_c} = \left(\frac{y}{R}\right)^{\frac{1}{n}}} \tag{4.89}$$

where n varies between 6 and 10: for Re numbers less than 100,000 $n = 6$, and for Re numbers larger than 100,000, $7 \leq n \leq 10$.

Example 4.2.7 Calculate the ratio between the mean (or bulk) velocity of flow through pipe, $\overline{U}$, and the maximum velocity in the cross section U_c for $n = 6, 7, 8, 9$, and 10.

Solution. The bulk velocity is

$$\overline{U} := \frac{\int_0^R 2\pi r u(r)\, dr}{\int_0^R 2\pi r\, dr} = \frac{2\int_0^R r u(r)\, dr}{R^2} \tag{4.78}$$

where $r = R - y$. Dividing both sides of (4.78) by U_c and substituting (4.89) yields

$$\frac{\overline{U}}{U_c} = \frac{2n^2}{(n+1)(2n+1)}. \tag{4.90}$$

The variation of $\overline{U}/U_c$ as a function of n is given in Table 4.3.

Table 4.3 Variation of ratio of mean to maximum velocity as a function of n.

n	6	7	8	9	10
$\overline{U}/U_c$	0.7912	0.8167	0.8366	0.8526	0.8658

Using experimental results, the skin friction coefficient in a pipe can be approximated as [McAdams, 1954, p. 155]

$$c_f = \frac{\tau_w}{\frac{1}{2}\rho\overline{U}^2} = 0.046 \left(\frac{\overline{U}D}{\nu}\right)^{-\frac{1}{5}} \tag{4.91}$$

for $\text{Re} \in (5000, 200,000)$.

The skin friction coefficient c_f of (4.91) can be written as a function of the radius R as opposed to the diameter D:

$$c_f = \frac{\tau_w}{\frac{1}{2}\rho\overline{U}^2} = 0.040 \left(\frac{\overline{U}R}{\nu}\right)^{-\frac{1}{5}}. \tag{4.92}$$

For fully developed flows, that is, flows whose velocity profile does not change along the pipe, a good approximation of the ratio between the bulk velocity $\overline{U}$ and the velocity in the center of the pipe is

$$\frac{\overline{U}}{U_c} = 0.8, \tag{4.93}$$

according to Table 4.3. The skin friction coefficient can be written as a function of U_c:

$$c_f = \frac{\tau_w}{\frac{1}{2}\rho(0.8U_c)^2} = 0.040 \left(\frac{0.8U_c R}{\nu}\right)^{-\frac{1}{5}}. \tag{4.94}$$

The skin friction coefficient redefined as a function of U_c then becomes:

$$c_{f\,U_c} = \frac{\tau_w}{\frac{1}{2}\rho U_c^2} = 0.027 \left(\frac{U_c R}{\nu}\right)^{-\frac{1}{5}}. \tag{4.95}$$

Note that c_f is different from $c_{f\,U_c}$ because the shear stress is normalized by different velocities.

The results presented above are valid for turbulent flows in round pipes with smooth walls. The wall roughness has little effect on laminar flows, but has a significant effect if the flow is turbulent. The wall roughness is accounted for by the average roughness height, k. Figure 4.10 shows the variation of the friction factor as a function of the relative roughness, k/D, and the Reynolds number based on the pipe diameter and bulk velocity, $\text{Re} = \overline{U}D/\nu$. This diagram is called the Moody diagram [Moody, 1944] and it was generated using the Colebrook formula [Colebrook, 1939]:

$$\frac{1}{\sqrt{f}} = -2\log\left(\frac{2.51}{\text{Re}\sqrt{f}} + \frac{k}{3.7D}\right), \tag{4.96}$$

which is an implicit formula for f. An explicit and accurate formula for the friction factor in turbulent pipe flows,

$$\frac{1}{\sqrt{f}} = -1.8\log\left[\frac{6.9}{\text{Re}} + \left(\frac{k}{3.7D}\right)^{1.11}\right], \tag{4.97}$$

was proposed by Haaland [1983].

Figure 4.10 Friction factors for turbulent flow in pipes with rough walls.

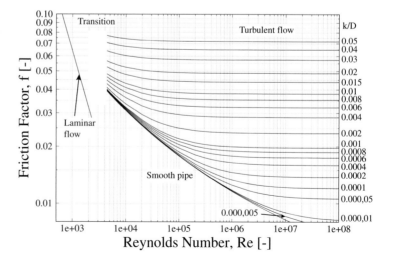

4.2.4.2 Turbulent Flow on a Flat Plate: Prandtl–Kármán Hypothesis

The Prandtl–Kármán hypothesis allows the extension of the relationships obtained for the turbulent flow in a pipe to the flow on a flat plate. The Prandtl–Kármán hypothesis states that the relationships obtained for the flow in a pipe can be applied to the flow on a flat plate if

1. U_c is replaced by the velocity outside the boundary layer U_∞;
2. the pipe radius R is replaced by the flat plate boundary layer thickness δ.

Example 4.2.8   Calculate the displacement thickness δ^* and the momentum thickness θ as functions of the boundary layer thickness δ for a turbulent boundary layer over a flat plate.

Solution
Using the definition of displacement thickness, δ^*, (4.3):

$$\delta^* = \int_0^h \left(1 - \frac{u}{U_\infty}\right) dy = \int_0^\delta \left(1 - \frac{u}{U_\infty}\right) dy + \int_\delta^h \left(1 - \frac{u}{U_\infty}\right) dy.$$

The integral between δ and h is approximately zero since beyond δ, $u \approx U_\infty$, such that

$$\delta^* = \int_0^\delta \left(1 - \frac{u}{U_\infty}\right) dy. \tag{4.98}$$

Using (4.89) and (4.98) yields

$$\delta^* = \int_0^\delta \left[1 - \left(\frac{y}{\delta}\right)^{\frac{1}{n}}\right] dy = \frac{\delta}{n+1}. \tag{4.99}$$

Using the definition of the momentum thickness, (4.5), in a similar way, one obtains

$$\theta = \int_0^\delta \left(\frac{y}{\delta}\right)^{\frac{1}{n}} \left[1 - \left(\frac{y}{\delta}\right)^{\frac{1}{n}}\right] dy = \delta \frac{n}{(n+1)(n+2)}. \tag{4.100}$$

Using equation (4.95) and the Prandtl–Kármán hypothesis, the skin friction coefficient c_f for a flat plate is

$$c_f = \frac{\tau_w}{\frac{1}{2}\rho U_\infty^2} = 0.027 \left(\frac{U_\infty \delta}{\nu}\right)^{-\frac{1}{5}}. \tag{4.101}$$

Assuming that the velocity variation in the boundary layer can be approximated using (4.89) with $n = 7$, then the momentum thickness is $\theta = \delta\, 7/72$ (see (4.100)), such that the skin friction coefficient becomes

$$c_f = \frac{\tau_w}{\frac{1}{2}\rho U_\infty^2} = 0.017 \left(\frac{U_\infty \theta}{\nu}\right)^{-\frac{1}{5}}. \tag{4.102}$$

For the flow on a flat plate, $U_\infty = \texttt{const}$, such that from (4.32),

$$\frac{d\theta}{dx} = \frac{\tau_w}{\rho U_\infty^2}. \tag{4.103}$$

Substituting the right-hand side of (4.103) using (4.102) yields

$$\frac{d\theta}{dx} = \frac{0.017}{2} \left(\frac{U_\infty \theta}{\nu} \right)^{-\frac{1}{5}}. \tag{4.104}$$

Integrating equation (4.104) yields

$$\theta(x) = 0.022 \left(\frac{U_\infty x}{\nu} \right)^{-\frac{1}{6}} x. \tag{4.105}$$

Once the momentum thickness $\theta(x)$ is known, the boundary layer thickness δ can be calculated using (4.100). The value of the boundary layer thickness is

$$\delta = 0.23 \left(\frac{U_\infty x}{\nu} \right)^{-\frac{1}{6}} x. \tag{4.106}$$

Example 4.2.9   What are the values of the boundary layer thickness, δ, displacement thickness, δ^*, momentum thickness, θ, and skin friction coefficient, c_f, at 0.2 m from the leading edge of a jet-engine inlet, for $U_\infty = 200$ m/s, if the flow is: (a) laminar and (b) turbulent. Assume that kinematic viscosity is $\nu = 14 \cdot 10^{-6}$ m^2/s.

Solution
Let us first determine the expressions of these variables as a function of x.

(a) Laminar

$$\delta = 5\sqrt{\frac{\nu x}{U_\infty}} = 5\sqrt{\frac{14 \cdot 10^{-6}}{200}}\sqrt{x} = 1.32 \cdot 10^{-3}\sqrt{x} \text{ m}$$

$$\delta^* = 1.72\sqrt{\frac{\nu x}{U_\infty}} = 0.455 \cdot 10^{-3}\sqrt{x} \text{ m}$$

$$\theta = 0.664\sqrt{\frac{\nu x}{U_\infty}} = 0.176 \cdot 10^{-3}\sqrt{x} \text{ m}$$

$$c_f = \frac{0.664}{\sqrt{\text{Re}_x}} = \frac{0.664}{\sqrt{\frac{U_\infty x}{\nu}}} = \frac{0.664}{\sqrt{\frac{200}{14 \cdot 10^{-6}}}} \cdot \frac{1}{\sqrt{x}} = 0.176 \cdot 10^{-3}\frac{1}{\sqrt{x}}.$$

(b) Turbulent

$$\delta = 0.23 \left(\frac{U_\infty x}{\nu} \right)^{-\frac{1}{6}} x = 0.23 \left(\frac{200}{14 \cdot 10^{-6}} \right)^{-\frac{1}{6}} x^{\frac{5}{6}} = 0.0148 x^{\frac{5}{6}} \text{ m}$$

$$\delta^* = \int_0^\delta \left(1 - \frac{u}{U_\infty} \right) dy = \int_0^\delta \left[1 - \left(\frac{y}{\delta} \right)^{\frac{1}{7}} \right] dy = \delta \int_0^1 (1 - z^{\frac{1}{7}}) \, dz = \frac{\delta}{8}$$

$$\delta^* = 1.85 \cdot 10^{-3} x^{\frac{5}{6}} \text{ m}$$

$$\theta = \delta \frac{7}{72} = 1.44 \cdot 10^{-3} x^{\frac{5}{6}} \text{ m}$$

$$c_f = 0.027 \left(\frac{U_\infty \delta}{\nu} \right)^{-\frac{1}{5}} = 0.027 \left(\frac{200}{14 \cdot 10^{-6}} \right)^{-\frac{1}{5}} (0.0148 \cdot x^{\frac{5}{6}})^{-\frac{1}{5}} = 0.00232 x^{-\frac{1}{6}}.$$

Let us now substitute for $x = 0.2$ m and compare the laminar and turbulent values.

$$\delta_l|_{x=0.2m} = 1.32 \times 10^{-3} \sqrt{0.2} = 0.6 \cdot 10^{-3} \text{m}$$

$$\delta_t|_{x=0.2m} = 0.0148 \cdot (0.2)^{\frac{5}{6}} = 3.9 \cdot 10^{-3} \text{m}$$

$$\delta_l^*|_{x=0.2m} = 0.455 \cdot 10^{-3} \sqrt{0.2} = 0.2 \cdot 10^{-3} \text{m}$$

$$\delta_t^*|_{x=0.2m} = 1.85 \cdot 10^{-3} (0.2)^{\frac{5}{6}} = 0.48 \cdot 10^{-3} \text{m}$$

$$\theta_l|_{x=0.2m} = 0.176 \cdot 10^{-3} \sqrt{0.2} = 0.08 \cdot 10^{-3} \text{m}$$

$$\theta_t|_{x=0.2m} = 1.44 \cdot 10^{-3} (0.2)^{\frac{5}{6}} = 0.38 \cdot 10^{-3} \text{m}$$

$$c_{f_l}|_{x=0.2m} = 0.176 \cdot 10^{-3} \frac{1}{\sqrt{0.2}} = 0.39 \cdot 10^{-3}$$

$$c_{f_t}|_{x=0.2m} = 0.00232 \cdot (0.2)^{-\frac{1}{6}} = 3 \cdot 10^{-3}.$$

The above results show that all turbulent parameters exceed the laminar counterparts.

4.3   Thermal Boundary Layer

There are several components of a jet propulsion system where a temperature difference exists between a solid wall and the fluid flowing past it. Examples of such components include burners, turbines, and exit nozzles. In these cases, the flow of heat is superimposed on the motion of the fluid. The heat transfer between the solid wall and the flow occurs by conduction and radiation. Conduction accelerated by the fluid motion is called convection, where the wall heat transfer rate per unit area, q''_w, is proportional to the temperature difference $T_w - T_\infty$, so that [Holman, 2010, p. 10]

$$q''_w = h_f (T_w - T_\infty) \tag{4.107}$$

where T_w is the wall temperature, T_∞ is the fluid temperature, and h_f is the heat transfer film coefficient in W/(m^2 K).

The heat transfer through radiation [Kreith, 1973, p. 11]

$$q'' = k_r (T_w^4 - T_\infty^4)$$

will be neglected herein because the coefficient k_r is very small, of the order of 10^{-8} W/(m^2 K^4). Radiation heat transfer becomes important when the temperature difference is large.

To determine the temperature distribution, the equations of fluid motion must be combined with those of heat transfer. Two cases can occur: (1) a cold fluid flows over a hot body, and (2) a hot fluid flows over a cold body, as shown in Fig. 4.11. In either case, it is expected that the temperature distribution will be similar to that of the velocity distribution, that is, it will vary most in the proximity of the body over a thin layer. By analogy with the flow phenomena, the thin layer in the neighborhood of the body where the major part of the transition from the temperature of the body to the temperature of the fluid occurs is called the *thermal boundary layer*, δ_T.

To determine the interaction between the flow phenomena and the thermal phenomena, let us establish the energy balance for a fluid element in motion shown in Fig. 4.12.

(a) $T_w > T_\infty$ (b) $T_w < T_\infty$

Figure 4.11 Temperature and velocity variation in boundary layer on a flat plate: (a) $T_w > T_\infty$ and (b) $T_w < T_\infty$.

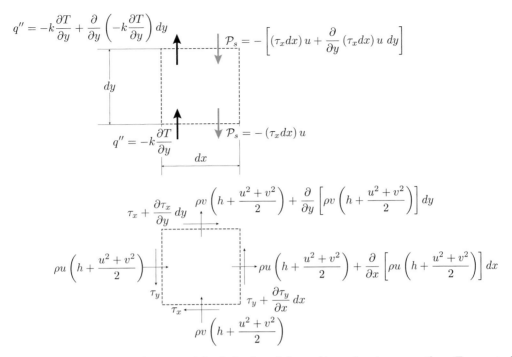

Figure 4.12 Small control volumes used for derivation of thermal boundary layer equations. Two control volumes had to be used to include all terms.

Let us start by writing the energy equation:

$$\dot{Q} = P_{shaft} + P_{shear} + \int_{\tau} \frac{\partial}{\partial t}(\rho e)\, d\tau + \int_{\sigma \equiv \partial \tau} \rho\left(h + \frac{V^2}{2} + gz\right) \vec{V} \cdot \vec{n}\, d\sigma \tag{2.63}$$

where the velocity $\vec{V}$ has the components (u, v).

To simplify this equation, let us make the following assumptions:

1. The shaft power, P_{shaft}, is zero, because there is no shaft in the control volume.
2. Steady flow, therefore the third term of the right hand side is zero.
3. Heat transfer is normal to the wall, that is, the heat transfer in the x direction is neglected. Therefore, according to Fourier's law of heat conduction, the heat flux q [J/(m² s)] per unit area A and time is

$$\frac{1}{A}\frac{dQ}{dt} = \frac{\dot{Q}}{A} = q'' = -k\frac{\partial T}{\partial y} \tag{4.108}$$

where k [W/(m K)] is the thermal conductivity.

4. Shear work is neglected on the surfaces normal to the wall (since $v \ll u$, $\tau_y \ll \tau_x$).

5. Potential energy term gz is neglected being much smaller than the enthalpy h and the kinetic energy term $V^2/2$.

The energy conservation equation (2.63) for the control volume shown in Fig. 4.12 becomes

$$\frac{\partial}{\partial y}\left(k\frac{\partial T}{\partial y}\right) dy\, dx = -\frac{\partial}{\partial y}(\tau_x u)\, dx\, dy + \frac{\partial}{\partial x}\left[\rho u\left(h + \frac{u^2 + v^2}{2}\right)\right] dx\, dy +$$
$$\frac{\partial}{\partial y}\left[\rho v\left(h + \frac{u^2 + v^2}{2}\right)\right] dy\, dx \qquad (4.109)$$

where

$$\frac{\partial}{\partial y}\left(k\frac{\partial T}{\partial y}\right) dy\, dx$$

is the net heat flux,

$$-\frac{\partial}{\partial y}(\tau_x u)\, dx\, dy$$

is the shear power, and

$$\frac{\partial}{\partial x}\left[\rho u\left(h + \frac{u^2 + v^2}{2}\right)\right] dx\, dy + \frac{\partial}{\partial y}\left[\rho v\left(h + \frac{u^2 + v^2}{2}\right)\right] dy\, dx$$

is the net enthalpy flux.

Equation (4.109) can be rearranged as

$$\rho u\frac{\partial}{\partial x}\left(h + \frac{u^2 + v^2}{2}\right) + \rho v\frac{\partial}{\partial y}\left(h + \frac{u^2 + v^2}{2}\right) + \left(h + \frac{u^2 + v^2}{2}\right)\left[\frac{\partial}{\partial x}(\rho u) + \frac{\partial}{\partial y}(\rho v)\right]$$
$$= \frac{\partial}{\partial y}\left(k\frac{\partial T}{\partial y}\right) + \frac{\partial}{\partial y}(\tau_x u), \qquad (4.110)$$

and since the third term of the left-hand side is zero because of the mass conservation equation, 4.110 reduces to

$$\rho u\frac{\partial}{\partial x}\left(h + \frac{u^2 + v^2}{2}\right) + \rho v\frac{\partial}{\partial y}\left(h + \frac{u^2 + v^2}{2}\right) = \frac{\partial}{\partial y}\left(k\frac{\partial T}{\partial y}\right) + \frac{\partial}{\partial y}(\tau_x u). \qquad (4.111)$$

The small terms $\frac{\partial}{\partial x}\left(\frac{v^2}{2}\right)$ and $\frac{\partial}{\partial y}\left(\frac{v^2}{2}\right)$ are neglected relative to $\frac{\partial}{\partial x}\left(\frac{u^2}{2}\right)$ and $\frac{\partial}{\partial y}\left(\frac{u^2}{2}\right)$ because $v \ll u$. Furthermore, using the relationship (2.76) between enthalpy and temperature reduces (4.111) to

$$\rho u c_p\frac{\partial T}{\partial x} + \rho v c_p\frac{\partial T}{\partial y} + \rho u\frac{\partial}{\partial x}\left(\frac{u^2}{2}\right) + \rho v\frac{\partial}{\partial y}\left(\frac{u^2}{2}\right) = \frac{\partial}{\partial y}\left(k\frac{\partial T}{\partial y}\right) + \frac{\partial}{\partial y}(\tau_x u) \qquad (4.112)$$

and then

$$\rho u c_p\frac{\partial T}{\partial x} + \rho v c_p\frac{\partial T}{\partial y} + \rho u\left(u\frac{\partial u}{\partial x} + v\frac{\partial u}{\partial y}\right) = \frac{\partial}{\partial y}\left(k\frac{\partial T}{\partial y}\right) + \frac{\partial}{\partial y}(\tau_x u). \qquad (4.113)$$

Using the x-momentum conservation equation (4.15) in the third term of the right-hand side yields

$$\rho u c_p \frac{\partial T}{\partial x} + \rho v c_p \frac{\partial T}{\partial y} - u\frac{dp}{dx} + u\frac{\partial \tau_x}{\partial y} = \frac{\partial}{\partial y}\left(k\frac{\partial T}{\partial y}\right) + \frac{\partial}{\partial y}(\tau_x u),$$ (4.114)

which can be rearranged using (4.9) as

$$\rho u c_p \frac{\partial T}{\partial x} + \rho v c_p \frac{\partial T}{\partial y} = u\frac{dp}{dx} + \frac{\partial}{\partial y}\left(k\frac{\partial T}{\partial y}\right) + \mu\left(\frac{\partial u}{\partial y}\right)^2.$$ (4.115)

If one considers the flow over a flat plate, then dp/dx is zero. The term $\mu\left(\frac{\partial u}{\partial y}\right)^2$ represents the viscous dissipation, that is, the work done on the fluid in the control volume. In many cases, this term can be neglected since it is much smaller than the other terms of (4.115). If one assumes k is independent of y, then (4.115) becomes

$$u\frac{\partial T}{\partial x} + v\frac{\partial T}{\partial y} = \frac{k}{\rho c_p}\frac{\partial^2 T}{\partial y^2}.$$ (4.116)

Recall the x-momentum conservation equation for incompressible flow over a flat plate:

$$u\frac{\partial u}{\partial x} + v\frac{\partial u}{\partial y} = \nu\frac{\partial^2 u}{\partial y^2}.$$ (4.33)

Equations (4.116) and (4.33) are similar if

$$k/(\rho c_p) = \nu.$$ (4.117)

The group ρc_p represents the thermal inertia per unit volume. $k/\rho c_p$ is defined as thermal diffusivity, α [m^2/s]:

$$\boxed{\alpha := \frac{k}{\rho c_p}}$$ Thermal diffusivity. (4.118)

Using the definition of the thermal diffusivity, (4.117) can be written as $\alpha = \nu$, which is equivalent to $\mu c_p/k = 1$. The ratio between the kinematic viscosity (or momentum diffusivity), ν, and the thermal diffusivity, α, is a dimensionless number defined as the Prandtl number. Using the definition of thermal diffusivity (4.118), the Prandtl number can be written as:

$$\boxed{\mathrm{Pr} := \frac{\nu}{\alpha} = \frac{\mu c_p}{k}}$$ Prandtl number. (4.119)

If $\mathrm{Pr} < 1$, then the thickness of the viscous boundary layer, δ, is smaller than the thickness of the thermal boundary layer, δ_T. Figure 4.13a shows the variation of the viscous and thermal boundary layers for $\mathrm{Pr} < 1$ and $T_w > T_\infty$. Figure 4.13b shows the variation of the viscous and thermal boundary layers for $\mathrm{Pr} > 1$ and $T_w < T_\infty$.

Ludwig Prandtl, a German scientist, is often referred to as the father of modern aerodynamics. Prandtl contributed to the development of rigorous mathematical analyses in aerodynamics (Fig. 4.14). Prandtl introduced the notion of the boundary layer in 1904 in the

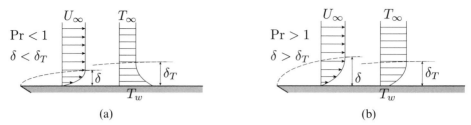

Figure 4.14 Ludwig Prandtl (1875–1953), courtesy of DLR, CC-BY 3.0.

seminal paper "Fluid flow in very little friction". He developed the theory of thin airfoils and the lifting-line theories. Prandtl and his student Theodor Meyer developed the Prandtl–Meyer expansion fans and the theory of supersonic shock waves in 1908. With Adolf Busemann, another student of his, Prandtl developed a method for designing supersonic nozzles in 1929. Further developments of supersonic theory were done by Prandtl and his student Theodore von Kármán. Prandtl was a professor at the Technical University in Hannover between 1901 and 1904 and subsequently director of the Institute for Technical Physics at the University of Göttingen.

The Prandtl number is a measure of the ratio between the viscous diffusion rate and the thermal diffusion rate. Note that the Prandtl number does not contain a length scale in its definition and is dependent only on the fluid and the state. For this reason, the Prandtl number can often be found in property tables as a function of temperature and pressure. Table 4.4 shows the variation of the Prandtl number for several substances.

The similarity condition (4.117) is satisfied if $Pr = 1$. As shown in Table 4.4, the values of the Prandtl number for most gases are close to unity. As a result, the solution of velocity u from the momentum equation (4.33) is identical to the solution of temperature T from the energy equation (4.116) if both u and T are written as variables that have the same boundary conditions. This last requirement can be satisfied by choosing variables u/U_∞ and $(T_w - T)/(T_w - T_\infty)$. Both of them vary from 0 at the wall to 1 at the edge of the boundary layer. Consequently, if $Pr = 1$, the variables u/U_∞ and $(T_w - T)/(T_w - T_\infty)$ have the same variation, and the thickness of the momentum boundary layer, δ, is equal to the thickness of the thermal boundary layers δ_T. Note that U_∞, T_∞, and T_w are constants.

Let us calculate the ratio between the wall shear stress and the wall heat flux:

$$\frac{\tau_w}{q_w''} = \frac{\mu(\frac{\partial u}{\partial y})|_{y=0}}{-k(\frac{\partial T}{\partial y})|_{y=0}} = \frac{\mu}{k} \cdot \frac{U_\infty \frac{\partial}{\partial y}(\frac{u}{U_\infty})|_{y=0}}{\frac{\partial}{\partial y}(\frac{T_w-T}{T_w-T_\infty})|_{y=0}} \cdot \frac{1}{T_w - T_\infty}. \tag{4.120}$$

If $Pr = 1$, then $\frac{\partial}{\partial y}\left(\frac{u}{U_\infty}\right)\Big|_{y=0} = \frac{\partial}{\partial y}\left(\frac{T_w-T}{T_w-T_\infty}\right)\Big|_{y=0}$ such that (4.120) yields

$$\frac{\tau_w}{q_w''} = \frac{\mu U_\infty}{k(T_w - T_\infty)}. \tag{4.121}$$

Using (4.121), let us write the wall shear stress in dimensionless form as

$$\frac{\tau_w}{\rho U_\infty^2} = \frac{\mu}{\rho U_\infty x} \cdot \frac{q_w'' x}{(T_w - T_\infty)k} = \frac{1}{Re_x} \cdot \frac{q_w'' x}{(T_w - T_\infty)k}.$$

Substituting the convection heat flux at wall, q_w'', given by (4.107) yields

$$\frac{\tau_w}{\rho U_\infty^2} = \frac{1}{Re_x} \cdot \frac{h_f(T_w - T_\infty)x}{(T_w - T_\infty)k} = \frac{h_f x}{k} \frac{1}{Re_x}. \tag{4.122}$$

The right-hand side of (4.122) is a dimensionless term. Since Re_x is dimensionless, $h_f x/k$ must also be dimensionless. The dimensionless term $h_f x/k$ is called the Nusselt[1] number:

$$\boxed{Nu_x := \frac{h_f x}{k}} \quad \text{Nusselt number.} \tag{4.123}$$

Combining (4.122), (4.123) and the definition of skin friction coefficient, c_f, (4.55) yields

$$Nu_x = \frac{1}{2}c_f Re_x. \tag{4.124}$$

Note that (4.124) is valid only for $Pr = 1$.

---

[1] Ernst Kraft Wilhelm Nusselt (1882–1957), a German engineer

Table 4.4 Prandtl number for different substances.

Substance	Pressure [bar]	Temperature [K]	Pr [-]	Reference
Mercury (saturated liquid)	6.9×10^{-8}	260	0.0316	[Perry and Green, 1997, p. 2-250]
Mercury (saturated liquid)	5.5×10^{-5}	340	0.0205	
Mercury (saturated liquid)	65.74	1000	0.0076	
Hydrogen	1	250	0.69	[Kakac et al., 1987, p. 22.17]
	1	350	0.69	
	1	400	0.68	
	1	700	0.68	
Nitrogen	1	77.33	0.8	[Bejan, 1993, p. 647 Jacobsen]
	1	700	0.71	
Dry air	1	93	0.92	[Bejan, 1993, p. 646]
	1	173	0.75	
	1	223	0.72	
	1	303	0.72	
	1	333	0.70	
	1	373	0.70	
	1	473	0.68	
	1	573	0.68	
	1	773	0.70	
	1	1273	0.72	
Steam	1	373.15	0.98	[Kakac et al., 1987, p. 22.34]
	1	2000	0.45	
Kerosene	1	293	25.7	[Bejan, 1993, p. 639 Vargaftik]
	1	333	18.3	
	1	473	8.5	
Gasoline	1	293	9.4	[Bejan, 1993, p. 639 Vargaftik]
	1	333	7.4	
	1	473	4.2	
Refrigerant 12 (CCl_2F_2) (gas)	1	300	0.798	[Kakac et al., 1987, p. 22.25]
Refrigerant 12 (gas)	1	500	0.674	
Refrigerant 12 (saturated liquid)	0.1	200.1	5.01	[Kakac et al., 1987, p. 22.24]
Refrigerant 12 (saturated liquid)	6	295.2	2.95	
Refrigerant 12 (saturated liquid)	30	367.2	4.20	
Water	1	273	13.44	[Bejan, 1993, p. 637-8]
	1	293	7.07	
	1	323	3.57	
	1	373	1.78	
Unused engine oil	-*	260	144500	[Kakac et al., 1987, p. 22.38]
	-	280	27200	
	-	300	6450	
	-	320	1990	
	-	360	395	
	-	400	155	

* indicates unspecified pressure

If the Prandtl number is not equal to 1, the thicknesses of the momentum and thermal boundary layers are different. Therefore, the Prandtl number provides a measure of the relative thickness of the momentum and thermal boundary layers. If the Prandtl number is less than 1, then thermal diffusivity is dominant.

4.3.1   Laminar Thermal Boundary Layer

For laminar flow over a flat plate, the Blasius solution produced

$$\frac{1}{2}c_f = 0.332\frac{1}{\sqrt{\text{Re}_x}} \tag{4.54}$$

such that

$$\text{Nu}_x = 0.332\sqrt{\text{Re}_x}.$$

If Pr is different from 1, then for laminar flow over a flat plate the Nu_x variation will be proportional to $\sqrt[3]{\text{Pr}}$. If the temperature of the wall is constant and $\text{Pr} > 0.5$, then

$$\boxed{\text{Nu}_x = 0.332\sqrt{\text{Re}_x}\sqrt[3]{\text{Pr}}}\quad \text{Flat plate, laminar flow, } \text{Pr} > 0.5\text{, and } T_w = \texttt{const}.$$

If the wall heat flux is constant, then

$$\boxed{\text{Nu}_x = 0.458\sqrt{\text{Re}_x}\sqrt[3]{\text{Pr}}}\quad \text{Flat plate, laminar flow, and } q''_w = \texttt{const}.$$

4.3.2   Turbulent Thermal Boundary Layer

The turbulent flow enhances thermal conductivity, k. Therefore, similar to the approach used for turbulent shear stresses where eddy viscosity ν_T enhanced the laminar viscosity ν, the heat transfer flux can be written as

$$q'' = -(k + \alpha_T \rho c_p)\frac{dT}{dy}$$

where α_T is *energy transfer diffusivity*.

Using the definition of thermal diffusivity α,

$$\alpha := \frac{k}{\rho c_p}, \tag{4.118}$$

the heat transfer flux can be written as

$$q'' = -\rho c_p (\alpha + \alpha_T)\frac{dT}{dy}.$$

4.3.2.1   Heat Transfer-Skin Friction Analogy

The main premise of the heat transfer-skin friction analogy [von Karman, 1939] for turbulent boundary layers is that *turbulence affects momentum and heat transfer in a similar fashion.*

Therefore, for fully developed flows in pipes, the variation of the skin friction and heat transfer with radius are similar:

$$\frac{\tau}{\tau_w} = \frac{r}{R} \quad \text{and} \quad \frac{q''}{q''_w} = \frac{r}{R},$$

such that

$$\frac{\tau}{q''} = \frac{\tau_w}{q''_w}. \tag{4.125}$$

In addition, the velocity and temperature profiles are similar, such that

$$\frac{d}{dy}\left(\frac{u}{\overline{U}}\right) = \frac{d}{dy}\left(\frac{T_w - T}{T_w - \overline{T}}\right) \tag{4.126}$$

where $\overline{U}, \overline{T}$ are the bulk velocity and temperature.

Let us calculate the ratio of the shear stress and heat flux:

$$\frac{\tau}{q''} = \frac{\rho(\nu + \nu_T)\dfrac{du}{dy}}{-\rho\,c_p\,(\alpha + \alpha_T)\dfrac{dT}{dy}} = \frac{\rho(\nu + \nu_T)\overline{U}\dfrac{d}{dy}\left(\dfrac{u}{\overline{U}}\right)}{\rho c_p(\alpha + \alpha_T)(T_w - \overline{T})\dfrac{d}{dy}\left(\dfrac{T_w - T}{T_w - \overline{T}}\right)} = \frac{(\nu + \nu_T)\overline{U}}{(\alpha + \alpha_T)c_p(T_w - \overline{T})}.$$

Since $\nu_T \gg \nu$ and $\alpha_T \gg \alpha$,

$$\frac{\tau}{q''} = \frac{\nu_T \overline{U}}{c_p \alpha_T (T_w - \overline{T})}, \tag{4.127}$$

and using (4.125) in (4.127) yields

$$\frac{\tau_w}{q''_w} = \frac{\nu_T \overline{U}}{c_p \alpha_T (T_w - \overline{T})}, \tag{4.128}$$

which can rearranged as

$$\frac{q''_w}{T_w - \overline{T}} \cdot \frac{1}{\rho c_p \overline{U}} = \frac{\alpha_T}{\nu_T} \cdot \frac{\tau_w}{\rho \overline{U}^2}.$$

Using the definition of the skin friction coefficient (4.55) and the heat convection flux (4.107) yields

$$\underbrace{\frac{h_f}{\rho c_p \overline{U}}}_{\text{heat transfer}} = \underbrace{\frac{\alpha_T}{\nu_T} \cdot \frac{c_f}{2}}_{\text{friction losses}} \tag{4.129}$$

where the ratio $\nu_T/\alpha_T \in (0.9, 1.7)$. The left-hand-side term is a measure of heat transfer, and the right-hand-side term is a measure of friction losses.

The dimensionless term on the left-hand side of (4.129) is called the Stanton[2] number:

$$\boxed{St := \frac{h_f}{\rho\, c_p\, \overline{U}}} \quad \text{Stanton Number.} \tag{4.130}$$

Experimental data obtained for long smooth tubes lead to the following relation:

$$\frac{h_f D}{k_b} = 0.023 \left(\frac{\rho \overline{U} D}{\mu}\right)_b^{0.8} \left(\frac{\mu\, c_p}{k}\right)_b^{0.33}, \tag{4.131}$$

where the subscript b denotes evaluation at the bulk mean fluid temperature. Multiplying (4.131) by $\left(\frac{\mu}{\rho \overline{U} D}\right)_b \cdot \left(\frac{k_b}{\mu c_p}\right)_b$ yields:

$$\frac{h_f}{\rho c_p \overline{U}} = 0.023 \left(\frac{\overline{U} D}{\nu}\right)_b^{-0.2} \left(\frac{\mu\, c_p}{k}\right)_b^{-0.67}. \tag{4.132}$$

Using (4.91),

$$c_f = 0.046 \left(\frac{\overline{U} D}{\nu}\right)_b^{-0.2}$$

and (4.130), (4.132) can be written as

$$St = \frac{1}{2} c_f \cdot Pr^{-0.67} \tag{4.133}$$

or

$$\frac{1}{2} c_f = St \cdot Pr^{0.67}. \tag{4.134}$$

Using the definitions of the Reynolds number (4.39) based on $\overline{U}$, the Prandtl number (4.119), and the Nusselt number (4.123), the Stanton number (4.130) can be written as a combination of these three dimensionless numbers, that is,

$$St = \frac{Nu_x}{Re_x Pr}. \tag{4.135}$$

4.3.2.2 Chilton–Colburn Analogy

The purpose of the Chilton–Colburn analogy is to extend the results of the thermal boundary layer in a pipe to a flat plate. This strategy is similar to the Prandtl–Kármán analogy that extended the results of the viscous boundary layer in a pipe to a flat plate.

---

[2] Sir Thomas Edward Stanton (1865–1931), a British engineer and student of Osborne Reynolds

Using the Chilton–Colburn analogy, (4.134) can be extended to a flat plate by substituting $\overline{U}$ by U_∞:

$$\boxed{\frac{1}{2}c_f = \frac{h_f}{\rho\, c_p\, U_\infty}\mathrm{Pr}^{0.67}}. \tag{4.136}$$

Example 4.3.1 Calculate the pressure drop, Δp, due to friction, and the fluid temperature variation in a pipe that is 24 diameters long. The diameter of the pipe is $D = 0.28$ m. The inlet air temperature is $\overline{T} = 600$ K and the inlet pressure is $p = 0.8$ MPa. The velocity of the air in the pipe is $\overline{U} = 220$ m/s. The pipe wall temperature is $T_w = 1150$ K and the dynamic viscosity is $\mu = 28.6 \times 10^{-6}$ Ns/m^2. Hint: Assume that the velocity profile at entrance is fully developed and that Prandtl number is 0.7.

Solution

$$\Delta p \overset{(4.73)}{=} \tau_w \frac{\pi LD}{\frac{\pi}{4}D^2}$$

$$\tau_w \overset{(4.55)}{=} \frac{1}{2}\rho\overline{U}^2 c_f \overset{(4.91)}{=} \frac{1}{2}\rho\overline{U}^2 0.046\,\mathrm{Re}^{-0.2}$$

$$\mathrm{Re} = \frac{\overline{U}D\rho}{\mu}$$

$$\rho = \frac{p}{RT} = \frac{800,000}{287 \times 600} = 4.65 \text{ kg/m}^3$$

$$\mathrm{Re} = \frac{220 \times 0.28 \times 4.65}{28.6 \times 10^{-6}} = 10.01 \times 10^6$$

$$\tau_w = \frac{1}{2}4.65 \times 220^2 \times 0.046 \times \left(10.01 \times 10^6\right)^{-0.2} = 206 \text{ Pa}$$

$$\Delta p = 206\frac{\pi\,24 \times 0.28 \times 0.28}{\frac{\pi}{4}0.28^2} = 19.78 \times 10^3 \text{ Pa}$$

$$\dot{Q} \overset{(4.107)}{=} h_f \pi\,DL(T_w - \overline{T}) = \rho\overline{U}\frac{\pi D^2}{4}c_p\Delta T \;\rightarrow\; \Delta T = h_f\frac{\pi\,DL(T_w - \overline{T})}{\rho\overline{U}\frac{\pi D^2}{4}c_p}$$

$$\Delta T = \frac{4L}{D}(T_w - \overline{T})\,\underbrace{\frac{h_f}{\rho\overline{U}c_p}}_{\mathrm{St}} \overset{(4.133)}{=} \frac{4L}{D}(T_w - \overline{T})\frac{1}{2}c_f\mathrm{Pr}^{-0.67}$$

$$c_f = 0.046 \, \text{Re}^{-0.2} = 0.046 \times \left(10.01 \times 10^6\right)^{-0.2} = 1.83 \times 10^{-3}$$

Start an iterative process for calculating ΔT.

First iteration

$$\Delta T = \frac{4 \times 24D}{D} \frac{1}{2} 1.83 \times 10^{-3} \times 0.7^{-0.67}(1150 - 600) = 61.4 \text{ K}$$

Assuming the temperature increase varies linearly along the pipe, from 0 at inlet to ΔT at length L, the average temperature $\overline{T}$ increases by $0.5 \, \Delta T$.

Second iteration

$$\Delta T = \frac{4 \times 24D}{D} \frac{1}{2} 1.83 \times 10^{-3} \times 0.7^{-0.67} \left[1150 - \left(600 + \frac{61.4}{2}\right)\right] = 57.9 \text{ K}$$

Third iteration

$$\Delta T = \frac{4 \times 24D}{D} \frac{1}{2} 1.83 \times 10^{-3} \times 0.7^{-0.67} \left[1150 - \left(600 + \frac{57.9}{2}\right)\right] = 58.1 \text{ K}$$

The iteration process is stopped since the change of the temperature variation between the last two iterations is considered small enough.

Problems

1. The free stream velocity over a compressor blade decreases linearly from 150 m/s to 135 m/s. The compressor blade length is 10 cm. One can assume that the kinematic viscosity is 14.6×10^{-6} m^2/s, the flow is laminar and the Thwaites formula,

$$\frac{\theta^2(x)}{\nu} = \frac{\theta_1^2}{\nu} \left(\frac{U_1}{U(x)}\right)^6 + \frac{0.45}{U_1} \left(\frac{U_1}{U(x)}\right)^6 \int\limits_{x_1}^{x} \left[\frac{U(\xi)}{U_1}\right]^5 d\xi,$$

is a good approximation of the momentum thickness.
   1. Calculate the integral of the Thwaites formula and determine the analytical expression for $\theta(x)$.
   2. Calculate the momentum thickness of the boundary layer at the trailing edge using the Thwaites formula.
   3. What is the momentum thickness if the free stream velocity is constant and equal to 150 m/s?
   4. Compare the two momentum thicknesses and explain the result.
2. Let us assume the free stream velocity on a compressor blade varies linearly. The chord of the airfoil is 0.1 m long. Consider two cases: (1) the free stream velocity at the mid chord is 80% of the value at the leading edge, and (2) the free stream velocity at the mid chord is 120% of the value at the leading edge.
   1. What is the free stream velocity variation for the two cases?
   2. What is the ratio between the momentum thicknesses at mid chord in the two cases?
   3. Comment the above result. Does this ratio make sense or not?

3. The air intake of a jet engine is shown in Fig. 4.15, where $A_1 = 0.271$ m^2, $A_2 = 0.254$ m^2, and $L = 0.2$ m. The mass flow rate of the jet engine is $\dot{m} = 50$ kg/s. The density of the air is assumed to be constant and equal to $\rho = 1.23$ kg/m^3. The flow in the air intake is turbulent, and the velocity variation can be approximated by $u/U = (y/\delta)^{1/7}$.

Figure 4.15 Engine air intake.

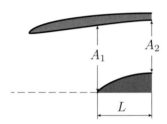

(a) Assuming that the flow along the walls of the air intake can be approximated as the flow over a flat plate, calculate boundary layer thickness, δ, displacement thickness, δ^*, momentum thickness, θ, and wall shear stress, τ_w, as a function of the distance from A_1 to A_2.

(b) Assuming that outside of the boundary layer the velocity in the air intake varies linearly, develop an algorithm to calculate displacement thickness, δ^*, momentum thickness, θ, and wall shear stress, τ_w, as functions of the distance from A_1 to A_2. Compare these results to the results obtained in part (a).

4. The front part of two-dimensional inlet diffuser in a jet engine has the dimensions shown in Fig. 4.16. The velocity at inlet is 120 m/s. The air can be assumed to be incompressible (the density is $\rho = 1.23$ kg/m^3).

1. Write a computer program to calculate the free stream velocity variation in the inlet assuming the flow is (a) laminar and (b) turbulent.
2. Plot the velocity variation along the inlet for the two cases.
3. Plot the boundary layer thickness for the two cases.
4. Calculate the viscous force acting on the inlet for the two cases.
5. Calculate the pressure force acting on the inlet for the two cases.

Hint: The acceptable error for the free stream velocity at the end of the diffuser is 1%.

Figure 4.16 Engine inlet diffuser.

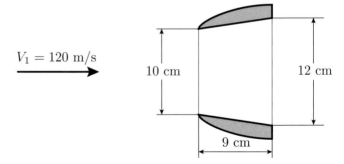

5. Compare the shape factor, H_{12}, of the laminar flow represented by the Blasius solution with the shape factor of a turbulent flow for which one can assume $n = 7$.

6. The velocity variation inside a turbulent boundary layer on a flat plate can be approximated by

$$\frac{u}{U} = \left(\frac{y}{\delta}\right)^{\frac{1}{n}}$$

where $6 \leq n \leq 10$, depending on the Reynolds number.

   1. Calculate the expressions of δ^*/δ and θ/δ as functions of n.
   2. Draw (approximately) the velocity profiles for a laminar flow and a turbulent flow with $n = 6$ and $n = 10$. For which value of n is the turbulent boundary layer closer to the laminar boundary layer? Explain the result.

7. The compressor airfoil can be approximated by a flat plate in the blade tip region. The chord of the blade is $c = 0.05$ m, the air density is 1.23 kg/m^3, the dynamic viscosity is $\mu = 18.5 \times 10^{-6}$ N s/m^2, and the velocity is $U_\infty = 150$ m/s.

   Calculate the boundary layer thickness, displacement thickness, momentum thickness, and wall shear stress at the trailing edge of the airfoil:

   1. for a laminar boundary layer;
   2. for a turbulent boundary layer (assume $n = 7$).

8. The compressor airfoil can be approximated by a flat plate in the blade tip region. The chord of the blade is $c = 0.1$ m, the air density is 1.23 kg/m^3, the dynamic viscosity is $\mu = 18.5 \times 10^{-6}$ N s/m^2, and the velocity is $U_\infty = 130$ m/s. Calculate the boundary layer thickness, displacement thickness, momentum thickness, and wall shear stress at the trailing edge of the airfoil: (a) for a laminar boundary layer; (b) for a turbulent boundary layer (assume $n = 7$).

9. A hypersonic vehicle flies at Mach 7 and altitude $H = 20$ km. At this altitude the atmospheric temperature is $T = 216.65$ K, the pressure is $p = 5530.7$ Pa, and the kinematic viscosity of air is $v = 1.5989 \times 10^{-4}$ m^2/s. The air heat capacity at constant pressure is $c_p = 1004$ J/(kg K). The flow on the hypersonic vehicle is turbulent, and the Prandtl number is 0.9. The air gas constant is $R = 287.16$ J/(kg K), and the ratio of specific heat capacities is $\gamma = 1.4$.

   The fin of the hypersonic vehicle has a chord of 1 m. The fin has a skin thickness of 8 mm and it is made out of a metalic material that has a thermal conductivity $k = 20$ W/(m K). The inside surface of the fin skin is cooled, and the heat convection coefficient is $h_f = 500 W/$(m^2 K).

   What should the temperature of the inside coolant be such that the heat flux through the fin surface is $q = 30$ kW/m^2?

10. The cooling oil of a jet engine flows through a steel pipe that has a 2 cm inner diameter. The thickness of the pipe is 3 mm. The thermal conductivity of the steel is $k = 20$ W/(K m). The oil flows with a velocity of 1 m/s and has a temperature of 400 K. The oil density is 850 kg/m^3, the heat capacity at constant pressure is

$c_{p_{oil}} = 2300$ J/(kg K), the kinematic viscosity is $\nu_{oil} = 50 \times 10^{-6}$ m²/s, and the Prandtl number is 155. The temperature at the exterior wall of the pipe is 320 K.

1. What is the temperature at the inner wall of the pipe?
2. What is the heat flux at the inner wall of the pipe?
3. What is the heat flux at the outer wall of the pipe?

11. Turbine blades are cooled with hot air from the compressor. The mass flow rate of this cooling air is 2% of the air mass flow rate of the jet engine. The temperature of the compressor air used for cooling is 500 K, the pressure is 4 bar, and the air mass flow rate of the jet engine is 2 kg/s (this is a small engine used for drones). The flow in the pipe is turbulent and the Prandtl number is 0.86. The inner diameter of the cooling air pipe is 1 cm. The thickness of the pipe is small such that one can assume the temperature across the pipe is constant. The temperature at the exterior of the pipe is 420 K. The dynamic viscosity of air at 293 K is 17.9×10^{-6} Pa s. The dynamic viscosity varies with temperature as $\mu(T_1)/\mu(T_2) = (T_1/T_2)^{0.8}$. The specific heat capacity of air at 500 K is 1006 J/(kg K).

1. What is the skin friction coefficient in the pipe?
2. What is the heat convection coefficient in the pipe?
3. What is the heat flux at the outer wall of the pipe?

Bibliography

B. S. Baldwin and T. J. Barth. A one-equation turbulence transport model for high Reynolds number wall-bounded flows. In *29th Aerospace Sciences Meeting*, volume AIAA91-0610, Reno, Nevada, January 1991. AIAA. 129

B. S. Baldwin and H. Lomax. Thin layer approximation and algebraic model for separated turbulent flow. In *16th Aerospace Sciences Meeting*, AIAA Paper 1978-257, Huntsville, Alabama, January 1978. AIAA. 129

A. Bejan. *Heat Transfer*. John Wiley & Sons, 1993. 146

H. Blasius. Grenzschichten in Flüssigkeiten mit kleiner Reibung. *Zeitschrift für angewandte Mathematik und Physik*, 56:1–37, 1908. 119, 121

H. Blasius. Das Ähnlichkeitsgesetz bei Reibungsvorgängen in Flüssigkeiten. *Forschungarbeiten auf dem Gebiete des Ingenieurwesens*, 131, 1913. 131

J. Boussinesq. Theorie de l'écoulement tourbillonant et tumultueux des liquides dans les lits rectilignes à grande section (tuyaux de conduite et cannaux découverts), quand cet écoulement s'est régularisé en un régime uniforme, c'est-à-dire, moyennement pareil à travers toutes les sections normales du lit. *Comptes Rendus de l'Académie des Sciences*, CXXII:1290–1295, 1896. 128

T. Cebeci and A. M. O. Smith. *Analysis of Turbulent Boundary Layers*. Academic Press, 1974. 129

P. Y. Chou. On velocity correlations and the solutions of the equations of turbulent fluctuations. *Quarterly of Applied Mathematics*, 3:38–54, 1945. 129

C. F. Colebrook. Turbulent flow in pipes, with particular reference to the transition region between the smooth and rough pipe laws. *Journal of the Institution of Civil Engineers*, 11(4):133–156, 1939. 136

J. R. d'Alembert. Essai d'une nouvelle théorie de la résistance des fluides. Technical report, Academie Royale des Sciences. Paris, France, 1752. 107

J. W. Deardorff. Three-dimensional numerical study of turbulence in an entraining mixed layer. *Boundary-Layer Meteorology*, 7:199–226, 1974. 129

C. duP. Donaldson and H. Rosenbaum. Calculation of the turbulent shear flows through closure of Reynolds equations by invariant modeling. Technical Report ARAP Report 127, Aeronautical Research Associates of Princeton, Princeton, NJ, 1968. 129

S. E. Haaland. Simple and explicit formulas for the friction factor in turbulent pipe flow. *Journal of Fluids Engineering*, 105(1):89–90, March 1983. 136

J. P. Holman. *Heat Transfer*. McGraw-Hill, tenth edition, 2010. 140

L. Howarth. On the solution of the laminar boundary layer equations. *Proceedings of the Royal Society of London. Series A, Mathematical and Physical Sciences*, 164:547–579, 1938. 123, 124

D. A. Johnson and L. S. King. A mathematically simple turbulence closure model for attached and separated turbulent boundary layers. *AIAA Journal*, 23(11):1684–1692, November 1985. 129

S. Kakac, R. K. Shah, and W. Aung. *Handbook of Single-Phase Convective Heat Transfer*. John Wiley & Sons, 1987. 146

John Kim, Parviz Moin, and Robert Moser. Turbulence statistics in fully developed channel flow at low Reynolds number. *Journal of Fluid Mechanics*, 177:133–166, 1987. doi: 10.1017/S0022112087000892. 130

A. N. Kolmogorov. Equations of turbulent motion of an incompressible fluid. *Izvestia Academy of Sciences, USSR; Physics*, 30(1 and 2):56–58, 1942. 129

F. Kreith. *Principles of Heat Transfer*. Intext Press, Inc., third edition, 1973. 140

B. E. Launder and D. B. Spalding. *Mathematical Models of Turbulence*. Academic Press, 1972. 129

B. E. Launder, G. J. Reece, and W. Rodi. Progress in the development of a Reynolds-stress turbulence closure. *Journal of Fluid Mechanics*, 68(3):537–566, 1975. 129

M. J. Lighthill. Physics of gas flow at very high speeds. *Nature*, 178(4529):343–5, August 1956. 107

W. H. McAdams. *Heat Transmission*. McGraw-Hill, third edition, 1954. 135

F. R. Menter. Two-equation eddy-viscosity turbulence models for engineering applications. *AIAA Journal*, 32(8):1598–1605, 1994. 129

P. Moin and K. Mahesh. Direct numerical simulation: A tool in turbulence research. *Annual Review of Fluid Mechanics*, 30(5):39–78, 1998. 129

L. F. Moody. Friction factors for pipe flow. *Transactions of the ASME*, 66(8):671–684, November 1944. 136

R. H. Perry and D. W. Green, editors. *Perry's Chemical Engineers' Handbook*. McGraw-Hill, seventh edition, 1997. 146

K. Pohlhausen. Zur Näherungsweisen Integration der Differentialgleichung der laminaren Grenischicht. *Zeitschrift fü angewandte Mathematik und Mechanik*, 1:252–290, 1921. 127

Jean Leonard Poiseuille. *Recherches expérimentales sur le mouvement des liquides dans les tubes de très-petits diamèters*. Imprimerie Royale, Paris, 1846. 132

L. Prandtl. Fluid motion with very small friction. In *Proceedings of the Third International Mathematical Congress*, Heidelberg, Germany, 1904. English Translation NACA TM 452 (1928). 107

L. Prandtl. Bericht über Untersuchungen zur ausgebildeten Turbulenz. *Zeitschrift für angewandte Mathematik und Mechanik*, 5:136–139, 1925. 129

L. Prandtl. Über ein neues Formelsystem für die ausgebildete Turbulenz. *Nac. Akad. Wiss. Göttingen, Math-Phys. Kl.*, pages 6–19, 1945. 129

J. Rotta. Statistische Theorie nichthomogener Turbulenz. *Zeitschrift für Physik*, 129(6):547–572, Nov 1951. ISSN 0044-3328. doi: 10.1007/BF01330059. URL https://doi.org/10.1007/BF01330059. 129

P. G. Saffman. A model for inhomogeneous turbulent flow. *Proceedings of the Royal Society A*, pages 417–433, 1970. 129

P. Sagaut. *Large Eddy Simulation for Incompressible Flows: An Introduction*. Springer, 2009. 129

H. Schlichting. *Boundary-Layer Theory*. McGraw-Hill, seventh edition, 1979. 128

J. Smagorinsky. General circulation experiments with the primitive equations I. The basic experiment. *Montly Weather Review*, 91(3):99–164, March 1963. 129

P. R. Spalart and S. R. Allmaras. A one-equation turbulence model for aerodynamic flows. In *30th Aerospace Sciences Meeting & Exhibit*, AIAA-92-0439, Reno, NV, January 1992. 129

B. Thwaites. Approximate calculation of the laminar boundary layer. *The Aeronautical Quarterly*, 1(3): 245–280, November 1949. 124, 127

T. von Karman. Analogy between fluid friction and heat transfer. *Engineering*, 148(3840):210–213, 1939. 133, 147

Frank M. White. *Viscous Fluid Flow*. McGraw Hill, 3rd edition, 2006. 126

D. C. Wilcox. Reassessment of the scale-determining equation for advanced turbulence models. *AIAA Journal*, 26(11):1299–1310, November 1988. 129

5 Introduction to Combustion

Combustion is the process that heats the working fluid in a jet engine. Combustion is a particular chemical reaction that has the following specific characteristics: (1) it is exothermic, (2) it is a fast oxidation of the combustion mixture, and (3) it is associated by light emission. The majority of chemical reactions take place in the *flame*, which is the region where the oxidation is visible.

The composition of the mixture prior to combustion consists of combustible constituents (or species), oxidant, and possible inert substances. The initial composition of the mixture species, called *reactants*, is changed by the combustion. The species that are produced at the end of combustion are called *products*. As a result of the heat release, the temperature of the combustion products is higher than that of the reactants.

This chapter presents some of the basic concepts needed to understand jet engine combustion. This information is needed to evaluate the performance and understand the design of engine burners.

5.1 Classification of Fuels

The fuels used for jet engines are different from the fuels used for aero piston engines because the characteristics of combustion in the two types of engines differ. All aviation fuels, however, must meet the requirements of the aircraft fuel system and distribution system. Although the focus of this text is on jet propulsion, fuels for aero piston engines will also be presented for comparison.

5.1.1 Types of Fuels

This section presents two types of fuels: fuels for aero piston engines and fuels for jet engines.

5.1.1.1 Fuels for Aero Piston Engines

The piston engines used in aviation include Otto-cycle and Diesel-cycle engines. The spark-ignition piston engines, such as the Otto-cycle engine, pose a severe constraint on the fuel properties in order to avoid detonation (or *knocking*). *Detonation* occurs when some of the unburned mixture in the combustion chamber explodes before the flame front initiated by the spark plug can reach it. As a result, pressure rises dramatically beyond design limits. Detonation must be avoided since it can cause vibrations and damage to engine parts.

Consequently, resistance to detonation is a major requirement of the fuel. This requirement can be achieved most effectively by using gasolines that inherently have low spontaneous ignitability. Detonation resistance is measured in terms of the *octane number* up to a point of 100, and of *performance number* thereafter.

The octane number is equal to the volumetric percentage of isooctane, i-C_8H_{18} (2, 2, 4-trimethylpentane), in a mixture with paraffin normal heptane, n-C_7H_{16}, that has the same knock resistance as the fuel tested. For fuels with a better knock resistance than isooctane reference, blends of isooctane and tetraethyl lead are used. The concentration of tetraethyl lead is converted empirically to an equivalent octane number extrapolated up to 120. The fuel knock resistance is measured at lean mixture, equivalent to cruising conditions, and at rich-mixture supercharge, equivalent to takeoff conditions. For example, the anti-knock grade ratings of 80/87 refer to 80 octane at lean mixture and 87 octane at rich-mixture supercharged conditions. Two commonly used types of fuel have anti-knock grade ratings of 100/130 and 115/145. For these fuels, the ratings indicate performance numbers for the values larger than 100.

The gasolines for aviation are blends of volatile hydrocarbon components ranging from approximately C_5 to C_{12}. The average properties of the mixture are equivalent to those of a fictitious molecule $C_{7.3}H_{15.3}$ of molar mass 103 kg/kmol. The hydrogen mass is 14.98%. The density of the mixture is approximately 720 kg/m^3 (0.72 kg/ℓ). The specifications for aviation gasolines are presented in Table 5.1.

The fuel used by Diesel engines is chemically identical to home heating oil.

Table 5.1 Aviation gasolines specifications.

Property	Limits			
Grade	80	100 Low-Lead	100	115
Aromatics, % vol., min.	-	-	-	5
Color	Red	Blue	Green	Purple
Freezing point, °C, max.	-60	-60	-60	-60
Specific energy, net, MJ/kg	43.5	43.5	43.5	44.0
Sulfur, total, % mass, max.	0.05	0.05	0.05	0.05
Tetraethyl lead content, as g Pb/ℓ, max.	0.14	0.56	0.85	1.28
Knock rating, lean mixture, motor method				
Octane number, min.	80	99.5	99.5	-
Performance number, min.	-	-	-	115
Knock rating, rich mixture, motor method				
Octane number, min.	87	-	-	-
Performance number, min.	-	130	130	145

5.1.1.2 Fuels for Jet Engines

Combustion in a jet engine is not plagued by any phenomena as detrimental as detonation in piston engines. As a result, jet engine fuel, also known as aviation turbine fuel or jet fuel, does not require highly refined blending of gasolines. Consequently, heavier fractions can be used for jet fuels. Jet fuel, however, must satisfy the aircraft fuel system requirements, as the aviation gasoline does. For this reason, fuels as heavy as gas oils are not acceptable because of their high viscosities and freezing points. The kerosene fraction of petroleum is currently considered as the most suitable compromise for jet fuel. The kerosene fraction burns cleanly and consequently minimizes problems with combustion deposits, metal temperature, corrosion, and particulate erosion.

Aviation kerosene is called Jet A-1 in the US and AVTUR in the UK. Aviation kerosene has average properties equivalent to that of a fictitious molecule $C_{12.5}H_{24.4}$ of molar mass 175 kg/kmol and hydrogen mass 14.06%. The color of aviation kerosene normally ranges from water white (colorless) to straw/pale yellow. The jet fuel should be clear and free from solid matter and undissolved water at normal ambient temperature.

Aviation kerosene is a blend of relatively involatile hydrocarbon components, similar in many respects to lamp kerosene. Cleanliness and several properties of the aviation kerosene that control freezing, smoke, atomization, and combustion efficiency are, however, much more strictly controlled than for lamp kerosene. Aviation kerosene consists of refined hydrocarbons derived from conventional sources including crude oil, natural gas liquid condensates, heavy oil, shale oil, and oil sands. Jet fuels are also allowed to contain synthetic hydrocarbons. The amount of synthetic hydrocarbons in jet fuels is currently not regulated.

Several variations of aviation kerosene are currently available, including Jet A, Jet A-1 (or AVTUR), JP-4 (or AVTAG), JP-5 (or AVCAT), JP-7 and JP-8. Jet A has the low freezing point of -40°C, while Jet A-1 has the low freezing point of -47°C. The temperature of fuel in an aircraft tank decreases at a rate proportional to the flight duration. For this reason, Jet A is used primarily for internal flights while Jet A-1 is used for international flights, which are longer duration, higher altitude flights.

JP-4 fuel was introduced as a blend of Jet A-1 and naphtha, to promote a wide-cut gasoline. JP-4 fuel has average properties equivalent to a fictitious molecule $C_{10}H_{20.3}$ of molar mass 141 kg/kmol and hydrogen mass of 14.56%.

JP-5 fuel was formulated to avoid potentially dangerous concentrations of vapor vented from fuel tanks on aircraft carriers. JP-5 is a lower volatility kerosene of high flash point. The *flash point* is defined as the lowest temperature at which the vapor-air mixture can sustain momentary combustion when a small test flame is introduced into the vapor space. Table 5.2 lists the physical characteristics for several aviation fuels.

Fuel Specifications

Some of the specifications of Jet A-1, JP-4, and JP-5 are shown in Table 5.3. These fuel specifications impose restrictions on fuel composition, volatility, fluidity, combustion, corrosion, stability, contaminants, and electrical conductivity. This section describes the

Table 5.2 Physical characteristics of aviation fuels.

Property	Kerosenes					Wide-cuts		
	JP-1	JP-5	Jet A	Jet A-1	JP-6	JP-3	JP-4	Jet B
Flash point [°C]	44	60	38-66		ns[1]	ns	ns	−20
Freezing point [°C]	−60	−48	−40	−47	−54	−60	−60	−51
Specific gravity[2] [-]	0.82	0.80-0.84	0.78		0.78-0.84	0.73-0.8	0.75-0.8	0.75

Table 5.3 Aviation kerosene specifications (from MIL-T-5624 for JP-4 and JP-5 and ASTM D-1655 for JET A-1).

Property	JP-4	JET A-1	JP-5
Composition			
Acidity, total, mg KOH/g, max.	0.015	0.015	0.015
Aromatics, % vol., max.	25.0	22.0	25.0
Olefins, % vol., max.	5.0	5.0	5.0
Sulfur, total, % mass, max.	0.30	0.30	0.30
Sulfur, mercaptan, % mass, max.	0.001	0.003	0.001
Volatility			
Flash point, °C min.	–	38	60
Density at 15°C, kg/l, min.	0.751	0.775	0.788
max.	0.802	0.830	0.845
Fluidity			
Freezing point, °C, max.	−58	−47	−46
Viscosity at -20°C, mm²/s, max.	–	8.0	8.8
Combustion			
Specific energy, net, MJ/kg, min.	42.8	42.8	42.6

types of specifications imposed for jet engine fuels and the motivation for enforcing these specifications.

Composition – Fuels specifications restrict the concentration of several components, including aromatics[3], alkenes[4], sulfur, and sulfur mercaptan. Fuel acidity is also controlled. Fuel acidity is a measure of the corrosive potential of the fuel in storage. Acidity is measured by dissolving

[1] not specified.

[2] or relative density, is the ratio of the density to the density of water at 4°C.

[3] Aromatics are hydrocarbons derived from crude oil and, in small quantities, from coal. The name aromatics is due to their distinctive perfumed smell. The aromatics can be monocyclics, with a general formula C_nH_{2n-6}, or polycyclics, with a general formula C_nH_{2n-6r}. n is the number of carbon atoms in the molecule, also known as the carbon number. r is the number of benzene rings. The main aromatics are benzene, toluene, naphthalene, and the xylenes.

[4] Alkenes (or olefins) are unsaturated hydrocarbons that contain one or more carbon-carbon double bonds. In the International Union of Pure and Applied Chemistry nomenclature, alkenes have the -ene suffix. The simplest alkenes have only one double bond. These mono-unsaturated alkenes, such as ethene, propene, or butene, form a homologous series with the general formula C_nH_{2n}.

the fuel in a mixture of toluene and isopropyl alcohol. The solution is then titrated with alcoholic potassium hydroxide to the orange-to-green end point of an added indicator that consists of p-naphthol-benzein dissolved in a toluene-isopropyl alcohol-water solution. The acidity is expressed as the mass of potassium hydroxide per mass of fuel.

The amount of aromatics and alkenes is limited in jet fuels. An increased concentration of aromatics raises the production of smoke and the deposition of carbon, and also influences the flame temperature and radiation in gas-turbine combustion chambers. A small amount of aromatics is desirable in order to ensure sufficient swell for both rubber and plastic seals in fuels systems. An increased concentration of alkenes deteriorates the stability of jet fuels in storage.

The amount of sulfur is limited in jet fuels because sulfur increases the formation of carbon in combustion chambers. Sulfur also increases the corrosivity of the cooled combustion products. In piston engine fuels, the sulfur severely inhibits the anti-detonation effectiveness of alkyl lead additives. Mercaptan sulfur is also limited, because it attacks certain metals and elastomers used in the fuel system.

Jet fuel may include additives to improve fuel performance, and handling, and maintenance. Fuel handling and maintenance additives include leak detection additives, electrical conductivity improver additives, and biocidal additives.

Additives for enhancing fuel performance include antioxidant additives, metal deactivator additives, and fuel system icing inhibitor additives. Antioxidants and metal deactivators are used to prevent the formation of oxidation deposits in the fuel system, to improve the oxidation stability of fuels in storage, and to counteract the effects of active metals (such as copper and copper-base alloys) in the fuel system.

Fuel is also used to lubricate the components of the aircraft/engine fuel system. The effectiveness of fuel as a lubricant is called *lubricity*. Fuels with low levels of lubricity can cause in-service problems, ranging from reductions of pump flow to mechanical failures resulting in engine shutdown. To avoid these problems, additives that improve lubricity are added to the fuel.

Volatility – Volatility is a property that is regulated by fuel specifications. *Volatility* indicates the ability of the fuel to vaporize. Fuel volatility is influenced by its distillation characteristics and vapor pressure.

Fuel specifications for volatility include initial boiling point, temperatures needed to evaporate 10, 20, 50 and 90% of the fuel volume, final boiling point, flash point vapor pressure + fuel density. The *initial boiling point* is the temperature at which the fuel must be heated to produce the first droplet of condensate. The *final boiling point* (or the end point) is the highest temperature reached during the distillation of a fuel sample. The final boiling point is limited for petroleum fuels to $370°C$ to avoid cracking of the heavier hydrocarbon molecules.

The distillation characteristics are also captured by the temperatures needed to evaporate certain percentages of fuel volume. Distillation temperature increases with fuel density because heavier fuels are less volatile. Distillation temperature provides information on the type of fuel blend. Distillation data also relate to other fuel properties, such as flash point and freezing point.

Vapor pressure is measured in the *Reid test* [Goodger and Vere, 1985, p. 42], by heating a fuel sample at 37.8°C in a cylindrical container and measuring the vapor absolute pressure. The pressure of fuel vapor drops abruptly as fuel density increases. As a result, Reid vapor pressure is not specified for heavier fuels such as Jet A-1 and JP-5. Reid vapor pressure provides an indication of venting loss and vapor lock in fuel systems.

Density, the last property of the volatile group of properties in Table 5.3, provides a broad indication of the type of fuel. The density has also a direct influence on droplet size in spray formation. The amount of fuel loaded on the aircraft is measured by volume, while the quantity of fuel needed for a certain mission is expressed on a mass basis. Consequently, the density value is needed to relate the fuel mass needed to the fuel volume loaded.

Fluidity – Fluidity is another property of the fuel regulated by the standard specifications. The fuel must remain liquid under all conditions of operation. Two properties describe fluidity of the fuel: freezing point and viscosity.

Freezing point is measured by continuously stirring and cooling the fuel until hydrocarbon crystals appear. The fuel container is then taken out of the refrigerant and keep in ambient air while continuously stirring. The temperature at which hydrocarbon crystals disappear is the freezing point. This temperature increases with density.

Viscosity increases with the decrease of temperature. Viscosity also increases with increase in density. The standard specifications limit the maximum value of viscosity at a low test temperature in order to permit adequate flow through the fuel system and adequate pressure at the spray nozzles. Viscosity also affects the lubrication ability of the fuel.

Fuel Combustion – Fuel combustion must cover the full range of engine conditions. Aviation kerosene combustion is characterized by several parameters, including heat of combustion, smoke point, and the content of naphthalenes.

Heat of combustion for aviation kerosene can be expressed as either higher heating value (*HHV*) or lower heating value (*LHV*), as presented in Section 5.2.6. The heat of combustion measures the energy released per unit mass, and has the units of J/kg. This description is appropriate for subsonic aircraft where the mass is a more stringent restriction than volume. For supersonic aircraft, where thin airfoil wing sections limit the volume of the fuel tanks located in the wings, it is beneficial to express the energy released per unit volume.

Figure 5.1 shows the energy released by several types of fuel vs. density. The energy released is expressed either per mass, that is, specific energy, or per volume, that is, energy density. The best choice of fuel for supersonic aircraft, based on the energy available per mass, is JP-5. The best choice of fuel for subsonic aircraft, based on the energy available per volume shown in Fig. 5.1, is JP-4 if one ignores the option of using gasoline.

Corrosion – Aircraft and engine fuel system corrosion is promoted by the various types of sulfur compounds present in the fuel. Although the composition specifications limit the

Figure 5.1 Fuel specific energy and energy density vs density at 15°C (adapted from [Goodger and Vere, 1985, p. 51].

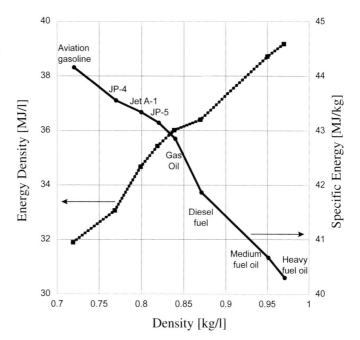

amount of sulfur present in fuel, sulfur cannot be totally eliminated. The level of corrosivity is measured by exposing either copper or silver strips to fuel samples in a test bomb container. The copper strip is exposed for two hours at 100°C, and the silver strip is exposed for four hours at 50°C. The corrosivity level of the fuel is assessed by comparing the metal strips against a colored reproduction of corrosion standard strips. The scales varies from 0 (no tarnish) to 4 (corrosion). The current requirements for fuel corrosivity limit corrosion to level 1 (slight tarnish). Copper corrosion may occur in bearings, valves and other fuel system parts that contain copper. Silver corrosion may occur on silver-faced slipper pads and some of the older types of engine fuel pumps.

Stability – As temperature increases above ambient values, gums are generated in the fuel as a result of the oxidation of the most unstable molecules. These gums could cause a blockage of the fuel system. For this reason, the fuel specifications regulate fuel stability. The oxidation stability is estimated by measuring the precipitate, soluble gum, and insoluble gum. Thermal oxidation stability is measured using a Jet Fuel Thermal Oxidation Tester in order to estimate the tendency to deposition of degradation products. This deposition tends to occur when fuel is exposed to high temperatures for long time.

Contaminants – Contaminants are defined as the components of the fuel that are non-combustible. These contaminants could originate in the crude oil or could be unintentionally added into the fuel during refinement and handling. Contaminants always reduce the performance of the fuel. Fuel specifications limit the amount of copper, gum, and water present in fuel.

5.1.2 Fuels Defined by Elements

The fuel can be defined either by its elements or by the chemical formula of its components. This section presents fuels defined by elements. The definition of the fuels based on their elements is the most general. It can cover both liquid and solid fuels.

To define fuel composition, let us introduce *mass fraction* and *mole fraction*. The mass fraction of a species i within a mixture is the ratio of the mass m_i of that species to the mass m of the mixture:

$$Y_i = m_i/m.$$

The mole fraction of species i is defined as the ratio between the number of moles v_i of species i to the total number of moles in the mixture, v:

$$X_i = v_i/v.$$

The sum of all mass fractions in the mixture is $\sum_{i=1}^{N} Y_i = 1$, where N is the total number of species. Similarly, the sum of all mole fractions in the mixture is $\sum_{i=1}^{N} X_i = 1$. The mole fraction and mass fraction are related by [Kee et al., 1996]

$$X_i = \frac{Y_i}{\mathcal{M}_i} \mathcal{M} \tag{5.1}$$

where $\mathcal{M}_i$ is the molecular mass of species i and $\mathcal{M}$ is the mean molecular mass of the mixture.

The main elements present in aviation fuel are carbon, hydrogen, sulfur, oxygen, nitrogen, and argon. The mass fractions of these elements are

$$Y_C = \frac{\text{kg C}}{\text{kg fuel}}, \quad Y_{H_2} = \frac{\text{kg H}_2}{\text{kg fuel}}, \quad Y_S = \frac{\text{kg S}}{\text{kg fuel}},$$

$$Y_{O_2} = \frac{\text{kg O}_2}{\text{kg fuel}}, \quad Y_{N_2} = \frac{\text{kg N}_2}{\text{kg fuel}}, \quad Y_{Ar} = \frac{\text{kg Ar}}{\text{kg fuel}}, \tag{5.2}$$

$$Y_r = \frac{\text{kg residuals}}{\text{kg fuel}}.$$

The sum of all these mass fractions equals 1, that is,

$$Y_C + Y_{H_2} + Y_S + Y_{O_2} + Y_{N_2} + Y_{Ar} + Y_r = 1.$$

The mass fraction of carbon in jet engine fuels is approximately $Y_C = 0.85$ and the fraction of hydrogen is approximately $Y_{H_2} = 0.15$. The same values of Y_C and Y_{H_2} can be found in fuels for car piston engines and airplane piston engines, as shown in Table 5.4. The difference between these fuels is the number of components, which varies from 700 for car piston engines to 8000 for jet engines.

Note that the composition by elements for all these fuels is the same. The number of components, however, varies. The fuel for car piston engines has the smallest number of components, because it is the first result of the distillation process. The fuels for airplane

Table 5.4 Composition of various type of fuels.

Engine type	Number of components (approx.)	Composition by elements
Car piston engine	700	$Y_C = 0.85; Y_{H_2} = 0.15$
Airplane piston engine	2,500	$Y_C = 0.85; Y_{H_2} = 0.15$
Jet engine	8,000	$Y_C = 0.85; Y_{H_2} = 0.15$

piston engines and jet engines are obtained later in the distillation process compared to the fuel for car piston engines, and for this reason they contain more components.

Fuel composition can also be written as the ratio between the number of moles (or kilomoles) per kilogram of fuel. Using (5.1), the number of kilomoles of fuel per kilogram of fuel is:

$$\sum_i \frac{Y_i}{\mathcal{M}_i} = \underbrace{\frac{Y_C}{\mathcal{M}_C} + \frac{Y_S}{\mathcal{M}_S} + \frac{Y_{H_2}}{\mathcal{M}_{H_2}}}_{\text{Combustible}} + \underbrace{\frac{Y_{O_2}}{\mathcal{M}_{O_2}}}_{\text{Oxidant}} + \underbrace{\frac{Y_{N_2}}{\mathcal{M}_{N_2}} + \frac{Y_{Ar}}{\mathcal{M}_{Ar}} + \frac{Y_r}{\mathcal{M}_r}}_{\text{Inert}} =$$

$$= \frac{Y_C}{12} + \frac{Y_S}{32} + \frac{Y_{H_2}}{2} + \frac{Y_{O_2}}{32} + \frac{Y_{N_2}}{28} + \frac{Y_{Ar}}{40} + \frac{Y_r}{M_r} = \mathcal{N} \frac{\text{kmol fuel}}{\text{kg fuel}}$$

where $\mathcal{M}_i$ is the molar mass of species i.

Let us assume that the fuel is combusted through the following reactions:

$$\frac{Y_C}{12} \text{ kmol C} + \frac{Y_C}{12} \text{ kmol O}_2 \rightarrow \frac{Y_C}{12} \text{ kmol CO}_2$$

$$\frac{Y_S}{32} \text{ kmol S} + \frac{Y_S}{32} \text{ kmol O}_2 \rightarrow \frac{Y_S}{32} \text{ kmol SO}_2 \tag{5.3}$$

$$\frac{Y_{H_2}}{2} \text{ kmol H}_2 + \frac{Y_{H_2}}{4} \text{ kmol O}_2 \rightarrow \frac{Y_{H_2}}{2} \text{ kmol H}_2\text{O}.$$

The minimum amount of oxygen, in kmol, needed for the complete combustion in these three oxidation reactions is

$$(minO)_{\text{kmol}} = \left(\frac{Y_C}{12} + \frac{Y_S}{32} + \frac{Y_{H_2}}{4} - \frac{Y_{O_2}}{32} \right) \frac{\text{kmol O}_2}{\text{kg fuel}} \tag{5.4}$$

where the amount of oxygen $Y_{O_2}/32$ was deducted since it is already present in the fuel and therefore it is not needed from the air. Complete combustion is also called *stoichiometric combustion*.

To find out the volume of air needed for stoichiometric combustion, one needs to know the composition, by volume, of dry air. The composition of dry air is shown in Table 5.5.

The volume of air, in kilomoles, needed for stoichiometric combustion is

$$(minL)_{\text{kmol}} = \frac{1}{0.2095} (minO)_{\text{kmol}} = 4.76(minO)_{\text{kmol}} \tag{5.5}$$

Table 5.5 Composition of dry air, by volume.

Species	Formula	Percentage [%]
Nitrogen	N_2	78.09
Oxygen	O_2	20.95
Argon	Ar	0.93
Carbon dioxide	CO_2	0.03

where $minL$ is the amount of air required for stoichiometric combustion per kg of fuel. The L notation comes from the German word Luft, which means air. The relationship between the volume, mass, and molar mass for any species s is given by

$$\boxed{(min\ s)_{kg} = \mathcal{M}_s \cdot (min\ s)_{kmol}}$$ (5.6)

where the molar mass $\mathcal{M}_s$ is in kg/kmol. Consequently, the mass of oxygen needed for stoichiometric combustion can be calculated as a function of the minimum number of kilomoles of oxygen and the molar mass

$$(minO)_{kg} = \mathcal{M}_{O_2} \cdot (minO)_{kmol}.$$ (5.7)

The minimum mass of oxygen is then obtained using equations (5.4) and (5.7):

$$(minO)_{kg} = 32 \cdot \left(\frac{Y_C}{12} + \frac{Y_{H_2}}{4} + \frac{Y_S - Y_{O_2}}{32} \right) \frac{\text{kg } O_2}{\text{kg fuel}}.$$

The volume of nitrogen, in kilomoles, needed for stoichiometric combustion can be calculated using the composition of dry air and neglecting the presence of argon and carbon dioxide

$$(minN)_{kmol} = \frac{1 - 0.2095}{0.2095} \cdot (minO)_{kmol} = 3.76(minO)_{kmol}.$$ (5.8)

The mass of nitrogen needed for stoichiometric combustion is

$$(minN)_{kg} \overset{(5.6)}{=} 28(minN)_{kmol} \overset{(5.8)}{=} 28 \cdot 3.76(minO)_{kmol} \overset{(5.7)}{=}$$ (5.9)

$$= \frac{28}{32} \cdot 3.76(minO)_{kg} = 3.29(minO)_{kg}.$$

The stoichiometric mass of air, neglecting the small contribution of argon and carbon dioxide, is

$$(minL)_{kg} = (minO)_{kg} + (minN)_{kg} \overset{(5.9)}{=} 4.29(minO)_{kg}.$$ (5.10)

Comparing (5.5) with (5.10), one notices that 4.76 times the volume of oxygen equals the volume of air, while only 4.29 times the mass of oxygen equals the mass of air. The difference is caused by the fact that the molecular mass of oxygen is larger than the molecular mass of nitrogen.

Recall that $minL$ is the amount of air, in kg, required for the stoichiometric combustion of one kg of fuel. L denotes the amount of air, in kg, supplied per kg of fuel. The ratio between

L, the amount of air per kg of fuel, and $minL$, the stoichiometric amount of air per kg of fuel, is called the *excess air coefficient*, λ:

$$\boxed{\lambda = \frac{L}{minL}}.$$
(5.11)

The *fuel-air mass ratio* is defined as

$$f := \dot{m}_f/\dot{m}_a$$
(5.12)

where $\dot{m}_f$ is the mass flow rate of fuel and $\dot{m}_a$ is the mass flow rate of air. Since $L = \dot{m}_a/\dot{m}_f$, the fuel-air mass ratio is

$$f := \dot{m}_f/\dot{m}_a = 1/L,$$

and using (5.11) yields

$$f = \frac{1}{\lambda\, minL}.$$
(5.13)

The *stoichiometric fuel-air mass ratio* is

$$f_{\text{stoich}} := (\dot{m}_f/\dot{m}_a)_{\text{stoich}},$$

which corresponds to the excess air $\lambda = 1$. Consequently, the stoichiometric fuel-air mass ratio is

$$f_{\text{stoich}} = (\dot{m}_f/\dot{m}_a)_{\text{stoich}} = 1/minL.$$

An alternate notation for the stoichiometric fuel-air mass ratio, f_{stoich}, is $f_{\lambda=1}$.

The ratio between the fuel-air mass ratio, f, and the stoichiometric fuel-air mass ratio, f_{stoich}, is the *equivalence ratio*, Φ

$$\boxed{\Phi := f/f_{\text{stoich}}.}$$

The link between the equivalence ratio and the excess air is

$$\Phi = f/f_{\text{stoich}} = minL/L = minL/(\lambda\, minL) = 1/\lambda.$$

A mixture that has more air than is needed for stoichiometric combustion is called a *lean mixture*. In this case, $\lambda > 1$ and $\Phi < 1$. A mixture that has less air than is needed for stoichiometric combustion is called a *rich mixture*. In this case, $\lambda < 1$ and $\Phi > 1$.

Example 5.1.1 Methane combustion is approximated using a global reaction with the stoichiometric equation

$$CH_4 + 2O_2 \rightarrow CO_2 + 2H_2O.$$
(5.14)

Calculate the mass fraction of methane knowing that the equivalence ratio is $\Phi = 0.6$, the molecular mass of air is $\mathcal{M}_{air} = 28.93$ kg/kmol, and the air composition is given in Table 5.5.

Solution

The fuel in this reaction is methane. Using (5.14), the stoichiometric ratio between the moles of fuel and oxygen is

$$\left(\frac{\text{mole fuel}}{\text{mole O}_2}\right)_{stoich} = \frac{1}{2}.$$

Therefore, using the volumetric air composition, the stoichiometric ratio between the moles of fuel and air is

$$\left(\frac{\text{mole fuel}}{\text{mole air}}\right)_{stoich} = \frac{1}{2/0.2095} = 0.1047.$$

The stoichiometric ratio between the mass of fuel and air is

$$\left(\frac{\dot{m}_f}{\dot{m}_{air}}\right)_{stoich} = \left(\frac{\text{mole fuel}}{\text{mole air}}\right)_{stoich}\frac{\mathcal{M}_f}{\mathcal{M}_{air}} = 0.1047\frac{16}{28.93} = 0.0579,$$

where the molecular mass of methane is 16 kg/kmol. The ratio between the mass of fuel and air for an equivalence ratio $\Phi = 0.6$ is

$$\frac{\dot{m}_f}{\dot{m}_{air}} = \Phi\left(\frac{\dot{m}_f}{\dot{m}_{air}}\right)_{stoich} = 0.6 \times 0.0579 = 0.0347.$$

The mass fraction of methane is

$$Y_{(CH_4)} = \frac{\dot{m}_f}{\dot{m}_f + \dot{m}_{air}} = \frac{1}{1 + 1/\frac{\dot{m}_f}{\dot{m}_{air}}} = 0.0336.$$

5.1.3   Fuels Defined by Chemical Formula

A second option for defining the fuels is by their chemical formula. Note, however, that this type of definition is not used for solid fuels.

The common components found in the fuels have the chemical formula C_nH_{2n+2}. These components include

$$\left.\begin{array}{ll} CH_4 & \text{- methane} \\ C_2H_6 & \text{- ethane} \\ C_3H_8 & \text{- propane} \\ C_4H_{10} & \text{- butane} \\ C_5H_{12} & \text{- pentane} \\ C_7H_{16} & \text{- heptane} \\ C_8H_{18} & \text{- octane} \end{array}\right\} C_nH_{2n+2}$$

Other compounds found in fuel, especially in rocket fuel, are

$$\left.\begin{array}{ll} CH_3OH & \text{- methanol} \\ C_2H_5OH & \text{- ethanol} \end{array}\right\} \text{Common for rockets}$$

Let's consider a fuel with the general formula $C_\alpha H_\beta O_\gamma$. The molar mass of this fuel is

$$M_{C_\alpha H_\beta O_\gamma} = 12\,\alpha + \beta + 16\,\gamma \ \ kg/kmol$$

The oxidation reaction of fuel with the general formula $C_\alpha H_\beta O_\gamma$ is

$$C_\alpha H_\beta O_\gamma + \left(\alpha + \frac{\beta}{4} - \frac{\gamma}{2}\right) O_2 \to \alpha CO_2 + \frac{\beta}{2} H_2O.$$

The amount of oxygen, in kmol, needed for stoichiometric combustion of a kmol of fuel is

$$(minO)_{kmol} = \left(\alpha + \frac{\beta}{4} - \frac{\gamma}{2}\right) \frac{kmol\ O_2}{kmol\ fuel}.$$

The amount of oxygen, in kg, needed for stoichiometric combustion of a kmol of fuel is obtained using equation (5.6)

$$(minO)_{kg} = M_{O_2}(minO)_{kmol} = 32(minO)_{kmol}\ \frac{kg\ O_2}{kmol\ fuel}.$$

To switch from fuel in kmol to fuel in kg, one applies equation (5.6) again,

$$(minO)_{kg} = \frac{32}{M_{C_\alpha H_\beta O_\gamma}}\ (minO)_{kmol}\ \frac{kg\ O_2}{kg\ fuel},$$

so that

$$(minO)_{kg} = \frac{32}{12\alpha + \beta + 16\gamma}\left(\alpha + \frac{\beta}{4} - \frac{\gamma}{2}\right) \frac{kg\ O_2}{kg\ fuel}.$$

The minimum amount of air needed to burn the $C_\alpha H_\beta O_\gamma$ fuel is determined using (5.10):

$$(minL)_{kg} = 4.29\ (minO)_{kg}\ \frac{kg\ air}{kg\ fuel}.$$

Example 5.1.2 The composition by volume of a fuel is:

$$
\begin{aligned}
X_{(CO)} &= 0.12 \\
X_{(H_2)} &= 0.43 \\
X_{(CH_4)} &= 0.33 \\
X_{(C_2H_4)} &= 0.05 \\
X_{(O_2)} &= 0.03 \\
X_{(N_2)} &= 0.02 \\
X_{(CO_2)} &= 0.02.
\end{aligned}
$$

How much air is necessary for stoichiometric combustion? What are the combustion products?

Solution

Let us write the chemical reactions that occur:

$$CO + \frac{1}{2} O_2 + \frac{1}{2} \times 3.76 \ N_2 \rightarrow CO_2 + \frac{1}{2} \times 3.76 \ N_2 \qquad | \times 0.12$$

$$H_2 + \frac{1}{2} O_2 + \frac{1}{2} \times 3.76 \ N_2 \rightarrow H_2O + \frac{1}{2} \times 3.76 \ N_2 \qquad | \times 0.43$$

$$CH_4 + 2 \ O_2 + 2 \times 3.76 \ N_2 \rightarrow CO_2 + 2 \ H_2O + 2 \times 3.76 \ N_2 \qquad | \times 0.33$$

$$C_2H_4 + 3 \ O_2 + 3 \times 3.76 \ N_2 \rightarrow 2 \ CO_2 + 2 \ H_2O + 3 \times 3.76 \ N_2 \qquad | \times 0.05.$$

Adding the above chemical equations yields

$$\underbrace{0.12 \ CO + 0.43 \ H_2 + 0.33 \ CH_4 + 0.05 \ C_2H_4 + 0.03 \ O_2 + 0.02 \ N_2 + 0.02 \ CO_2}_{\text{fuel}}$$

$$+ \underbrace{1.055 \ O_2 + 3.9668 \ N_2}_{\text{air}} \rightarrow \underbrace{(0.55 + 0.02) \ CO_2 + 1.19 \ H_2O + 4.1668 \ N_2}_{\text{combustion products}}.$$

5.2   Thermodynamics of Chemistry

In the thermodynamics of chemistry, one departs from considering pure substance alone. The thermodynamics of chemistry deals with systems which are not necessarily homogeneous in either composition or state. Parts of these systems may vary in composition in the course of a process. Such non-homogeneous systems can be best studied in terms of the homogeneous subsystems of which they comprise.

A *phase* is defined as a subsystem which is homogeneous in composition and state.

Example 5.2.1   Let us consider the combustion of solid carbon in air. The stoichiometric equation for this chemical reaction is

$$\underbrace{C}_{\text{solid phase}} + \underbrace{a \ \text{air}}_{\text{gaseous phase}} \rightarrow \underbrace{b \ O_2 + c \ N_2 + CO_2}_{\text{gaseous phase}} \qquad (5.15)$$

where a denotes the number of moles of air, and b and c denote the number of moles of products.

5.2.1   First Law of Thermodynamics Applied to Chemical Reactions

The first law of thermodynamics, introduced in Section 2.4.2, states that for any change of state in any system,

$$\delta Q = \delta W + dE. \qquad (2.48)$$

Let us determine particular forms of the law applied to chemical reactions. If one considers only changes of state which can be carried out without the aid of electricity, magnetism, without capillary forces, gravity or appreciable motion, then the variation of energy is equal to the variation of internal energy, $dE = dU$. As a result, $\delta Q = \delta W + dU$. The work is equal to $p\delta V$ if shear stresses are neglected, so that $\delta Q = p\delta V + dU$. If the change of state occurs at constant pressure, then

$$\delta Q = \delta(pV) + dU$$

and using the definition of enthalpy, $H = U + pV$, gives

$$\delta Q = dH$$

which after integration yields

$$Q_p = H_2 - H_1 = \Delta H \tag{5.16}$$

where subscript p reminds us that the heat transfer occurs at constant pressure, and H_2 and H_1 are the enthalpies at the final and initial states, respectively.

For a constant-volume reaction, where variations in potential and kinetic energy can be neglected as was done above, no external work is done, and the first law is

$$Q_v = \Delta U$$

where the v subscript indicates heat transfer at constant volume.

5.2.2 Thermochemical Laws

Thermochemistry is the branch of thermodynamics concerning the heat changes associated with chemical reactions. This section presents two fundamental thermochemical laws: the Lavoisier and Laplace law, and the law of constant heat summation (or Hess' law).

5.2.2.1 Lavoisier and Laplace Law

The law introduced by A. L. Lavoisier and P. S. Laplace in 1780 states *"The quantity of heat that must be supplied to decompose a compound into its elements is equal to the heat transferred when the compound is formed from its elements."*

In a more general sense, the law states that the heat transfer accompanying a chemical reaction in one direction is equal in magnitude but opposite in sign to the heat transfer associated with the same reaction in the reverse direction [Kuo, 1986, p.41].

Example 5.2.2   When the combustion of methane in oxygen occurs at a temperature of 298.16 K, the heat released during this exothermic reaction is 50,146 kJ for each kg of methane

$$CH_4(g) + 2O_2(g) \rightarrow CO_2(g) + 2H_2O(g) - 50,146 \text{ kJ}.$$

The negative sign assigned above to the 50,146 kJ indicates that this amount of heat was removed from the system in order to keep the temperature constant.

To obtain methane and oxygen from carbon dioxide and water from a reaction that occurs at a temperature of 298.16 K, 50,146 kJ need to be provided to the reactants for every kg of methane that results from the reaction

$$CO_2(g) + 2H_2O_{(g)} \rightarrow CH_4(g) + 2O_2(g) + 50,146 \text{ kJ}.$$

In these thermochemical equations, the subscript (g) denotes the gaseous phase.

5.2.2.2 Law of Constant Heat Summation (or Hess' Law)

The law of constant heat summation states "*In a given chemical reaction, the resultant heat change, at either constant pressure or constant volume, is the same whether the reaction takes place in one or several stages.*" This law was established empirically by G. H. Hess in 1840.

The law implies that the net heat of reaction depends only on the initial and final states. Consequently, the heats of reaction have an additive property [Williams, 1985]. One important outcome of Hess' law is that thermochemical equations can be added and subtracted like algebraic equations.

Example 5.2.3 Let us consider methane combustion at constant pressure. This reaction can be written using one global reaction:

$$CH_4(g) + 2O_2(g) \rightarrow CO_2(g) + 2H_2O(g) \qquad\qquad -802.337 \text{ kJ}. \qquad\qquad (5.17)$$

This is an exothermic reaction where 802.337 kJ need to be removed from the system for each mole of methane such that the temperature remains constant. In other words, the heat liberated during the reaction of a single mole of methane is 802.337 kJ.

The methane combustion can also be written as two global reactions:

$$CH_4(g) + 1.5O_2(g) \rightarrow CO(g) + 2H_2O(g) \qquad\qquad -519.352 \text{ kJ} \qquad\qquad (5.18)$$
$$CO(g) + 0.5O_2(g) \rightarrow CO_2(g) \qquad\qquad\qquad\qquad -282.985 \text{ kJ}. \qquad\qquad (5.19)$$

Both reactions (5.18) and (5.19) are exothermic. By comparing the one-reaction and the two-reaction mechanisms, it is apparent that the heat liberated is the same.

The benefit of Hess' law lies in the fact that it allows us to calculate enthalpy changes for some reactions that are difficult to measure, by using reactions whose enthalpy changes are easy to measure. This will be illustrated in the next section.

5.2.3 Standard Heats of Formation

Standard heats of formation of chemical compounds are of great practical value when studying chemical reactions. They provide an easy method of determining the enthalpy change occurring in any reaction. Before introducing the standard heat of formation, let us define the *standard state* of an element as the form that is stable at room temperature (298.15 K) and atmospheric pressure[5].

The *heat of formation* of a substance, $\Delta\mathcal{H}_f^{\circ}$ [J/mol], also called the *standard heat of formation, enthalpy of formation*, or *standard enthalpy of formation*, is defined as the heat released or absorbed when one mole of the substance is formed from its constituent elements, each substance being in its normal physical state (gas, liquid, or solid) at 298.15 K and 1 atm. The standard heat of formation can also be stated as the enthalpy ΔH_f° [J] of the substance in its standard state, in reference to its elements in their standard state at the same temperature [Kuo, 1986, p. 38].

The heat of formation of an element is arbitrarily assigned a value of zero, or has a standard heat of formation of zero.

Example 5.2.4 What are the heats of formation $\Delta\mathcal{H}_f^{\circ}$ for $C_8H_{18}(\ell)$ and $CO_2(\ell)$?
  From Table 5.6 one reads:

$$\Delta\mathcal{H}_{f,C_8H_{18}}^{\circ} = -249.9 \text{ kJ/mol}$$

$$\Delta\mathcal{H}_{f,CO_2}^{\circ} = -393.5 \text{ kJ/mol}.$$

Example 5.2.5 Using Hess' Law, calculate the enthalpy of formation of methane.

Solution
The direct reaction for the formation of methane,

$$C(s) + 2H_2(g) \rightarrow CH_4(g),$$

cannot conveniently be carried out. Instead, the enthalpy variations of the following reactions can be easily measured:

$$CH_4(g) + 2O_2(g) \rightarrow CO_2(g) + 2H_2O(\ell) \qquad -890.309 \text{ kJ}$$
$$C(s) + O_2(g) \rightarrow CO_2(g) \qquad -393.510 \text{ kJ}$$
$$2H_2 + O_2(g) \rightarrow 2H_2O(\ell) \qquad -571.660 \text{ kJ}$$

---

[5] After 1982, the recommended value for the standard state pressure is 1 bar instead of 1 atm = 1.01325 bar.

Table 5.6 Tables of standard heats of formation. All substances are in the gaseous state, except for those indicated as liquid by (ℓ) and solid by (s).

Substance	Formula	$\Delta \mathcal{H}_f^{\circ}$ [J/mol]
Argon	Ar	0
Carbon (graphite)	C(s)	0
Carbon monoxide	CO	–110,527
Carbon dioxide	CO_2	–393,510
Methane	CH_4	–74,873
Ethane	C_2H_6	–84,685
Propane	C_3H_8	–103,849
n-Butane	$C_4H_{10}(\ell)$	–147,655
n-Heptane	$C_7H_{16}(\ell)$	–224,392
n-Octane	$C_8H_{18}(\ell)$	–249,957
n-Decane	$C_{10}H_{22}(\ell)$	–294,366
Hydrogen atom	H	217,999
Hydrogen	H_2	0
Water	H_2O	–241,818
Water	$H_2O(\ell)$	–285,830
Helium	He	0
Nitrogen atom	N	472,683
Nitrogen	N_2	0
Nitrogen oxide	NO	90,291
Nitrogen dioxide	NO_2	33,095
Ammonia	NH_3	–45,898
Oxygen atom	O	249,173
Oxygen	O_2	0
Hydroxyl	OH	38,987
Sulfur	S	276,980
Sulfur dioxide	SO_2	–296,842
Sulfur trioxide	SO_3	–395,765

where the heat released is reported per mole. Deducting the first equation from the sum of the last two equations yields

$$C(s) + 2H_2(g) \rightarrow CH_4(g) - 74.861 \text{ kJ}$$

which is the value given in Table 5.6.

5.2.4 Heats of Reaction

The heat change occurring during a chemical reaction is in general an indefinite quantity because it depends on the path of the process. If the process is carried out at either constant pressure or constant volume, the heat change has a definite value that is determined only by the initial and final states of system, as shown in Section 5.2.1. For this reason the heat changes of chemical reactions are measured under either constant pressure or constant volume conditions.

Let us consider a single arbitrary chemical reaction

$$\sum_{s=1}^{N_s} v_s' \mathfrak{M}_s \rightarrow \sum_{s=1}^{N_s} v_s'' \mathfrak{M}_s \tag{5.20}$$

where v_s' are the number of moles of species s appearing as a reactant, v_s'' are the number of moles of species s appearing as a product, $\mathfrak{M}_s$ denotes the chemical symbol of species s, and N_s is the total number of species. Species that are not reactants have $v_s' = 0$, and species that are not products have $v_s'' = 0$

To define the heat of reaction, let us consider a closed system containing a number of moles v_s' of N_s different species at a given temperature T and pressure p. If these species undergo an isobaric process, then according to (5.16),

$$Q = \Delta H = H_{\text{products}} - H_{\text{reactants}} = \sum_{s=1}^{N_s} (v_s'' - v_s') \mathcal{H}_{\mathfrak{M}_s, T} \tag{5.21}$$

where $\mathcal{H}_{\mathfrak{M}_s, T}$ [J/mol] is the enthalpy of species s. If the isobaric process has the same initial and final temperatures, then the heat liberated by the system, Q, is the *heat of reaction* or *enthalpy of reaction*, ΔH_r, for this process [Kuo, 1986, p. 46]. Consequently, the enthalpy of reaction, ΔH_r, is defined as the heat transferred, per mole, for a specific chemical reaction when the reactants and products are at the same temperature, and the pressure remains constant.

The decision to define the heat of reaction as (+Q), instead of (-Q), is arbitrary. Some textbooks define the heat of reaction as the heat released (-Q) as opposed to the heat absorbed (+Q). Since in combustion most reactions are exothermic, one could argue that it is more convenient to define the heat of reaction as (-Q) so as to work with positive heats of reaction [Williams, 1985, p. 538].

For the arbitrary chemical reaction (5.20), the heat of reaction at standard state is

$$\Delta H_{r, T_0} = \sum_{s=1}^{N_s} v_s'' \Delta \mathcal{H}_{f, \mathfrak{M}_s}^{\circ} - \sum_{s=1}^{N_s} v_s' \Delta \mathcal{H}_{f, \mathfrak{M}_s}^{\circ} \tag{5.22}$$

where T_0 is the standard temperature 298.15 K.

Example 5.2.6 Calculate the standard heat of reaction for *n*-decane, $C_{10}H_{22}$.

Solution
The chemical reaction for *n*-decane combustion is

$$C_{10}H_{22} + 15.5O_2 \rightarrow 10CO_2 + 11H_2O(g). \tag{5.23}$$

Neglecting the dissociation in the final products, the standard heat of reaction is obtained using (5.22) and the heats of formation listed in Table 5.6:

$$\Delta H_{r, T_0} = 10 \times (-393,510) + 11 \times (-241,818) - [-294,336 - 0] =$$
$$= -6,300,762 \text{ J per mole of } C_{10}H_{22}.$$

To calculate the heat of reaction at an arbitrary temperature T, the variation of enthalpy between temperature T and T_0 must be included [Kuo, 1986, p. 49]:

$$\mathcal{H}_{\mathfrak{M}_s, T} = \Delta \mathcal{H}^{\circ}_{f, \mathfrak{M}_s} + \int_{T_0}^{T} C_{p, \mathfrak{M}_s} dT \tag{5.24}$$

so that

$$\Delta H_{r, T} = \sum_{s=1}^{N_s} v''_s \left(\Delta \mathcal{H}^{\circ}_{f, \mathfrak{M}_s} + \int_{T_0}^{T} C_{p, \mathfrak{M}_s} dT \right) - \sum_{s=1}^{N_s} v'_s \left(\Delta \mathcal{H}^{\circ}_{f, \mathfrak{M}_s} + \int_{T_0}^{T} C_{p, \mathfrak{M}_s} dT \right) \tag{5.25}$$

where $C_{p, \mathfrak{M}_s}$ [J/(mol K)] is the molar heat capacity at constant pressure of species $\mathfrak{M}_s$. Since the molar heat capacity varies with temperature, several options are available for calculating $\Delta \mathcal{H}^{\circ}_{T, \mathfrak{M}_s} - \Delta \mathcal{H}^{\circ}_{T_0, \mathfrak{M}_s} = \int_{T_0}^{T} C_{p, \mathfrak{M}_s} dT$: (1) use tables for variation of $C_{p, \mathfrak{M}_s}$ [Anonymous, 1947], (2) use polynomials to approximate the variation of $C_{p, \mathfrak{M}_s}$ with temperature [McBride et al., 1993], and (3) use tables for variation of $\Delta \mathcal{H}^{\circ}_{T, \mathfrak{M}_s} - \Delta \mathcal{H}^{\circ}_{T_0, \mathfrak{M}_s}$ with temperature [Anonynous, 1971].

Example 5.2.7 Calculate $\int_{400 \text{ K}}^{700 \text{ K}} C_{p, CH_4} dT$ using the enthalpy variation given in Table 5.7 and the polynomial approximation [McBride et al., 1993]

$$C_p/\mathcal{R} = 5.14987613 - 1.36709788 \times 10^{-2} T + 4.91800599 \times 10^{-5} T^2 -$$
$$- 4.84743026 \times 10^{-8} T^3 + 1.66693956 \times 10^{-11} T^4 \tag{5.26}$$

where the universal gas constant $\mathcal{R} = 8.3145$ kJ/(kmol K).

Solution

Using Table 5.7 yields

$$\Delta \mathcal{H}^{\circ}_{T=700K} - \Delta \mathcal{H}^{\circ}_{T=400K} = 28,761 - 13,909 = 14,852 \text{ kJ/kmol.}$$

Integrating the polynomial approximation (5.26) from $T_1 = 400$ K to $T_2 = 700$ K yields

$$\int_{T_1}^{T_2} C_p dT = \mathcal{R} \left(5.14987613 T - 1.36709788 \times 10^{-2} T^2/2 + 4.91800599 \times 10^{-5} T^3/3 - \right.$$
$$\left. -4.84743026 \times 10^{-8} T^4/4 + 1.66693956 \times 10^{-11} T^5/5 \right) |_{T_1}^{T_2} = 14,881 \text{ kJ/kmol.}$$

Table 5.7 Tables of molar heat capacity and enthalpy variation for methane.

Temperature [K]	C_p [kJ/(kmol K)]	$\Delta\mathcal{H}_T^\circ - \Delta\mathcal{H}_0^\circ$ [kJ/kmol]
298.16	35.639	10,033
300	35.708	10,100
400	40.500	13,909
500	46.342	18,271
600	52.227	23,227
700	57.794	28,761
800	62.932	34,830
900	67.601	41,385
1000	71.795	48,387
1100	75.529	55,755
1200	78.833	63,499
1300	81.744	71,577
1400	84.305	79,907
1500	86.556	88,446

Figure 5.2 Methane heat capacity at constant pressure vs. temperature: tabulated vs. polynomial approximation values.

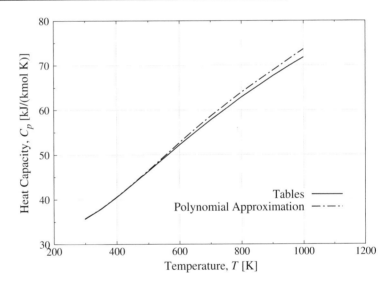

The error introduced by using the polynomial approximation is

$$\epsilon = (14,881 - 14,852)/14,852 = 0.0019$$

or 0.19%. The polynomial expressions, although approximate, are convenient for numerical computation. Figure 5.2 shows a comparison between the tabulated values of heat capacities of methane and the polynomial approximation of these values. It is apparent that as the temperature increases, the differences increase. For this reason, a second polynomial is used to approximate C_p for temperatures higher than 1000 K.

Example 5.2.8 Calculate the standard heat of reaction for n-decane, $C_{10}H_{22}$, at a temperature of 1300 K using the linear approximations of the molar heat capacities at constant pressure given in Table 5.8.

Table 5.8 Coefficients of linear approximation of molar heat capacities, $C_p = a + b \times T$ (from [Flack, 2005]).

Compound	a [kJ/(kmol K)]	b [kJ/(kmol K^2)]
CO	27.4	0.0058
CO_2	28.8	0.0280
H_2O	30.5	0.0103
O_2	27.0	0.0079
N_2	27.6	0.0051
$C_8H_{18}(\ell)$	254	0
$C_{10}H_{22}(\ell)$	296	0

Solution

Using the chemical reaction for n-decane combustion (5.23) and neglecting the dissociation in the final products, the standard heat of reaction is obtained using (5.25), the heats of formation listed in Table 5.6, and the linear approximation of the heat capacities listed in Table 5.8:

$$\Delta H_{r,T} = 10 \times \left[(-393,510) + a_{CO_2} \times (1300 - 298) + b_{CO_2} \times (1300^2 - 298^2)/2 \right] +$$
$$+ 11 \times \left[(-241,818) + a_{H_2O} \times (1300 - 298) + b_{H_2O} \times (1300^2 - 298^2)/2 \right] -$$
$$- 1 \times \left[-294,336 + a_{C_{10}H_{22}} \times (1300 - 298) + b_{C_{10}H_{22}} \times (1300^2 - 298^2)/2 - \right]$$
$$- 15.5 \times \left[0 + a_{O_2} \times (1300 - 298) + b_{O_2} \times (1300^2 - 298^2)/2 \right].$$

Using the standard heat of reaction $\Delta H_{r,T_0}$ calculated in example 5.2.6 yields

$$\Delta H_{r,T} = -6,300,732 + (1300 - 298) \times [10a_{CO_2} + 11a_{H_2O} - a_{C_{10}H_{22}} - 15.5a_{O_2}] +$$
$$+ (1300^2 - 298^2)/2 \times [10b_{CO_2} + 11b_{H_2O} - b_{C_{10}H_{22}} - 15.5b_{O_2}] =$$
$$= -6,300,732 + (1300 - 298) \times [10 \times 28.8 + 11 \times 30.5 - 296 - 15.5 \times 27.0] +$$
$$+ (1300^2 - 298^2)/2 \times [10 \times 0.0280 + 11 \times 0.0103 - 0 - 15.5 \times 0.0079] =$$
$$= -6,175,070 \text{ kJ per mole of } C_{10}H_{22}.$$

5.2.5 Heat of Combustion

The heat of combustion, or enthalpy of combustion, ΔH_c, is defined as the heat transferred when 1 mole of a substance is completely burned at constant pressure, the products being cooled to the initial temperature of the fuel.

5.2.6 Higher and Lower Heating Values

The *higher heating value* (*HHV*) is defined as the heat transferred when 1 kg of fuel is completely combusted, the temperature of the reactants and products being the same and all water formed from combustion being liquid.

The *lower heating value* (*LHV*) is defined as the heat transferred when 1 kg of fuel is completely combusted, the temperature of the reactants and products being the same and all water formed from combustion being vapor.

The difference between the *HHV* and *LHV* is

$$HHV - LHV = m_{H_2O} h_{\ell v},$$

where $h_{\ell v}$ is *latent heat of vaporization* and m_{H_2O} is the mass of water resulting from the combustion of 1 kg of fuel. The value of the latent heat of vaporization of water is $h_{\ell v} = 2258$ kJ/kg.

Table 5.9 shows the lower heating values and higher heating values for several fuels. Note that all the fuels listed in Table 5.9 have lower heating values higher than the standard fuel.

Table 5.9 Higher and lower heating values for various substances at 298.15 K.

Compound	Formula	*LHV* [kJ/kg]	*HHV* [kJ/kg]
Methane	$CH_4(g)$	50,012	55,499
Ethane	$C_2H_6(g)$	44,521	48,638
Propane	$C_3H_8(\ell)$	46,351	50,343
n-Butane	$C_4H_{10}(\ell)$	45,342	49,128
n-Heptane	$C_7H_{16}(\ell)$	44,555	48,069
n-Octane	$C_8H_{18}(\ell)$	44,422	47,890
n-Nonane	$C_9H_{20}(\ell)$	44,315	
n-Decane	$C_{10}H_{22}(\ell)$	44,231	
Jet A	$CH_{1.94}(\ell)$	43,252	47,000
JP-4	$CH_{2.0}(\ell)$	43,486	
JP-5	$CH_{1.92}(\ell)$	43,017	

Table 5.10 Molecular mass.

Name	Symbol	Molecular mass [kg/kmol]
Hydrogen	H	1.008
Carbon	C	12.011
Nitrogen	N	14.007
Oxygen	O	15.999
Sulfur	S	32.06
Argon	Ar	39.95

Table 5.10 shows the molecular mass for several species that are used in this chapter. These molecular masses are needed to calculate the molecular masses of different compounds present in jet engine combustion.

Example 5.2.9 The lower heating value of n-octane, C_8H_{18}, is 44,422 kJ/kg. The latent heat of vaporization of water is 2258 kJ/kg. Calculate the higher heating value of n-octane.

Solution
The n-octane combustion

$$C_8H_{18} + 12.5O_2 \rightarrow 8\,CO_2 + 9\,H_2O$$

shows that 9 moles of H_2O result for every mole of C_8H_{18}. The molecular mass of C_8H_{18}, calculated using the values of Table 5.10 rounded to integers, is $8 \times 12 + 18 \times 1 = 114$ kg/kmol, and the molecular mass of water is 18 kg/kmol. Consequently, $9 \times 18 / 114 = 1.421$ kg of water result for every kilogram of C_8H_{18}. Therefore, the higher heating value of C_8H_{18} is

$$HHV = LHV + m_{H_2O}h_{\ell v} = 44,422 + 1.421 \times 2258 = 47,631 \text{ kJ/kg}.$$

The calculated value is a good approximation of the value reported in Table 5.9.

5.2.7   Adiabatic Flame Temperature

The *adiabatic flame temperature* is defined as the temperature that results when a fuel is burned adiabatically during a constant-pressure process without changes in kinetic or potential energy. The adiabatic flame temperature is the maximum temperature that can be achieved for the given reactants. Dissociation, radiation, incomplete combustion, and excess air lower the flame temperature.

The adiabatic flame temperature in a gas turbine is controlled with the amount of *excess air*, λ. The adiabatic flame temperature is adjusted so that it does not exceed the maximum permissible temperature in the combustor and the turbine.

The adiabatic flame temperature is calculated using (5.21) where Q is zero:

$$H_{\text{products}} - H_{\text{reactants}} = Q = 0,$$

which can be written as

$$H_{\text{reactants}} = H_{\text{products}}. \tag{5.27}$$

Finding the adiabatic flame temperature is typically an iterative process, unless the heat capacities of the species are approximated as polynomial functions of temperature. The following example illustrates how to calculate the adiabatic flame temperature when the heat capacities are approximated as linear functions of temperature.

Example 5.2.10   Liquid n-octane (C_8H_{18}) is burned at constant pressure assuming complete combustion. The fuel is supplied at temperature $T_1 = 400$ K and the air is supplied at temperature $T_2 = 700$ K. The composition of dry air, by volume, is given in Table 5.5. The argon and carbon dioxide compounds can be neglected. Determine adiabatic flame temperatures for the following cases: (1) stoichiometric combustion, $\lambda = 1$ and (2) excess air, $\lambda = 3$.

Solution
Neglecting the presence of argon and carbon dioxide compounds, for every mole of oxygen, there are $(1 - 0.2095)/0.2095 = 3.76$ moles of nitrogen.

For the general case when excess air is λ, the chemical reaction is

$$C_8H_{18} + 12.5\lambda\, O_2 + 12.5 \times 3.76\lambda\, N_2 \rightarrow 8\, CO_2 + 9\, H_2O + 12.5(\lambda - 1)\, O_2 + 12.5 \times 3.76\lambda\, N_2.$$

The enthalpy of n-octane at temperature T is

$$\mathcal{H}_{C_8H_{18},T} = \Delta \mathcal{H}^{\circ}_{f,C_8H_{18}} + \int_{T_0}^{T} C_{p,C_8H_{18}} dT$$

where $T_0 = 298.15$ K. If the temperature variation of heat capacity is approximated using the values from Table 5.8, the enthalpy of n-octane at temperature T becomes

$$\mathcal{H}_{C_8H_{18},T} = \Delta \mathcal{H}^{\circ}_{f,C_8H_{18}} + a_{C_8H_{18}}(T - T_0) + 0.5 b_{C_8H_{18}}(T^2 - T_0^2).$$

The enthalpies for the other species can be written in a similar fashion so that (5.27) becomes

$$
\Delta \mathcal{H}^\circ_{f, C_8H_{18}} + a_{C_8H_{18}}(T_1 - T_0) + 0.5 b_{C_8H_{18}}(T_1^2 - T_0^2) +
$$
$$
+ 12.5\lambda \left(\Delta \mathcal{H}^\circ_{f, O_2} + a_{O_2}(T_2 - T_0) + 0.5 b_{O_2}(T_2^2 - T_0^2) \right) +
$$
$$
+ 47\lambda \left(\Delta \mathcal{H}^\circ_{f, N_2} + a_{N_2}(T_2 - T_0) + 0.5 b_{N_2}(T_2^2 - T_0^2) \right) =
$$
$$
= 8 \left(\Delta \mathcal{H}^\circ_{f, CO_2} + a_{CO_2}(T_3 - T_0) + 0.5 b_{CO_2}(T_3^2 - T_0^2) \right) +
$$
$$
+ 9 \left(\Delta \mathcal{H}^\circ_{f, H_2O} + a_{H_2O}(T_3 - T_0) + 0.5 b_{H_2O}(T_3^2 - T_0^2) \right) +
$$
$$
+ 12.5(\lambda - 1) \left(\Delta \mathcal{H}^\circ_{f, O_2} + a_{O_2}(T_3 - T_0) + 0.5 b_{O_2}(T_3^2 - T_0^2) \right) +
$$
$$
+ 47\lambda \left(\Delta \mathcal{H}^\circ_{f, N_2} + a_{N_2}(T_3 - T_0) + 0.5 b_{N_2}(T_3^2 - T_0^2) \right)
$$

where T_3 is the adiabatic flame temperature. The above equation is a quadratic equation of T_3 that can be written as

$$
T_3^2 \, 0.5 \, (8 b_{CO_2} + 9 b_{H_2O} + 47\lambda b_{N_2} + 12.5(\lambda - 1) b_{O_2}) + \tag{5.28}
$$
$$
+ T_3 \, (8 a_{CO_2} + 9 a_{H_2O} + 47\lambda a_{N_2} + 12.5(\lambda - 1) \, a_{O_2}) + r = 0,
$$

where

$$
r = 8 \Delta \mathcal{H}^\circ_{f, CO_2} + 9 \Delta \mathcal{H}^\circ_{f, H_2O} - \Delta \mathcal{H}^\circ_{f, C_8H_{18}} - 12.5 \Delta \mathcal{H}^\circ_{f, O_2} -
$$
$$
- T_0(8 a_{CO_2} + 9 a_{H_2O} + 12.5(\lambda - 1) a_{O_2}) - T_0^2 \, 0.5 \, (8 b_{CO_2} + 9 b_{H_2O} + 12.5(\lambda - 1) b_{O_2}) -
$$
$$
- a_{C_8H_{18}}(T_1 - T_0) - 0.5 b_{C_8H_{18}}(T_1^2 - T_0^2) - 12.5\lambda \left(a_{O_2}(T_2 - T_0) + 0.5 b_{O_2}(T_2^2 - T_0^2) \right) -
$$
$$
- 47\lambda(a_{N_2} T_2 + 0.5 b_{N_2} T_2^2).
$$

Substituting in (5.28) the enthalpies of formation from Table 5.6 and the coefficients a and b from Table 5.8 yields an adiabatic flame temperature $T_3 = 2544.7$ K for $\lambda = 1$, and $T_3 = 1477.8$ K for $\lambda = 3$.

Example 5.2.11 Liquid n-decane ($C_{10}H_{22}$) is burned at constant pressure assuming complete combustion. The fuel is supplied at temperature $T_1 = 450$ K and the air is supplied at temperature $T_2 = 650$ K. The composition of dry air, by volume, is given in Table 5.5. The argon and carbon dioxide compounds can be neglected. Determine the adiabatic flame temperature if excess air $\lambda = 2.9$.

Solution
As shown in the previous example, for every mole of oxygen in dry air there are 3.76 moles of nitrogen. Therefore, the chemical reaction for n-decane combustion is

$$C_{10}H_{22} + 15.5\lambda\, O_2 + 15.5 \times 3.76\lambda\, N_2 \rightarrow 10\, CO_2 + 11\, H_2O + 15.5(\lambda - 1)\, O_2 + \\ + 15.5 \times 3.76\lambda\, N_2,$$

which for $\lambda = 2.9$ yields

$$C_{10}H_{22} + 44.95\, O_2 + 169.012\, N_2 \rightarrow 10\, CO_2 + 11\, H_2O + 29.45\, O_2 + 169.012\, N_2.$$

Using (5.24) for the enthalpies of each species in the above reaction, (5.27) yields

$$
\begin{aligned}
\Delta\mathcal{H}^{\circ}_{f,C_{10}H_{22}} &+ a_{C_{10}H_{22}}(T_1 - T_0) + 0.5 b_{C_{10}H_{22}}(T_1^2 - T_0^2) + \\
+ 44.95\, &\left(\Delta\mathcal{H}^{\circ}_{f,O_2} + a_{O_2}(T_2 - T_0) + 0.5 b_{O_2}(T_2^2 - T_0^2)\right) + \\
+ 169.012\, &\left(\Delta\mathcal{H}^{\circ}_{f,N_2} + a_{N_2}(T_2 - T_0) + 0.5 b_{N_2}(T_2^2 - T_0^2)\right) = \\
= 10\, &\left(\Delta\mathcal{H}^{\circ}_{f,CO_2} + a_{CO_2}(T_3 - T_0) + 0.5 b_{CO_2}(T_3^2 - T_0^2)\right) + \\
+ 11\, &\left(\Delta\mathcal{H}^{\circ}_{f,H_2O} + a_{H_2O}(T_3 - T_0) + 0.5 b_{H_2O}(T_3^2 - T_0^2)\right) + \\
+ 29.45\, &\left(\Delta\mathcal{H}^{\circ}_{f,O_2} + a_{O_2}(T_3 - T_0) + 0.5 b_{O_2}(T_3^2 - T_0^2)\right) + \\
+ 169.012\, &\left(\Delta\mathcal{H}^{\circ}_{f,N_2} + a_{N_2}(T_3 - T_0) + 0.5 b_{N_2}(T_3^2 - T_0^2)\right).
\end{aligned}
\tag{5.29}
$$

Substituting in (5.29) the enthalpies of formation from Table 5.6 and the coefficients a and b from Table 5.8 yields an adiabatic flame temperature $T_3 = 1463.57$ K for $\lambda = 2.9$.

5.3 Standard Fuel

The type of fuel used in an engine has a significant impact on the engine performance. To allow a fair comparison between engines, it is necessary to present their performances obtained using the same fuel. For this reason, the standard fuel was introduced. The composition of standard fuel was assumed to be 86.08% C, 13.92% H by mass [Hodge, 1955, p. 60]. The lower heating value (LHV) of standard fuel is $LHV = 43,500$ kJ/kg. The amount of air needed for the stoichiometric combustion of one kilogram of standard fuel is 14.66 kg, that is, $minL = 14.66$ kg air/kg fuel or $minL = 0.506$ kmol air/kg fuel.

Once the turbine inlet temperature, T_{03}, is specified, the engine designer must determine the excess air, λ, needed to achieve that temperature. To find the excess air using standard fuel, energy conservation is used between the inlet in the burner, state 02, and the exit of the burner, state 03, assuming the heat and work transfer are zero, so that:

$$\dot{m}_a \cdot h_{02} + \dot{m}_f \cdot (h_f + \xi_{comb} \cdot LHV) = \dot{m}_g \cdot h_{03} \tag{5.30}$$

where

$\dot{m}_a$ - mass flow rate of air entering the combustor

$\dot{m}_f$ - mass flow rate of fuel

$\dot{m}_g$ - mass flow rate of gases leaving the combustor: $\dot{m}_g = \dot{m}_a + \dot{m}_f$

ξ_{comb} - combustion efficiency.

Typical values of combustion efficiency are $0.97 > \xi_{comb} > 0.9$. The enthalpy of the fuel, h_f, is much smaller than the lower heating value, LHV, and will be neglected. Using

$$f = \frac{\dot{m}_f}{\dot{m}_a} = \frac{1}{L} = \frac{1}{\lambda \cdot minL}, \tag{5.11}$$

(5.30) becomes:

$$h_{02} + \frac{\xi_{comb} \cdot LHV}{\lambda \cdot minL} = \left(1 + \frac{1}{\lambda \cdot minL}\right) \cdot h_{03}. \tag{5.31}$$

The enthalpy at the exit of the combustor, h_{03}, is a function of the excess air, λ. The enthalpy of combustion products for λ excess air is a weighted function of the enthalpy of the air and the enthalpy of the combustion products for stoichiometric combustion:

$$h_\lambda = \frac{1 + minL}{1 + \lambda \cdot minL} \cdot h_{\lambda=1} + \frac{(\lambda - 1)minL}{1 + \lambda \cdot minL} \cdot h_a, \tag{5.32}$$

where

h_λ - enthalpy of combustion products for λ excess air

$h_{\lambda=1}$ - enthalpy of combustion products for stoichiometric combustion

h_a - enthalpy of air.

The two weighting functions are

$$r = \frac{1 + minL}{1 + \lambda\, minL} \quad \text{and} \quad q = \frac{(\lambda - 1)\, minL}{1 + \lambda\, minL}, \tag{5.33}$$

and, as expected, the sum of the weighting functions is 1, that is, $r + q = 1$.

Problem 5.3.1   Prove Eq. (5.32).

Solution
For stoichiometric conditions, (5.27) yields

$(\dot{m}_a + \dot{m}_f)_{stoich} h_{\lambda=1} = \dot{m}_{a\ stoich} h_a + \dot{m}_{f\ stoich} h_f.$

Divide all terms of above equation by $\dot{m}_{f\,stoich}$ and use the definition of $minL = \dot{m}_{a\,stoich}/\dot{m}_{f\,stoich}$ so that

$$(min\ L + 1)h_{\lambda=1} = min\ L\ h_a + h_f. \tag{5.34}$$

For an arbitrary value of λ, (5.27) yields

$$(\dot{m}_a + \dot{m}_f)h_\lambda = \dot{m}_a h_a + \dot{m}_f h_f. \tag{5.35}$$

Substitute h_f from (5.34) into (5.35):

$$(\dot{m}_a + \dot{m}_f)h_\lambda = \dot{m}_a h_a + \dot{m}_f \left[(min\ L + 1)\,h_{\lambda=1} - h_a min\ L\right] \tag{5.36}$$

and divide by $\dot{m}_a$

$$(1+f)h_\lambda = h_a + f\left[(min\ L + 1)\,h_{\lambda=1} - h_a min\ L\right]. \tag{5.37}$$

Using

$$f = 1/(\lambda min\ L) \tag{5.13}$$

yields

$$\left(1 + \frac{1}{\lambda min\ L}\right)h_\lambda = h_a + \frac{1}{\lambda min\ L}\left[(min\ L + 1)\,h_{\lambda=1} - h_a min\ L\right] \tag{5.38}$$

which can be rearranged as Eq. (5.32).

Let us find the enthalpy of the combustion products knowing the temperature and excess air.

Example 5.3.1 The excess air in a combustor is $\lambda = 2.72$ and the temperature is $T = 1400$ K. What is the enthalpy?

Solution
Read from the tables of thermodynamic properties of air and stoichiometric combustion products $h_a = 1515.2$ kJ/kg and $h_{\lambda=1} = 1642.6$ kJ/kg. Calculate the weighting functions

$$r = (1 + 14.66)/(1 + 2.72 \times 14.66) = 0.383$$

and

$$q = (1.72 \times 14.66)/(1 + 2.72 \times 14.66) = 0.617.$$

Using (5.32),

$$h_{\lambda=2.72} = 0.383 \times 1642.6 + 0.617 \times 1515.2 = 1564 \text{ kJ/kg.}$$

We can also find the temperature of the combustion products when the enthalpy and excess air are known, but this requires an iterative process, as shown in the following example.

Example 5.3.2 The excess air in a combustor is $\lambda = 2.984$ and the enthalpy of the combustion products is $h_\lambda = 1225$ kJ/kg. What is the temperature?

Solution
The temperature is the root of the equation

$$f(T) = h_\lambda - r\, h_{\lambda=1}(T) - q\, h_a(T) = 0$$

where r and q are the weightings (5.33)

$$r = (1 + 14.66)/(1 + 2.984 \times 14.66) = 0.350$$
$$q = (1.984 \times 14.66)/(1 + 2.984 \times 14.66) = 0.650.$$

We start with a guess of the temperature T, read the enthalpies for air and stoichiometric combustion products from the thermodynamic properties tables, and calculate $f(T)$. A good initial guess is obtained by reading the air and stoichiometric combustion gases temperatures corresponding to the enthalpy $h_\lambda = 1225$ kJ/kg and weighting them:

$$T = r T_{\lambda=1} + q T_a = 0.350 \times 1082.22 + 0.650 \times 1155.05 = 1129.56 \text{ K}.$$

The results are shown in Table 5.11.

Table 5.11 Enthalpy and function $f(T)$ variation with temperature.

Iteration	Temperature K	Enthalpy $h_{\lambda=1}$ kJ/kg	Enthalpy h_a kJ/kg	Function $f(T)$ kJ/kg
1	1129.56	1285.91	1195.21	−1.95
2	1125	1279.97	1189.91	3.57
3	1127.95	1283.81	1193.34	−0.0045

Note that the temperature at iteration 2 was guessed so as to bracket the solution. The temperature at iteration 3 was calculated as a linear interpolation of the first two iterations, that is,

$$T_{\text{iter}=3} = 1125 + (1129.56 - 1125)\frac{3.57}{3.57 + 1.95} = 1127.95 \text{ K}.$$

The solution was considered converged after iteration 3.

Using the expression of the enthalpy of mixture (5.32) in the energy balance equation (5.31) yields an analytical expression of the excess air, λ.

Problem 5.3.2   Find the analytical expression of the excess air as a function of the stagnation enthalpy at states 2 and 3.

Solution

$$\lambda = \frac{h_{03\lambda=1}(1 + minL) - \xi_{comb}LHV - h_{03a}minL}{minL(h_{02} - h_{03a})}.$$

(5.39)

The internal energy of combustion products with λ excess air, u_λ, can be calculated as a function of the internal energy of air, u_a, and the internal energy of the products of stoichiometric combustion, $u_{\lambda=1}$, using

$$u_\lambda = \frac{1 + minL}{1 + \lambda\, minL} u_{\lambda=1} + \frac{(\lambda - 1)\, minL}{1 + \lambda\, minL} u_a.$$

(5.40)

Similarly to the enthalpy expression (5.32), the internal energy is calculated by weighting the values of the internal energy for air and the stoichiometric products.

The entropy of the mixture is calculated using an expression similar to (5.32), except for a correction term, $\Delta s'$:

$$s_\lambda = \frac{1 + minL}{1 + \lambda minL} s_{\lambda=1} + \frac{(\lambda - 1)minL}{1 + \lambda minL} s_a + \Delta s'.$$

(5.41)

The entropy correction term, $\Delta s'$, is [Pimsner, 1983, p. 29]

$$\Delta s' = \mathcal{R}[r(v_{CO_2} + v_{H_2O}) \ln \tfrac{1}{r} + q\, v_{O_2}^a \ln \tfrac{1}{q}$$
$$- r \tfrac{1}{v_{N_2}} \ln \tfrac{1}{v_{N_2}} - q \tfrac{1}{v_{N_2}^a} \ln \tfrac{1}{v_{N_2}^a} + (rv_{N_2} + qv_{N_2}^a) \ln \tfrac{1}{rv_{N_2} + qv_{N_2}^a}]$$

(5.42)

where

v_{CO_2} - number of CO_2 kmol in 1 kmol of gases

v_{H_2O} - number of H_2O kmol in 1 kmol of gases

v_{N_2} - number of N_2 kmol in 1 kmol of gases

$v_{O_2}^a$ - number of O_2 kmol in 1 kmol of air: $v_{O_2}^a = 0.2095$

$v_{N_2}^a$ - number of N_2 kmol in 1 kmol of air (including Ar): $v_{N_2}^a = 0.7905$

and $\mathcal{R}$ is the universal gas constant $\mathcal{R} = 8.3145$ kJ/(kmol K). For standard fuel, the gas constant of stoichiometric combustion products is equal to the gas constant of air, and therefore the weightings r and q are given by (5.33).

Since the composition of standard fuel is 86.08% C, 13.92% H, then

$$v_{CO_2} = \frac{\frac{0.8608}{12}}{\frac{0.8608}{12} + \frac{0.1392}{2} + \frac{0.7905}{0.2095}\left(\frac{0.8608}{12} + \frac{0.1392}{4}\right)} = 0.132$$

$$\nu_{H_2O} = \cfrac{\frac{0.1392}{2}}{\frac{0.8608}{12} + \frac{0.1392}{2} + \frac{0.7905}{0.2095}\left(\frac{0.8608}{12} + \frac{0.1392}{4}\right)} = 0.128$$

$$\nu_{N_2} = \cfrac{\frac{0.7905}{0.2095}\left(\frac{0.8608}{12} + \frac{0.1392}{4}\right)}{\frac{0.8608}{12} + \frac{0.1392}{2} + \frac{0.7905}{0.2095}\left(\frac{0.8608}{12} + \frac{0.1392}{4}\right)} = 0.740.$$

Consequently, the entropy correction is a function of the excess air only and does not depend on the thermodynamic parameters of the working fluid.

Example 5.3.3 The excess air is $\lambda = 2.72$, enthalpy $h = 1503.1$ kJ/kg, and pressure $p = 5.136$ bar. What is the entropy s?

Solution
For the given enthalpy, read entropy from the air tables:

$$s_{a,\ p=1} = 8.3532 \text{ kJ/(kg K)}$$

and read entropy from the stoichiometric tables:

$$s_{\lambda=1,\ p=1} = 8.6149 \text{ kJ/(kg K)}.$$

Calculate the entropy of the mixture at pressure of 1 bar using equation (5.41):

$$s_\lambda = \frac{1 + minL}{1 + \lambda\, minL}\, s_{\lambda=1} + \frac{(\lambda - 1)\, minL}{1 + \lambda\, minL}\, s_a + \Delta s'.$$

The weighting functions r and q are

$$r = \frac{1 + minL}{1 + \lambda minL} = \frac{1 + 14.66}{1 + 2.72 \times 14.66} = 0.383$$

and

$$q = \frac{(\lambda - 1)\, minL}{1 + \lambda\, minL} = \frac{1.72 \times 14.66}{1 + 2.72 \times 14.66} = 0.617.$$

The correction $\Delta s'$ is computed using equation (5.42):

$$\Delta s' = 8.3145\left[0.383 \times (0.132 + 0.128)\ln\frac{1}{0.383} + 0.617 \times 0.2095 \times \ln\frac{1}{0.617} - \right.$$

$$- 0.383\frac{1}{0.740}\ln\frac{1}{0.740} - 0.617\frac{1}{0.7905}\ln\frac{1}{0.7905} +$$

$$\left. +(0.383 \times 0.740 + 0.617 \times 0.7905)\ln\frac{1}{0.383 \times 0.740 + 0.617 \times 0.7905}\right] = 0.1584 \text{ kJ/(kg K)}$$

$$s_{\lambda,\ p=1} = 0.383 \times 8.6149 + 0.617 \times 8.3532 + 0.1584 = 8.6118 \text{ kJ/(kg K)}$$

$$s_{\lambda,\ p=5.136} = s_{\lambda,\ p=1} - R\ln 5.136 = 8.6118 - 0.28716\ln 5.136 = 8.1420 \text{ kJ/(kg K)}.$$

Finding the enthalpy of a mixture given the value of entropy requires an iterative process, as illustrated in the following example.

Example 5.3.4   The excess air is $\lambda = 2.984$, the pressure is $p = 1.023$ bar, and the entropy is $s = 7.8160$ kJ/(kg K). What is the enthalpy assuming the molecular mass of the mixture is approximately equal to the molecular mass of air?

Solution
The temperature of the mixture must be calculated first. To use the thermodynamic tables, let us find the entropy of a dummy state that has a pressure of 1 bar:

$$s = s_{p=1} - R \ln p$$

such that

$$s_{p=1} = s + R \ln p = 7.8160 + 0.28716 \times \ln 1.023 = 7.8226 \text{ kJ/(kg K)}.$$

The next step is to determine the root of the function

$$f(T) = s_{p=1} - rs_{\lambda=1}(T) - qs_a(T) \tag{5.43}$$

where r and q are the weighting functions

$$r = \frac{1 + minL}{1 + \lambda \, minL} = 0.35 \quad \text{and} \quad q = \frac{(\lambda - 1) \, minL}{1 + \lambda \, minL} = 0.65.$$

To bracket the root of $f(T)$ given in (5.43), let us determine the values of the temperature corresponding to the entropy $s_{p=1}$ for two cases: air and stoichiometric gases, respectively. Given the entropy of 7.8226 kJ/(kg K), the thermodynamic tables for air and stoichiometric combustion products yield temperatures of 879.3 K and 685.9 K, respectively. A good estimate of an initial value for calculating the root of (5.43) can be obtained from $T_{\text{init}} = rT_{\lambda=1} + qT_a = 811.6$ K. The results of the iterative process are shown in Table 5.12.

Table 5.12 Variation of entropy and function $f(T)$ with temperature. Note that the temperature of iteration 1 corresponds to entropy 7.8226 kJ/(kg K) for air and the temperature of iteration 2 corresponds to entropy 7.8226 kJ/(kg K) for $\lambda = 1$.

Iteration	Temperature T K	Entropy $s_{\lambda=1}$ kJ/(kg K)	Entropy s_a kJ/(kg K)	Function $f(T)$ kJ/(kg K)
1	685.9	7.8226	7.5509	0.1766
2	879.3	8.1197	7.8226	−0.1040
3	811.6	8.0222	7.7337	−0.0120
4	803.6	8.0103	7.7228	−0.0008
5	803.06	8.0094	7.7220	0.0000

Once the temperature of the mixture is known, the enthalpy can be calculated using (5.32):

$$h = rh_{\lambda=1} + qh_a = 0.35 \times 877.2 + 0.65 \times 825.2 = 843.4 \text{ kJ/kg}.$$

Example 5.3.5 The temperature of the air entering the combustor is $T_{02,a} = 650$ K. What should the excess air be such that the temperature in the combustor is $T_{03} = 1463.57$ K?

Solution
Given the temperature of the combustion products T_{03}, the enthalpies for air and stoichiometric combustion products are obtained from tables of Appendices B and C: $h_{03,a} = 1591.70$ kJ/kg and $h_{03,\lambda=1} = 1728.46$ kJ/kg. Given the air temperature $T_{02,a}$, the enthalpy is obtained from the air table of Appendix B: $h_{02,a} = 659.73$ kJ/kg.

The excess air is obtained from (5.39) assuming that combustion efficiency $\xi_{comb} = 100\%$:

$$\lambda = \frac{h_{03\lambda=1}(1 + minL) - \xi_{comb} LHV - h_{03a} minL}{minL(h_{02} - h_{03a})} =$$

$$= \frac{1728.46 \times 15.66 - 43,500 - 1591.70 \times 14.66}{14.66(659.73 - 1591.70)} = 2.91.$$

The excess air obtained using standard fuel, $\lambda = 2.91$ is almost identical to the excess air value, $\lambda = 2.9$, used in example 5.2.11 to obtain the adiabatic flame temperature of 1463.57 K. Note that the enthalpy of the fuel is neglected in this example, according to (5.39). The enthalpy of the fuel at 450 K can be calculated assuming the fuel is n-decane and using Tables 5.6 and 5.8:

$$\mathcal{H}_{C_{10}H_{22}, T=450 \text{ K}} = \Delta\mathcal{H}^{\circ}_{f, C_{10}H_{22}} + a(T - T_0) = -294.366 + 296 \times (450 - 298.15) =$$
$$= 44,697 \text{ kJ/kmol}.$$

Using the molecular mass of n-decane $\mathcal{M}_{C_{10}H_{22}} = 10 \times 12 + 22 = 142$ kg/kmol, the enthalpy of n-decane is

$$h_{C_{10}H_{22}, T=450 \text{ K}} = 44,697/142 = 314.8 \text{ kJ/kg}.$$

It is apparent that the fuel enthalpy, $h_{C_{10}H_{22}, T=450 \text{ K}}$, is much smaller than LHV, and therefore can be neglected in (5.39).

Problems

1. Calculate the mass fraction of propane (C_3H_8) knowing that the excess air is $\lambda = 3$, the molecular mass of air is $\mathcal{M}_{air} = 28.93$ kg/kmol, and the air composition by volume is 20.95% O_2 and 79.05% N_2.
   Hint: Write propane combustion using a one-step global reaction.

2. Calculate the mass fraction of octane knowing that the excess air is $\lambda = 2.6$ and the molecular mass of air is $\mathcal{M}_{air} = 28.93$ kg/kmol.

   Hint: Write octane combustion using a one-step global reaction.

3. The temperature at exit of a compressor is 600 K. The efficiency of the combustion is 97%. The enthalpy at the inlet of the turbine is 1515 kJ/kg.

   1. What is the excess air, λ?
   2. What is the temperature at the inlet of the turbine?

      Note: The lower heating value of the standard fuel is 43,500 kJ/kg and $minL = 14.66$.

4. Kerosene entering a combustor has a total temperature of 450 K. The air entering the combustor from a compressor has a total temperature of 650 K. The kerosene can be approximated as liquid n-decane, $C_{10}H_{22}$. Calculate the total temperature of the combustion products if

   1. The fuel burns at stoichiometric conditions.
   2. The fuel burns with an excess air of $\lambda = 2.9$.

   Assume dissociation of the compounds is negligible.

5. The lower heating value of n-octane, C_8H_{18}, is 44,422 kJ/kg.

   1. Write the thermochemical equation of the n-octane combustion.
   2. What is ratio between the mass of water and n-octane?
   3. What is the higher heating value if the latent heat of vaporization of water is 2258 kJ/kg?

6. The temperature of the air entering a combustor is $T_{02} = 700$ K. What should the excess air be so that the temperature at exit of the burner is $T_{03} = 1500$ K?

Bibliography

Anonymous. Tables of selected values of chemical thermodynamic properties. Technical report, NBS, 1947. 176

Anonymous. JANAF Thermochemical tables. Technical Report NSRDS-NBS 37, U.S. Office of Standard Reference Data, June 1971. 176

R. D. Flack. *Fundamentals of Jet Propulsion with Applications*. Cambridge University Press, 2005. 178

E. Goodger and R. Vere. *Aviation Fuels Technology*. Macmillan, 1985. 162, 163

J. Hodge. *Cycles and Performance Estimation*. Academic Press, 1955. 183

R. J. Kee, F. M. Rupley, E. Meeks, and J. A. Miller. Chemkin iii: A FORTRAN chemical kinetics package for the analysis of gas-phase chemical and plasma kinetics. Technical report SAND 89-8009. Technical Report SAND96-8216, Sandia National Laboratories, 1996. 164

K. K. Kuo. *Principles of Combustion*. John Wiley & Sons, 1986. 171, 173, 175, 176

B. J. McBride, S. Gordon, and M. A. Reno. Coefficients for calculating thermodynamic and transport properties of individual species. Technical Report NASA TM 4513, NASA, 1993. 176

V. Pimsner. *Motoare Aeroreactoare*. Editura Didactica si Pedagogica, 1983. 187

F. A. Williams. *Combustion Theory*. Perseus Books, second edition, 1985. 172, 175

Part II

Air-Breathing Engines

6   Thermodynamics of Air-Breathing Engines

6.1   Introduction

This chapter presents a thermodynamic analysis of various types of jet engines. The general thrust equation, introduced based on simple reasoning in Section 1.4, is derived using a rigorous approach based on mass and momentum conservation equations. The performance parameters needed to evaluate propulsion systems are presented next. The Brayton cycle, the ideal cycle of a jet engine, is then discussed. The assumptions of the Brayton cycle are gradually relaxed, and the real cycles of turbojet, turbofan, turboprop/turboshaft and ramjet engines are subsequently presented.

6.2   Thrust Equation

To derive the thrust equation for the jet engine shown in Fig. 6.1, let us consider two control volumes: a control volume that bounds the flow going into the engine, marked with a dotted line, and a control volume that includes both the flow going into and around the engine, marked with a dashed line. Note that the dashed line wraps around the jet engine.

Mass conservation of the flow inside the engine yields

$$\underbrace{\rho_e u_e A_e}_{\dot{m}_e} = \underbrace{\rho u A_i}_{\dot{m}_a} + \dot{m}_f \tag{6.1}$$

where $\dot{m}_f$ is the mass flow rate of fuel entering in the engine. The mass flow rate of air going through the side boundaries, in this case the upper and lower boundaries of Fig. 6.1, is denoted by $\dot{m}_s$. The mass conservation equation applied to the control volume around the engine yields

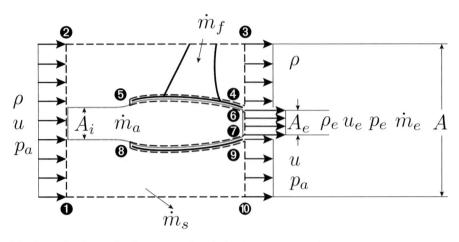

Figure 6.1 Control volumes for thrust equation derivation.

$$\rho_e u_e A_e + \rho u(A - A_e) + \dot{m}_s = \rho u A + \dot{m}_f$$

which simplifies to

$$\dot{m}_s = \dot{m}_f + \rho u A_e - \rho_e u_e A_e. \tag{6.2}$$

Substituting $\dot{m}_f$ from (6.1) into (6.2) yields

$$\dot{m}_s = \rho u(A_e - A_i). \tag{6.3}$$

Momentum conservation in the x-direction applied to a control volume around the engine yields

$$\int_{\sigma \equiv \partial \tau} \rho u \hat{i}(\vec{V} \cdot \vec{n}) \, d\sigma \overset{(2.31)}{=} \hat{i} F_{Vx} + \hat{i} F_{Sx} \tag{6.4}$$

where F_{Vx} and F_{Sx} are the x-components of the volume forces and surface forces, respectively. The volume forces F_{Vx} can be neglected by comparison to the other terms of the equation. The surface forces F_{Sx} are split into pressure forces and viscous forces, as discussed in Section 2.3.2.3. The pressure forces acting on the inlet 1-2 and outlet 3-10 yield $A_e(p_a - p_e)$. The sum of the pressure and viscous forces acting on engine surfaces 4-5-6 and 7-8-9 accounts for the thrust, $\vec{\mathcal{T}}$. Note that the forces acting on the *engine* surfaces are equal to the forces acting on the *control volume* surfaces 4-5-6 and 7-8-9, but have opposite direction, so that (6.4) yields

$$\int_{\sigma \equiv \partial \tau} \rho u \hat{i}(\vec{V} \cdot \vec{n}) \, d\sigma = \hat{i} A_e(p_a - p_e) - \vec{\mathcal{T}}. \tag{6.5}$$

The left-hand-side term is made up of the outgoing fluxes $u_e \dot{m}_e + \rho u(A - A_e)u + u \dot{m}_s$ and incoming fluxes $-u \dot{m}_a - \rho u(A - A_i)u$, such that (6.5) yields

$$\hat{i}\left[u_e \dot{m}_e + \rho u(A - A_e)u + u \dot{m}_s - u \dot{m}_a - \rho u(A - A_i)u\right] = \hat{i} A_e(p_a - p_e) - \vec{\mathcal{T}}. \tag{6.6}$$

Substituting $\dot{m}_s$ from (6.3) into (6.6) yields the expression of the thrust vector

$$\vec{T} = -\hat{i}\left[\dot{m}_e u_e - \dot{m}_a u + A_e(p_e - p_a)\right],$$

which shows that the thrust acts from right to left and pushes the engine forward. The magnitude of the thrust is

$$\boxed{T = \dot{m}_e u_e - \dot{m}_a u + A_e(p_e - p_a)}. \tag{6.7}$$

The first two terms of the thrust are momentum terms while the last is an area-pressure term. The term $\dot{m}_e u_e$ is called the *momentum thrust*. This is the component of the thrust due to the momentum of the gases leaving the engine. The second momentum term, $-\dot{m}_a u$, is the inlet momentum. This is called the *ram drag* and represents the thrust penalty for bringing air into the engine. The third term $A_e(p_e - p_a)$ acts on the exit nozzle. The momentum thrust and the area-pressure term represent the *gross thrust*.

If the flow in the exit nozzle of the engine is subsonic, then $p_e = p_a$, and the term $A_e(p_e - p_a)$ is zero. If the flow is supersonic in the exit nozzle and the flow does not expand to ambient pressure, then the difference between pressures p_e and p_a could be significant, and the term $A_e(p_e - p_a)$ should not be neglected.

The thrust equation (6.7) can be rearranged as

$$T = \dot{m}_e u_e + A_e p_e - (\dot{m}_a u + A_e p_a), \tag{6.8}$$

a form that highlights that the thrust is equal to the difference between the impulse functions (3.47) at exit and inlet of the engine.

Using (6.1) and the mass flow rate of fuel to air ratio, f, defined in (5.12), Eq. (6.7) becomes

$$\boxed{T = \dot{m}_a\left[(1+f)u_e - u\right] + A_e(p_e - p_a)} \tag{6.9}$$

The thrust equation (6.7) was derived for a jet engine having a single exhaust stream. In the case of engines with two distinct streams, that is, turbofan and turboprop engines, the thrust expression must account for and differentiate between the mass flow rate that passes through or around the combustion chamber. In addition, the term $A_e(p_e - p_a)$ can be neglected since the gases in the exit nozzle expands to atmospheric pressure. Using the subscript C for the cold mass flow rate, the mass flow rate that bypasses the combustion chamber, and H for the hot mass flow rate, the mass flow rate that passes through the combustion chamber, the thrust equation for turbofan and turboprop engines is

$$T = (\dot{m}_{aH} + \dot{m}_f)u_{eH} - \dot{m}_{aH}u + \dot{m}_{aC}(u_{eC} - u). \tag{6.10}$$

The mass flow rate of fuel to air ratio in this case is defined as

$$f := \frac{\dot{m}_f}{\dot{m}_{aH}} \tag{6.11}$$

such that (6.10) becomes

$$T = \dot{m}_{aH}\left[(1+f)u_{eH} - u\right] + \dot{m}_{aC}(u_{eC} - u). \tag{6.12}$$

6.3 Engine Performance

Jet engines are complex systems whose performance is assessed by multiple parameters. This section presents the parameters that define various aspects of jet engine efficiency.

6.3.1 Propulsion Efficiency

Propulsion efficiency is defined as the ratio between the thrust power, $\mathcal{T}u$, and the rate of production of propellant kinetic energy, $\dot{m}_a\left[(1+f)u_e^2/2 - u^2/2\right]$:

$$\boxed{\eta_p := \frac{\mathcal{T}u}{\dot{m}_a\left[(1+f)u_e^2/2 - u^2/2\right]}.} \tag{6.13}$$

Propulsion efficiency measures how efficiently the rate of production of propellant kinetic energy is transformed into thrust power.

To get better insight into the parameters that affect propulsion efficiency, let us simplify the expression (6.13) by noting that $f \ll 1$ and $A_e(p_e - p_a) \ll \dot{m}_a\left[(1+f)u_e - u\right]$, so that [Bathie, 1996]

$$\eta_p \approx \frac{\dot{m}_a(u_e - u)u}{\dot{m}_a\left(\frac{u_e^2}{2} - \frac{u^2}{2}\right)} = 2\frac{u}{u_e + u} = 2\frac{u/u_e}{1 + u/u_e}.$$

The variation of the propulsion efficiency vs. velocity ratio u/u_e is shown in Fig. 6.2. The maximum propulsion efficiency occurs when $u = u_e$. However, as velocity ratio increases, the dimensionless thrust decreases to the point that at $u = u_e$, the dimensionless thrust, $\mathcal{T}/(\dot{m}_a u_e)$, is zero. Consequently, a compromise value of u/u_e must be found between the propulsion efficiency and the thrust.

6.3.2 Thermal Efficiency

For turbojet, turbofan, and ramjet engines, thermal efficiency is defined as the ratio between the rate of production of propellant kinetic energy and the rate of heat released by burning fuel:

Figure 6.2 Propulsion efficiency and dimensionless thrust vs. velocity ratio u/u_e.

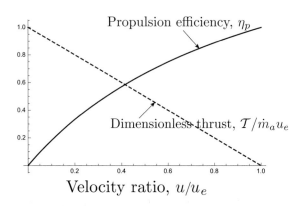

$$\eta_{th} := \frac{\dot{m}_a \left[(1+f)\frac{u_e^2}{2} - \frac{u^2}{2}\right]}{\dot{m}_f \, LHV} \tag{6.14}$$

where LHV is the lower heating value, that is, the heat of reaction of the fuel.

For turboprop and turboshaft engines, the thermal efficiency is defined as the ratio between the engine shaft power and the rate of heat released by burning fuel:

$$\eta_{th} := \frac{\mathcal{P}_s}{\dot{m}_f \, LHV} \tag{6.15}$$

where $\mathcal{P}_s$ is the engine shaft power.

6.3.3 Propeller Efficiency

For turboprop engines, the shaft power produced by the engine is converted into thrust power by a propeller. The ratio between the thrust power due to the propeller, $\mathcal{T}_{pr}u$, and the shaft power, $\mathcal{P}_s$, is defined as propeller efficiency:

$$\eta_{pr} := \frac{\mathcal{T}_{pr} \, u}{\mathcal{P}_s} \tag{6.16}$$

where $\mathcal{T}_{pr}$ is the thrust due to the propeller.

An alternative definition takes into account the fact that turboprop engines derive a certain amount of thrust from the gases exhausting the exit nozzle [Hill and Peterson, 1992]. In this case, the propulsion efficiency is defined as the ratio between the entire thrust power, $\mathcal{T}u$, and the equivalent shaft power, $\mathcal{P}_{es}$:

$$\eta_{pr} := \frac{\mathcal{T} \, u}{\mathcal{P}_{es}}. \tag{6.17}$$

The equivalent shaft power takes into account the fact that the turboprop/turboshaft engines produce both shaft power and thrust. Therefore, equivalent shaft power is defined as

$$\mathcal{P}_{es} = \mathcal{P}_s + \mathcal{T}_e u$$

where $\mathcal{T}_e$ is the exhaust thrust. Section 6.8 will introduce a method for relating thrust and power.

6.3.4 Overall Efficiency

For turbojet, turbofan, and ramjet engines, the overall efficiency is defined as the product of propulsion efficiency and thermal efficiency,

$$\eta_o := \eta_p \, \eta_{th}, \tag{6.18}$$

and using (6.13) and (6.14) yields

$$\eta_o = \frac{T\,u}{\dot{m}_f\,LHV}. \tag{6.19}$$

For turboprop and turboshaft engines, overall efficiency is defined as the product of propeller efficiency and thermal efficiency,

$$\eta_o := \eta_{pr}\eta_{th},$$

so that using (6.15) and (6.16) yields

$$\eta_o = \frac{T_{pr}\,u}{\dot{m}_f\,LHV}.$$

Let us determine the conditions that maximize the overall efficiency for a turbojet engine. For this, we will simplify (6.19) by neglecting the $A_e(p_e - p_a)$ term compared to the $\dot{m}_e u_e - \dot{m}_a u$ term, and f compared to 1, so that

$$\eta_o \stackrel{\text{neglect} A_e(p_e-p_a)}{=} \frac{(\dot{m}_e u_e - \dot{m}_a u)u}{\dot{m}_f LHV} \stackrel{f \ll 1}{=} \frac{(u_e - u)u}{f LHV}.$$

Consequently for given u_e, the overall efficiency η_o is a maximum if $u = u_e/2$, that is, when the velocity of the exhaust gases leaving the exhaust nozzle is double the speed of the aircraft.

For a turbofan engine, the average exit velocity is (see Example 2.4.1)

$$\bar{u}_e = \frac{\dot{m}_{aH}(1+f)u_{eH} + \dot{m}_{aC}u_{eC}}{\dot{m}_{aH}(1+f) + \dot{m}_{aC}}, \tag{6.20}$$

where f, the ratio between the mass flow rate of fuel, $\dot{m}_f$, and the mass flow rate of air, $\dot{m}_{aH}$, was defined in (6.11). Consequently, the overall efficiency for a turbofan engine is

$$\eta_o = \left(1 + \frac{\dot{m}_{aC}}{\dot{m}_{aH}}\right)\frac{(\bar{u}_e - u)u}{f LHV}, \tag{6.21}$$

and the maximum overall efficiency occurs when $u = \bar{u}_e/2$.

6.3.5   Takeoff Thrust

Takeoff thrust, one of the most important engine performance values, describes the capacity to provide static and low speed thrust so that the aircraft can takeoff. The takeoff thrust for a turbojet is obtained from (6.9) by setting $u = 0$, neglecting f relative to unity, and neglecting the pressure term $A_e(p_e - p_a)$ compared to the $\dot{m}_a u_e$ term:

$$T|_{u=0} = \dot{m}_a u_e. \tag{6.22}$$

Using (6.14), the thermal efficiency at takeoff is

$$\eta_{th}|_{u=0} = \frac{\dot{m}_a \frac{u_e^2}{2}}{\dot{m}_f LHV}$$

such that the takeoff thrust can be written as

$$T|_{u=0} = \frac{2\,\eta_{\text{th}}|_{u=0}\,LHV\,\dot{m}_{\text{f}}}{u_{\text{e}}}. \tag{6.23}$$

It is apparent from (6.23) that the takeoff thrust increases with thermal efficiency, fuel lower heating value, and mass flow rate of fuel. It is interesting that the takeoff thrust increases as the exhaust gas velocity u_{e} decreases. This idea will be further explored when discussing turbofan engines.

6.3.6 Aircraft Range – Breguet Equation

The efficiency of the propulsion system affects how far an airplane can fly on a given mass of fuel. To determine the aircraft range, let us ignore the climb and descent and consider only the case of the airplane flying steady and level. In this case, the thrust of the engine should equal the drag of the airplane, and the weight of the airplane should equal the lift. Consequently

$$T = D = L\left(\frac{D}{L}\right) = \frac{mg}{L/D} \tag{6.24}$$

where m is the instantaneous vehicle mass, L is the lift, and D is the drag. Using (6.24) in the definition of overall efficiency (6.19) yields

$$\eta_{\text{o}} = \frac{mgu}{L/D}\frac{1}{\dot{m}_{\text{f}}LHV}$$

such that the mass flow rate of fuel is

$$\dot{m}_{\text{f}} = \frac{mgu}{\frac{L}{D}\eta_{\text{o}}LHV}. \tag{6.25}$$

Assuming that the mass of the airplane only varies over time due to fuel consumption yields

$$\dot{m}_{\text{f}} = -\frac{dm}{dt}$$

where, since the mass decreases over time and therefore the term dm/dt is negative, the minus sign makes the right-hand-side term positive, in agreement with the left-hand side, which is also positive.

Since it was assumed that the airplane flies with constant speed, then

$$\dot{m}_{\text{f}} = -u\frac{dm}{ds}. \tag{6.26}$$

Combining (6.25) and (6.26) yields

$$\frac{mgu}{\frac{L}{D}\eta_{\text{o}}LHV} = -u\frac{dm}{ds}$$

so that

$$ds = -\frac{dm}{m}\frac{\frac{L}{D}\eta_{\text{o}}LHV}{g}. \tag{6.27}$$

Integrating (6.27) yields the *Breguet range equation*

$$
\boxed{s = \frac{L}{D} \frac{\eta_o LHV}{g} \ln \frac{m_1}{m_2}}
\tag{6.28}
$$

where m_1 is the initial (or takeoff) mass, and m_2 is the final (or landing) mass. It is apparent that the range depends on the takeoff and landing mass, on the airplane L/D, on the overall efficiency of the propulsion system, and on the LHV of the fuel.

The overall efficiency, η_o, increases with the airplane Mach number. For subsonic flows, L/D is approximately constant. Consequently $\eta_o L/D$ continually increases with Mach number as long as the airplane flies at subsonic speeds. If the airplane flies at supersonic speeds, L/D drops significantly and therefore the range decreases. To recover the $\eta_o L/D$ value of high subsonic flows, the airplane must fly at a Mach number typically higher than 3 so that the decrease in L/D is compensated for by the increase in the overall efficiency.

Substituting the expression of η_o from (6.19) in the Breguet range equation (6.28) yields

$$
s = \frac{L}{D} \frac{LHV}{g} \frac{Tu}{\dot{m}_f LHV} \ln \frac{m_1}{m_2},
$$

and for a given airplane speed u, the range s is proportional to $T/\dot{m}_f$, that is

$$
s \sim \frac{T}{\dot{m}_f}.
$$

The inverse of $T/\dot{m}_f$ turns out to be an important engine performance parameter.

6.3.7   Thrust Specific Fuel Consumption (TSFC)

The inverse of $T/\dot{m}_f$ is defined as the thrust specific fuel consumption (TSFC), or simply specific fuel consumption

$$
\boxed{\text{TSFC} := \frac{\dot{m}_f}{T}}.
\tag{6.29}
$$

If one neglects the term $A_e(p_e - p_a)$ in the expression of thrust, the TSFC for a turbojet is

$$
\text{TSFC} = \frac{\dot{m}_f}{\dot{m}_a \left[(1+f)u_e - u \right]}.
$$

Typical values of TSFC are given in Table 6.1 for turbojet, turbofan, and ramjet engines. It is apparent that the most economical is the turbofan engine. This is why the vast majority of commercial airplanes are powered by turbofan engines.

Table 6.1 Range of thrust specific fuel consumption values for jet engines in kg/(N hour).

Turbofan	$0.03 - 0.05$
Turbojet	$0.075 - 0.11$
Ramjet	$0.17 - 0.26$

Example 6.3.1 Calculate the mass flow rate of fuel on a Boeing 747 powered by four turbofan engines. Each engine produces 44,000 lbf of thrust and has a TSFC = 0.035 kg/(N hour).

Solution

$$T = 44,000 \text{ lbf} = 44,000 \times 0.45 \text{ kgf} = 19,800 \times 9.81 \text{ N} = 194,238 \text{ N}$$

$$\dot{m}_\text{f} = T \times \text{TSFC} = 194,238 \times 0.035 = 6798 \text{ kg/hour}$$

For four engines, $\dot{m}_\text{f} = 27,193$ kg/hour.

6.3.8 Brake Specific Fuel Consumption (BSFC)

For turboshaft and turboprop engines, power, as opposed to thrust, is the relevant performance parameter. For this reason, the fuel consumption per kW is germane. Since in practice the power of an engine is measured with a brake, the fuel consumption for turboprop and turboshaft engines is called brake specific fuel consumption (BSFC) and is defined as

$$\boxed{\text{BSFC} := \frac{\dot{m}_\text{f}}{\mathcal{P}_\text{s}}}.$$

6.3.9 Equivalent Brake Specific Fuel Consumption (EBSFC)

On an airplane powered by a turboprop engine, part of the thrust is produced by the propeller and part is produced by the jet. Consequently, when assessing fuel efficiency, the fuel mass flow rate should be divided by equivalent shaft power. As a result, the equivalent brake specific fuel consumption (EBSFC) is defined by replacing the shaft power $\mathcal{P}_\text{s}$ by the equivalent shaft power $\mathcal{P}_\text{es}$, which also includes the power generated by the thrust at speed u:

$$\text{EBSFC} := \frac{\dot{m}_\text{f}}{\mathcal{P}_\text{es}} = \frac{\dot{m}_\text{f}}{\mathcal{P}_\text{s} + \mathcal{T}_\text{e}u}.$$

Typical values of EBSFC are in the range 0.27 to 0.36 kg/kWh.

6.3.10 Specific Thrust

By definition, the specific thrust is $\mathcal{T}_\text{sp} = T/\dot{m}_\text{a}$. With the thrust given by (6.7), the specific thrust becomes

$$\mathcal{T}_\text{sp} = \frac{T}{\dot{m}_\text{a}} = \frac{u_\text{e}\dot{m}_\text{e} - u\dot{m}_\text{a} + A_\text{e}(p_\text{e} - p_\text{a})}{\dot{m}_\text{a}} = u_\text{e}(1+f) - u + \frac{A_\text{e}}{\dot{m}_\text{a}}(p_\text{e} - p_\text{a}). \tag{6.30}$$

The last term can be rewritten as

$$\frac{A_\text{e}}{\dot{m}_\text{a}}(p_\text{e} - p_\text{a}) = \frac{A_\text{e}}{\dot{m}_\text{e}}(1+f)p_\text{e}\left(1 - \frac{p_\text{a}}{p_\text{e}}\right) = u_\text{e}(1+f)\frac{1}{\gamma M_e^2}\left(1 - \frac{p_\text{a}}{p_\text{e}}\right) \tag{6.31}$$

so that an alternative form of the specific thrust is

$$T_{sp} = u_e(1+f) - u + u_e(1+f)\frac{1}{\gamma M_e^2}\left(1 - \frac{p_a}{p_e}\right).$$

By defining an *equivalent exit velocity*, $u_{e_{eq}}$:

$$u_{e_{eq}} = u_e\left[1 + \frac{1}{\gamma M_e^2}\left(1 - \frac{p_a}{p_e}\right)\right], \tag{6.32}$$

the specific thrust becomes

$$T_{sp} = (1+f)u_{e_{eq}} - u. \tag{6.33}$$

Note that the specific thrust has the dimensions of velocity, and its value is proportional to the velocity at exit nozzle.

The takeoff specific thrust is obtained by setting the velocity u to zero:

$$T_{sp_0} = (1+f)u_{e_{eq}}. \tag{6.34}$$

6.4 Brayton Cycle

The Brayton cycle is the ideal cycle of the gas turbine engine. George Brayton (1830–1892) was an American mechanical engineer who built the first constant-pressure engine. The Brayton cycle is occasionally called the Joule cycle because of the theoretical contributions of James P. Joule (1818–1889), an English physicist and mathematician [Archer and Saarlas, 1996, p. 137].

The Brayton cycle is an *irreversible* cycle because the heat transfer does not occur at constant temperature. The Brayton cycle is a *constant-pressure* cycle because the heat addition and rejection take place at constant pressure. The Brayton cycle is an *open* cycle because the working fluid is discarded at the end of expansion and replaced with a fresh charge.

The following assumptions are made for the Brayton cycle:

1. The working fluid operating in the engine is air.
2. The mass flow rate is constant, that is, one neglects (i) the addition of mass flow rate of fuel, (ii) the fluid removed from compressor for cooling, and (iii) the fluid added in the turbine for cooling.
3. The heat addition and heat rejection occur at constant pressure, that is, these processes are isobar.
4. The compression and expansion processes are adiabatic and reversible, that is, these processes are isentropic.

The Brayton cycle consists of two isentropic processes and two isobaric processes of heating and cooling. Assuming that the heat rejection occurs at atmospheric pressure, the

Figure 6.3 (a) Brayton, (b) Otto, and (c) Carnot cycles. Circled symbols **s, p, v, T** denote isentropic, isobaric, isochoric, and isothermal processes, respectively. Symbol ⟳ denotes heat addition or rejection.

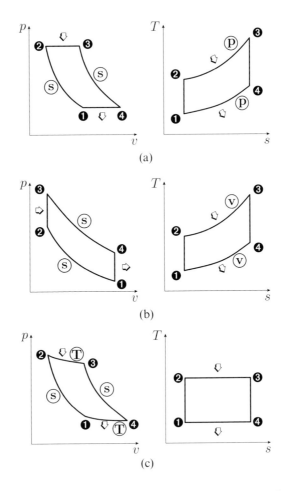

(a)

(b)

(c)

Brayton cycle is specified by two parameters: (i) the stagnation pressure at exit from the compressor, and (ii) the stagnation temperature at exit from the burner.

Engine cycles are typically represented by enthalpy–entropy $(h-s)$ diagrams, called *Mollier diagrams* and by pressure-volume $(p-v)$ diagrams, called *indicator diagrams*. The $(h-s)$ diagrams can be substituted by temperature–entropy $(T-s)$ diagrams if the specific heat c_p is constant. Figure 6.3 uses $p-v$ and $T-s$ diagrams to compare the irreversible Brayton and Otto cycles and the reversible Carnot cycle. The difference between the three cycles is the way heat addition and rejection occurs. In the Carnot cycle, the heat transfer occurs at constant temperature; in the Brayton cycle heat transfer occurs at constant pressure, while in the Otto cycle heat transfer occurs at constant volume.

Using (2.52) in the $T-s$ diagram, the heat addition in the Carnot cycle is the area below the 2–3 process and the heat rejection is the area below the 4-1 process. The work of the engine is obtained by using (2.45) and is a measure of the area 1-2-3-4-1. The thermal efficiency (2.46) yields

$$\eta_{th} = 1 - T_1/T_2.$$

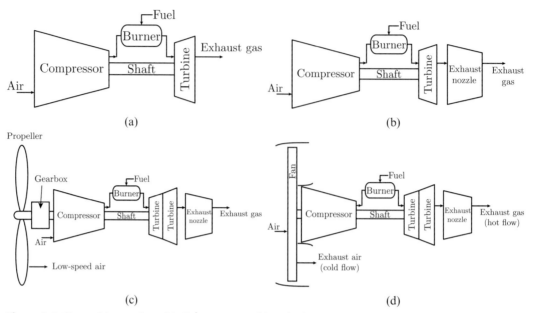

Figure 6.4 Gas turbine engines. (a) Gas generator, (b) turbo jet, (c) turboprop, and (d) turbofan.

It is apparent by comparing the $T - s$ diagrams of the three cycles that the Carnot cycle has the highest thermal efficiency. Indeed, the Carnot cycle determines the limit of transforming heat transfer into work transfer. While it is disappointing that the Brayton cycle, the ideal cycle of the jet engine, has a lower efficiency that the Carnot cycle, the jet engine components can be designed to have a high aerodynamic efficiency, therefore compensating for the lower efficiency of the thermodynamic cycle.

The schematic of gas turbine engines operating under the Brayton cycle is shown in Fig. 6.4. The common element present in all these engines is the basic gas generator. Different gas turbine engines result from adding different components to the gas generator. For all turbine engines shown in Fig. 6.4, a mass flow rate of air enters the compressor where the pressure is increased. The high-pressure air is mixed with fuel, and the temperature of the mixture is increased in the combustion chamber. The combustion products then enter the turbine where they expand. The pressure and temperature of the gases decrease while a part of their energy is captured by the turbine. The turbine, which is connected to the compressor through a shaft, provides the power for operating the compressor. As the gases leaving the engine are cooled, fresh air enters the compressor and the cycle repeats.

In the following paragraphs we will evaluate the values of the specific work in the compressor and turbine, and the heat added in the combustor. Let us consider the schematic diagram of a turbojet engine shown in Fig. 6.5. Throughout this book the state at inlet in the compressor will be denoted by 1, following the notation convention used by Bathie [1996] and Archer and Saarlas [1996]. An alternative is to denote this state by 2, as done by Hill and Peterson [1992] and Flack [2005].

Figure 6.5 State locations in a turbojet engine.

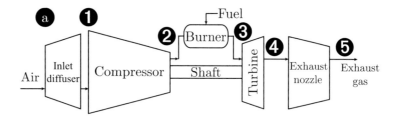

Let us start by integrating the first law of thermodynamics (3.10) across the compressor:

$$q_{1-2} - w_{1-2} = h_2 + \frac{V_2^2}{2} - \left(h_1 + \frac{V_1^2}{2}\right) \overset{(3.12)}{=} h_{02} - h_{01}. \tag{6.35}$$

In the Brayton cycle shown in Fig. 6.3, the compression process between states 1 and 2 is isentropic, that is, adiabatic and reversible. Because the process is adiabatic, (6.35) reduces to

$$-w_{1-2} = h_{02} - h_{01}. \tag{6.36}$$

Since $h_{02} > h_{01}$, the specific work $w_{1-2} < 0$, which is in agreement with the heat engine sign convention. For simplicity, the work in the compressor, w_C, will denote the absolute value of w_{1-2}, that is

$$w_C = |w_{1-2}|,$$

and therefore

$$w_C = h_{02} - h_{01}.$$

Since the expansion process in the Brayton cycle is isentropic, following a similar reasoning as above, the specific work is

$$-w_{3-4} = h_{04} - h_{03},$$

and since $h_{03} > h_{04}$, the specific work $w_{3-4} > 0$, which is in agreement with the heat engine sign convention. The specific work w_{3-4} is the specific work done in the turbine, w_T, so that

$$w_T = h_{03} - h_{04}. \tag{6.37}$$

Heat addition at constant pressure occurs in the combustor, between states 2 and 3 in Fig. 6.3. During this process the work transfer is zero because there are no moving parts in the combustor. As a result, the heat transfer added in the combustor is

$$q_{2-3} = h_{03} - h_{02}.$$

The notation for the heat transfer added in the combustor q_{2-3} is q_{in}, that is

$$q_{in} = h_{03} - h_{02}. \tag{6.38}$$

No work is being done between states 04 and 05 because typically the nozzle does not move. Consequently, $h_{04} = h_{05}$, and therefore one can also write $w_T = h_{03} - h_{05}$. Note, however, that state 05 is not commonly used in the cycle analysis of gas turbine engines.

The *net work* of the gas turbine engine is defined as

$$w_{net} := h_{03} - h_5 - (h_{02} - h_{01}).$$ (6.39)

Note that state 5 in this definition is a static state, as opposed to a stagnation state.

Cycle thermal efficiency is defined as

$$\eta_{th} := \frac{w_{net}}{q_{in}}.$$ (6.40)

Cycle thermal efficiency is an important indicator of the efficiency of a gas turbine engine. Cycle thermal efficiency measures how much of the heat added into the combustor has been transformed into net work.

The expressions of the specific work in the compressor and turbine, the heat added in the combustor, as well as the definitions of the net work and cycle thermal efficiency introduced with the Brayton cycle, are also valid for the jet engine real cycle.

To determine the real cycle, two additional efficiencies are needed: the *compressor efficiency*, η_C, and the *turbine efficiency*, η_T. The compressor efficiency is defined as the ratio between the ideal and the real work in the compressor shown in Fig. 6.6a:

$$\eta_C = \frac{w_{C_i}}{w_C}.$$ (6.41)

The turbine efficiency is defined as the ratio between the real and the ideal work in the turbine shown in Fig. 6.6b:

$$\eta_T = \frac{w_T}{w_{T_i}}.$$ (6.42)

Note that as the compressor efficiency improves, the work required to compress the fluid to a certain pressure decreases. On the other hand, as the turbine efficiency improves, the work the turbine extracts from the fluid increases.

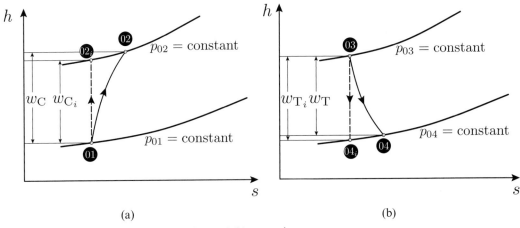

Figure 6.6 Real and ideal (a) compression and (b) expansion.

Using the notation of Fig. 6.6, the compressor and turbine efficiencies become

$$\eta_C = \frac{w_{C_i}}{w_C} = \frac{h_{02i} - h_{01}}{h_{02} - h_{01}} \qquad (6.43)$$

and

$$\eta_T = \frac{w_T}{w_{T_i}} = \frac{h_{03} - h_{04}}{h_{03} - h_{04i}}. \qquad (6.44)$$

An important parameter for the compressor is the compression pressure ratio, π_{0C}, defined as

$$\pi_{0C} = \frac{p_{02}}{p_{01}}. \qquad (6.45)$$

Similarly, the expansion pressure ratio in the turbine, π_{0T} is defined as

$$\pi_{0T} = \frac{p_{03}}{p_{04}}. \qquad (6.46)$$

6.5 Gas Generator

6.5.1 Introduction

The gas generator consists of a compressor, combustor, and turbine. These three components are present in any gas turbine engine and therefore constitute the core engine. This section will explore the ideal cycle of a gas generator whose working fluid is air, therefore ignoring the fuel injection in the combustor and the presence of combustion products in the combustor and turbine. In a first step, we will assume that the air properties do not vary with temperature, and the ideal cycle of a gas generator with air standard, constant properties will be predicted. The variation of the thermodynamic properties of air will be discussed and a method for predicting these properties will be introduced that uses the thermodynamic properties tables. Subsequently, the ideal cycle of a gas generator with air standard, variable properties will be predicted.

6.5.2 Gas Generator Ideal Cycle with Air Standard, Constant Properties

The ideal cycle of a gas generator in which the working fluid is assumed to be air with constant properties is a useful approximation that reveals simple relationships between some of the engine parameters. These relationships can be used as a starting point in the cycle analysis of a gas turbine engine. The purpose of this section is to derive such relations.

In the case of the gas generator, state 4 is identical with state 5 since there is no exhaust nozzle downstream of the turbine. It is assumed that the pressure at exit from the turbine is equal to the pressure at inlet in the compressor. If one assumes that the kinetic energy of gases at state 4 is negligible, that is, all the energy of the gases has been extracted by the turbine that drives the compressor, then in this case, *and only in this case*,

$$h_5 = h_4 = h_{04}. \qquad (6.47)$$

Consequently using (6.47), the net work (6.39) can written as

$$w_{net} = h_{03} - h_{04} - (h_{02} - h_{01}) = w_T - w_C. \tag{6.48}$$

For the particular case of the gas generator, the cycle thermal efficiency (6.40) becomes

$$\eta_{th} = \frac{w_{net}}{q_{in}} = \frac{c_p(T_{03} - T_{04} - T_{02} + T_{01})}{c_p(T_{03} - T_{02})} = 1 - \frac{T_{04} - T_{01}}{T_{03} - T_{02}}. \tag{6.49}$$

To obtain a simplified expression of the gas turbine parameters, let's assume that the variation of the velocities squared is negligible. This assumption will be made only for the case of the ideal gas generator and it will not be applied to any other engines. If one assumes that the variations of the velocities $V_4^2 - V_1^2$ and $V_3^2 - V_2^2$ are negligible, the cycle thermal efficiency becomes

$$\eta_{th} = 1 - \frac{T_4 - T_1}{T_3 - T_2}. \tag{6.50}$$

Since processes 1-2 and 3-4 are isentropic and processes 2-3 and 4-1 are isobaric:

$$\frac{p_2}{p_1} = \left(\frac{T_2}{T_1}\right)^{\frac{\gamma}{\gamma-1}} = \frac{p_3}{p_4} = \left(\frac{T_3}{T_4}\right)^{\frac{\gamma}{\gamma-1}} \tag{6.51}$$

so that

$$T_2/T_1 = T_3/T_4. \tag{6.52}$$

The thermal efficiency from (6.50) can be rewritten as

$$\eta_{th} = 1 - \frac{(T_4/T_1 - 1)T_1}{(T_3/T_2 - 1)T_2},$$

and using (6.52) it becomes

$$\eta_{th} = 1 - \frac{T_1}{T_2}.$$

Using (6.51), the cycle thermal efficiency can also be written as a function of the pressure ratio:

$$\eta_{th} = 1 - \frac{1}{(p_2/p_1)^{\frac{\gamma-1}{\gamma}}}. \tag{6.53}$$

6.5.2.1 Maximum Net Work for Fixed T_1 and T_3

Let us calculate the maximum net work for an ideal gas generator assuming a fixed inlet temperature T_1 and fixed turbine inlet temperature T_3. Assuming that the variation of the velocity squared is small, the net work from (6.48) can be approximated as

$$w_{net} \simeq c_p (T_3 - T_4 - T_2 + T_1),$$

and using (6.52) yields

$$w_{net} = c_p \left(T_3 - \frac{T_1 T_3}{T_2} - T_2 + T_1\right).$$

Figure 6.7 Optimum pressure ratio p_2/p_1 vs. temperature ratio T_3/T_1 for maximum net work at different ratios of specific heat capacities, γ.

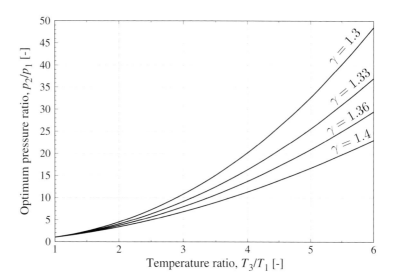

When temperatures T_1 and T_3 are fixed, the net work is only a function of temperature T_2, that is, $w_{net} = w_{net}(T_2)$. Consequently, the maximum net work is obtained from

$$\frac{\partial w_{net}}{\partial T_2} = 0$$

which yields

$$\frac{\partial w_{net}}{\partial T_2} = c_p \left(\frac{T_1 \, T_3}{T_2^2} - 1 \right) = 0.$$

As a result, the net work reaches a maximum when the temperature T_2 is

$$T_2 = \sqrt{T_1 \, T_3}, \tag{6.54}$$

and the maximum net work is

$$w_{net_{max}} = c_p \, (T_3 + T_1 - 2\sqrt{T_1 \, T_3}).$$

Note that when the maximum net work is achieved, the temperature $T_4 = T_2$ because

$$T_4 = T_1 \, T_3/T_2 = T_1 \, T_3/\sqrt{T_1 \, T_3} = \sqrt{T_1 \, T_3} = T_2. \tag{6.55}$$

In addition, the pressure ratio in the compressor is a function of temperature ratio between the turbine inlet temperature and the temperature at the inlet in the gas generator:

$$\frac{p_2}{p_1} = \left(\frac{T_2}{T_1} \right)^{\frac{\gamma}{\gamma-1}} = \left(\frac{\sqrt{T_1 T_3}}{T_1} \right)^{\frac{\gamma}{\gamma-1}} = \left(\frac{T_3}{T_1} \right)^{\frac{\gamma}{2(\gamma-1)}}. \tag{6.56}$$

The variation of optimum pressure ratio p_2/p_1 vs. temperature ratio T_3/T_1 is shown in Fig. 6.7 for ratios of specific heat capacities ranging from 1.3 to 1.4.

Example 6.5.1 A gas generator is modeled using an ideal cycle. One assumes that the working fluid is air with a constant specific heat capacity at constant pressure, $c_p = 1004.5$ J/(kg K), and a constant ratio of specific heat capacities, $\gamma = 1.4$. The gas generator has a compressor inlet temperature of 288 K and a turbine inlet temperature of 1500 K. The mass flow rate through the gas generator is $\dot{m} = 10$ kg/s.
Calculate:

1. the pressure ratio p_2/p_1 that gives maximum net work;
2. the compressor work, turbine work, heat added and cycle thermal efficiency for the pressure ratio determined in 1;
3. the power of the gas generator.

Solution

1. The temperature T_2 that maximizes the net work is

$$T_2 \overset{(6.54)}{=} \sqrt{T_1 \, T_3} = \sqrt{288 \times 1500} = 657.3 \text{ K}.$$

Assuming an isentropic process between states 1 and 2 yields the pressure ratio

$$\frac{p_2}{p_1} = \left(\frac{T_2}{T_1}\right)^{\frac{\gamma}{\gamma-1}} = \left(\frac{657.3}{288}\right)^{3.5} = 17.96.$$

2. The compressor work is obtained by neglecting the variation of kinetic energy between states 1 and 2

$$w_C = h_{02} - h_{01} \simeq h_2 - h_1 = c_p(T_2 - T_1) = 1004.5(657.3 - 288) =$$
$$= 370,929 \text{ J/kg}.$$

The turbine work is calculated by neglecting the variation of kinetic energy between states 3 and 4

$$w_T = h_{03} - h_{04} \simeq h_3 - h_4 = c_p(T_3 - T_4).$$

At maximum net work, $T_4 \overset{(6.55)}{=} T_2 = 657.3$ K, so that

$$w_T = c_p(T_3 - T_4) = 1004.5(1500 - 657.3) = 846,525 \text{ J/kg}$$

$$q_{in} = h_{03} - h_{02} \simeq h_3 - h_2 = c_p(T_3 - T_2) = 1004.5(1500 - 657.3) =$$
$$= 846,525 \text{ J/kg}.$$

Note that w_T is equal to q_{in} because $T_2 = T_4$. The thermal efficiency is

$$\eta_{th} = \frac{w_{net}}{q_{in}} = \frac{846,525 - 370,929}{846,525} = 0.562.$$

3. The power of the gas generator is

$$\mathcal{P} = w_{\text{net}} \times \dot{m} = (846,525 - 370,929) \times 10 = 4,755,962 \text{ W} = 4,755.962 \text{ kW}.$$

6.5.3   Gas Generator Ideal Cycle with Air Standard, Variable Properties

The next step in removing the simplifying assumptions used for calculating ideal or real cycles is to take into account the variation of air properties with temperature. To accomplish this we will consider that the air is a pure substance which, as discussed in Section 2.4.8.1, has only two independent static properties in the absence of magnetism, electricity, and capillarity. Note that temperature, T, and enthalpy, h, are not independent properties because, for the range of pressures and temperatures present in propulsion systems, enthalpy is a function of temperature only, as discussed in Section 2.4.8.1. Consequently a state cannot be determined by the pair (h, T).

The thermodynamic properties of air are specified either in tables (see appendix B) or with approximating formulae. The table gives the variation of enthalpy, internal energy, and entropy for temperatures varying between 213 K (-60°C) and 1773 K (1500°C), with a temperature increment of 10 K. The table was generated for a pressure of 1 bar, that is, 10^5 Pa. If the pressure is 1 bar, for a given temperature one can read from the table the values of enthalpy, internal energy and entropy. Alternatively, if the pressure is 1 bar and the enthalpy is given, one can read the values of temperature, internal energy, and entropy. Similarly, one can determine three out of the four parameters, as long as the pressure is 1 bar.

To use the table data at any other pressure that one might encounter in the engine, three approaches need to be considered while solving the ideal or real cycle. Note that in these cases, h can be substituted by T. In all cases, the extension of the table data will be done using the entropy variation equation (2.93).

Method H   Find h_1 when s_1, p_1 are given

If the pressure p_1 is 1 bar, then the value of enthalpy h_1 corresponding to the entropy s_1 can be read from the tables or calculated with the available formula valid for a pressure of 1 bar. When pressure p_1 is different from 1 bar, which is the case in most instances, the following methodology must be followed:

1. Calculate the entropy $s_{1'}$ of a fictitious state $1'$ shown in Fig. 6.8, which has enthalpy $h_{1'} = h_1$ and pressure $p_{1'} = 1$ bar, using (2.93) where $T_1 = T_{1'}$:

$$s_1 - s_{1'} = c_{\text{p}} \ln \frac{T_1}{T_{1'}} - R \ln \frac{p_1}{p_{1'}} \tag{6.57}$$

Since $h_{1'} = h_1$ then $T_{1'} = T_1$ so that (6.57) yields

$$s_{1'} = s_1 + R \ln p_1.$$

2. Read from tables enthalpy $h_{1'}$ corresponding to entropy $s_{1'}$. Note that $h_1 = h_{1'}$.

Figure 6.8 $h-s$ diagram.

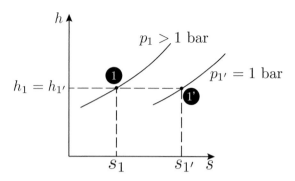

Example 6.5.2 What is the enthalpy of air at pressure $p_1 = 10$ bar and entropy $s_1 = 7.1$ kJ/(kg K)?

Solution

$$s_{1'} = s_1 + R \ln p_1 = 7.1 + 0.28716 \ln 10 = 7.7612 \text{ kJ/(kg K)}$$

For $s_{1'} = 7.7612$ kJ/(kg K) read from tables $h_{1'} = h_1 = 857.2$ kJ/kg.

Method S Find s_1 when h_1, p_1 are given.

As shown in Fig. 6.8, let us introduce a fictitious state $1'$ that has a pressure $p_{1'} = 1$ bar and the same enthalpy as state 1; then read from the tables the entropy $s_{1'}$ corresponding to $h_{1'}$:

$$h_1 = h_{1'} \overset{\text{table}}{\longrightarrow} s_{1'}.$$

Since $p_{1'} = 1$ bar, calculate s_1 from (2.93) where $T_1 = T_{1'}$:

$$s_1 = s_{1'} - R \ln p_1.$$

Example 6.5.3 What is the entropy of air at pressure $p_1 = 20$ bar and enthalpy $h_1 = 1200$ kJ/kg?

Solution

For enthalpy $h_1 = h_{1'}$, read from tables entropy at pressure 1 bar, $s_{1'} = 8.1123$ kJ/(kg K). Then calculate the entropy at pressure 20 bar:

$$s_1 = s_{1'} - R \ln p_1 = 8.1123 - 0.28716 \ln 20 = 7.2521 \text{ kJ/(kg K)}.$$

Method P Find p_1 when s_1, h_1 are given

Let us consider again Fig. 6.8, and since $h_1 = h_{1'}$, read from the tables the value of $s_{1'}$ at pressure $p_{1'} = 1$ bar. Using $T_1 = T_{1'}$ simplify (2.93) such that

$$s_1 - s_{1'} = -R \ln \frac{p_1}{p_{1'}}. \tag{6.58}$$

Since $p_{1'} = 1$ bar, (6.58) yields

$$p_1 = \exp\left(-\frac{s_1 - s_{1'}}{R}\right).$$

Example 6.5.4 What is the pressure of air at enthalpy $h_1 = 900$ kJ/kg and entropy $s_1 = 6.9078$ kJ/(kg K)?

Solution
For enthalpy $h_1 = h_{1'} = 900$ kJ/kg, read from the tables the entropy $s_{1'} = 7.8116$ kJ/(kg K). Then calculate pressure from

$$p_1 = \exp[-(s_1 - s_{1'})/R] = \exp[-(6.9078 - 7.8116)/0.28716] = 23.27 \text{ bar.}$$

Having presented the three methods H, S, and P for determining the dependent parameters, let us illustrate in the following example how to use them for calculating the ideal cycle of a gas generator.

Example 6.5.5 A gas generator is modeled using an ideal cycle with air standard, variable properties. At inlet in the compressor, the stagnation temperature is 288 K and the stagnation pressure is 1.01325 bar. The compressor pressure ratio is 17.96 and the turbine inlet temperature is $T_{03} = 1500$ K. The stagnation pressure at exit from the turbine is the same as the stagnation pressure at inlet in the compressor, 1.01325 bar. The mass flow rate of air in the engine is 10 kg/s.

Calculate the compressor work, w_C, turbine work, w_T, heat added in the combustor, q_{in}, cycle thermal efficiency, η_{th}, and power, $\mathcal{P}$.

Solution
Figure 6.9 shows the $h - s$ diagram of the ideal cycle of a gas generator where the working fluid is air with variable properties. The states shown in the diagram are the stagnation states.

- **State 01**
  Given the stagnation temperature $T_{01} = 288$ K, from the air tables we get the stagnation enthalpy $h_{01} = 288.15$ kJ/kg. Given the stagnation pressure $p_{01} = 1.01325$ bar, we then use the S method to calculate $s_{01} = 6.6565$ kJ/(kg K).

Figure 6.9 $h - s$ diagram of the gas generator ideal cycle.

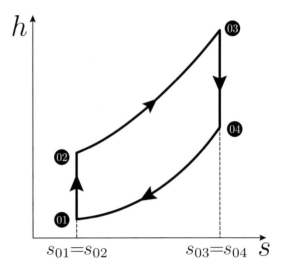

- **State 02**
  Since the process from 01 to 02 is isentropic,

  $$s_{02} = s_{01} = 6.6565 \text{ kJ/(kg K)}.$$

  The stagnation pressure is calculated using the compressor pressure ratio, π_{0C}, and the stagnation pressure at state 01:

  $$p_{02} = p_{01}\pi_{0C} = 18.20 \text{ bar}.$$

  Having determined the entropy and the pressure, the enthalpy is calculated using the H method which yields $h_{02} = 657.16 \text{ kJ/kg}$.
- **State 03**
  Stagnation temperature $T_{03} = 1500$ K is given at state 03. In addition, the process from 02 to 03 is isobaric, therefore the stagnation pressure $p_{03} = p_{02} = 18.20$ bar. Knowing the stagnation temperature, we read the stagnation enthalpy from the air tables and get $h_{03} = 1635.78 \text{ kJ/kg}$. Having determined the pressure and enthalpy, the entropy is calculated using the S method, which yields $s_{03} = 7.6119 \text{ kJ/(kg K)}$.
- **State 04**
  The stagnation pressure was specified to be $p_{04} = 1.01325$ bar. Since the process from 03 to 04 is isentropic, the stagnation entropy is $s_{04} = s_{03} = 7.6119 \text{ kJ/(kg K)}$. Using the H method, we obtain the stagnation enthalpy, $h_{04} = 743.74 \text{ kJ/kg}$.

  Having calculated the parameters of all states, let us calculate the compressor work

  $$w_C = h_{02} - h_{01} = 657.16 - 288.15 = 369.01 \text{ kJ/kg}$$

  the turbine work

  $$w_T = 1635.78 - 743.74 = 892.04 \text{ kJ/kg}$$

the net work

$$w_{net} = w_T - w_C = 892.04 - 369.01 = 523.03 \text{ kJ/kg}$$

the heat added in the burner

$$q_{in} = h_3 - h_2 = 1635.78 - 657.16 = 978.62 \text{ kJ/kg}$$

the thermal efficiency

$$\eta_{th} = \frac{w_{net}}{q_{in}} = \frac{523.03}{978.62} = 0.5345$$

and the power of the gas generator

$$\mathcal{P} = w_{net} \times \dot{m} = 523.03 \text{ kJ/kg} \times 10 \text{ kg/s} = 5230.3 \text{ kW}.$$

Note that although we used the same parameters as in Example 6.5.1, the temperature at state 02 was no longer equal to that at state 04 because the working fluid had variable properties.

6.6   Turbojet

6.6.1   Introduction

The turbojet consists of an inlet diffuser, a compressor, a combustor (or burner), a turbine, and an exit nozzle, as shown in Fig. 6.5. The inlet diffuser decreases the inlet velocity and increases the air pressure at inlet in the compressor. In the compressor the air density and pressure are increased. The temperature increases moderately. From the compressor, the air enters the combustor where fuel is injected. The fuel–air mixture burns and the temperature of the combustion products increases significantly. The gases expand in the turbine and the turbine extracts part of the energy of the gases. This energy is used to drive the compressor, the oil, fuel and hydraulics pumps, and the electrical generator. The compressor and turbine are on the same shaft. The hot gases continue to expand in the exit nozzle and leave the engine with high speed that produces thrust.

6.6.2   Turbojet Configurations

There are several types of turbojet engines, as shown in Fig. 6.10: (1) single-spool, (2) single-spool with afterburning, (3) two-spool, and (4) two-spool with afterburning. In the following paragraph we will provide examples of these types of turbojet engines. The Junkers Jumo 004, one of the first successful turbojets and the world's first production turbojet engine in operational use, was developed in Germany during the Second World War. The engine had an eight-stage axial compressor, six axial combustion chambers, and a one-stage turbine. It weighted 745 kg and produced a thrust of 8.8 kN at 8700 rpm. The engine was designed by

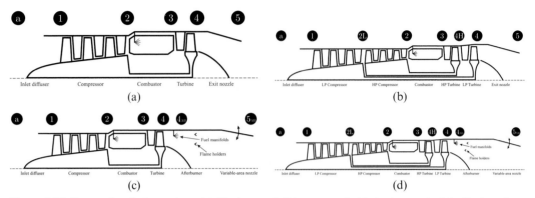

Figure 6.10 Types of turbojet engines. (a) Single-spool, (b) two-spool, (c) single-spool with afterburning, and (d) two-spool with afterburning

Figure 6.11 Westinghouse J30 turbojet (from US Navy Naval Aviation News, August 1947).

the Austrian engineer Anselm Franz (1900–1994) who after the Second World War moved to the United States as part of Operation Paperclip. Dr. Franz designed several turboshaft and turbofan engines at Lycoming, where he rose to Vice President: (i) the T53 used for Bell UH-1 Huey and AH-1 Cobra helicopters and the OV-1 Mohawk ground attack aircraft, (ii) the T55 used on the CH-47 Chinook, Bell 309, and Piper PA-48 Enforcer, and (iii) the AGT-1500 used on M1 Abrams tanks.

The first US-designed turbojet was the Westinghouse J30, which was designed and developed in only eight months between 1942 and 1943 (Fig. 6.11). The engine was the second axial-flow turbojet to run outside of Germany. The engine was first run on 19 March 1943 and was used on several airplanes, including the McDonnell F-4 Phantom. The engine had a maximum thrust of 6.2 kN (1,400 lbf) at 18,000 rpm, a dry weight of 367 kg (809 lb), a diameter of 48 cm (19 in), and a length of 2.6 m (100 in). The engine had a six-stage axial compressor with an overall pressure ratio 3:1, an annular stainless steel combustor, and a single-stage turbine with a turbine inlet temperature of 1089 K (1500° F). The mass flow rate was 13.61 kg/s (30 lb/s), the thrust specific fuel consumption was 0.13 kg/(N hr) (1.28 lb/(lbf hr)), and the thrust-to-weight ratio was 16.9 N/kg (1.72 lbf/lb).

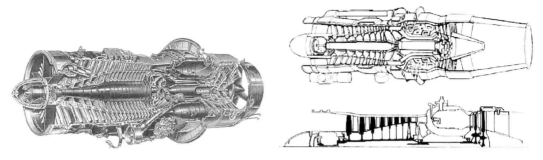

Figure 6.12 Rolls-Royce Viper turbojet (courtesy of Rolls-Royce).

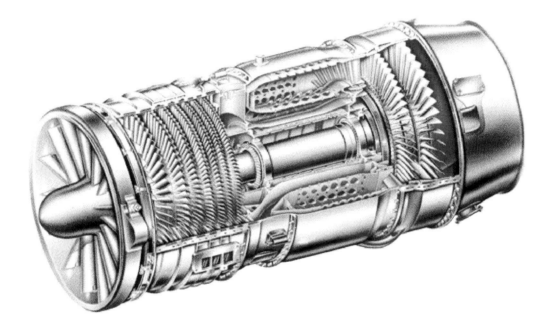

Figure 6.13 General Electric J85 turbojet (courtesy of GE Aviation).

Figure 6.12 shows the Rolls-Royce Viper turbojet engine whose first run was in April 1951. Numerous versions of the engine were developed, including an afterburning variant. The Viper ASV.12 version had a seven-stage axial compressor, an annular combustor with 24 burners, and a single-stage turbine. The maximum thrust is 12 kN at 13,800 rpm. The overall pressure ratio was 4.3:1, the mass flow rate was 20 kg/s, the thrust specific fuel consumption was 0.11 kg/(N hr), and the thrust-to-weight ratio was 4.9. Note that the engine shown in Fig. 6.12 is a version with an eight-stage axial compressor and two-stage turbine.

Figure 6.13 shows the General Electric J85 single-shaft turbojet engine. The engine had an eight-stage axial-flow compressor powered by a two-stage turbine and generated 13.1 kN of dry thrust, that is, thrust without afterburners. A version with afterburners generated up to 22 kN of wet thrust.

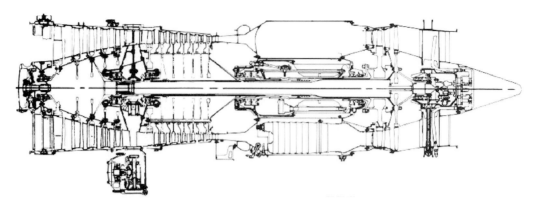

Figure 6.14 Pratt & Whitney J52 turbojet (reproduced with permission from Pratt & Whitney).

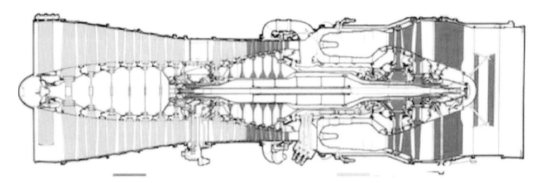

Figure 6.15 Rolls-Royce/SNECMA Olympus 593 turbojet (courtesy of Rolls-Royce).

An example of a two-spool, or twin-spool, turbojet engine was the Pratt & Whitney J52 shown in Fig. 6.14. The J52 had maximum thrust of 50 kN. The low-pressure spool had a five-stage axial compressor and a one-stage turbine, while the high-pressure spool had a seven-stage axial compressor and a one-stage turbine. The overall pressure ratio was 14.6:1 and the air mass flow rate was 64.9 kg/s (143 lb/s). The thrust specific fuel consumption was 0.079 kg/(N hr). The dry weight of the engine was 1052 kg and the thrust-to-weight ratio was 4.83. The engine had a length of 3 m and a diameter of 0.965 m.

A two-spool turbojet with afterburning, the Rolls-Royce-SNECMA Olympus 593, is shown in Fig. 6.15. The engine, which was used on Concorde, had a thrust of 142 kN dry and 169 kN wet (reheat). The low- and high-pressure axial-flow compressors had seven stages each. The overall pressure ratio was 15.5:1. The combustor was annular, with 16 burners. The low- and high-pressure turbines had one stage each. The thrust specific fuel consumption was 0.122 kg/(N hr) at cruise and 0.142 kg/(N hr) at sea level. The dry weight of the engine was 3175 kg (7000 lb) and the thrust-to-weight ratio was 5.4:1. The engine had a length of 4.039 m and a diameter of 1.212 m.

6.6.3 Turbojet Real Cycle with Air Standard

This section illustrates the methodology for calculating the real cycle of a turbojet assuming that the working fluid is air with variable properties. Both takeoff and in-flight conditions are considered.

6.6.3.1 Turbojet Real Cycle with Air Standard, Variable Properties

The steps needed to predict the real cycle of a turbojet whose working fluid is air with variable properties are similar to the steps followed for predicting the cycle of the gas generator. Therefore, two independent parameters that define each state must be identified and the third parameter is then calculated using one of the H, S, or P methods described in Section 6.5.3. The prediction of the real cycle requires that the efficiencies of the compressor and turbine be used to transition from the ideal to the real states.

Example 6.6.1 is presented to illustrate the methodology for calculating the real cycle of a turbojet with a working fluid that is air with variable properties.

Example 6.6.1 A turbojet operates in an engine test cell with a compressor inlet temperature of 287.16 K, a compressor efficiency of 85%, and a turbine efficiency of 88%. The efficiency of the inlet diffuser is 100%. The compressor pressure ratio is 36. The turbine inlet temperature is 1160 K, the compressor inlet pressure is 101.3 kPa, and the air enters the compressor at a rate of $\dot{m}_a = 1$ kg/s. There is a 4% pressure drop in the combustor. The exit nozzle efficiency is 97% and the exit static pressure is 1.5% above the compressor inlet static pressure. Assume the fluid in the turbojet is standard air.

1. Calculate the net power and cycle thermal efficiency.
2. What is the specific thrust of the engine?
3. Draw the real cycle.

Solution
The table of thermodynamic properties of air given in appendix B will be used to predict the states of the real cycle. Whenever these tables are used, it will be indicated by the word "table".

- **State a**
  The pressure and temperature are specified at state "a"

  $p_a = 101.3$ kPa $= 1.013$ bar

  $T_a = 287.16$ K $\overset{\text{table}}{\rightarrow} h_a = 287.3$ kJ/kg.

  Knowing the temperature and pressure, the entropy is calculated using the S method where

  $s_{a_{p=1\ \text{bar}}}(T_a) \overset{\text{table}}{=} 6.6573$ kJ/(kg K)

and

$$s_a = s_{a_{p=1 \text{ bar}}} - R \, \ln p_a = 6.6573 - 0.28716 \, \ln 1.013 = 6.6536 \text{ kJ/(kg K)}.$$

- **State 0a**

  The stagnation state "0a" is identical to the static state "a" since the velocity of the engine is zero, as it operates in an engine cell. Therefore

  $$T_{0a} = T_a = 287.16 \text{ K}$$
  $$h_{0a} = h_a = 287.3 \text{ kJ/kg}$$
  $$p_{0a} = p_a = 101.3 \text{ kPa} = 1.013 \text{ bar}$$
  $$s_{0a} = s_a = 6.6536 \text{ kJ/(kg K)}.$$

- **State 01**

  The stagnation state "01" is identical to the stagnation state "0a" since the inlet diffuser efficiency is 100%, that is, there are no stagnation pressure losses in the inlet. As a result

  $$T_{01} = T_{0a} = 287.16 \text{ K}$$
  $$h_{01} = h_{0a} = 287.3 \text{ kJ/kg}$$
  $$p_{01} = p_{0a} = 101.3 \text{ kPa} = 1.013 \text{ bar}$$
  $$s_{01} = s_{0a} = 6.6536 \text{ kJ/(kg K)}.$$

- **State 02i**

  The process between states 01 and 02i is isentropic:

  $$s_{02i} = s_{01} = 6.6536 \text{ kJ/(kg K)}.$$

  Using the compressor pressure ratio yields

  $$p_{02i} = p_{01} \, \pi_{0C} = 1.013 \times 36 = 36.468 \text{ bar}.$$

  Having determined the pressure and entropy, the enthalpy is calculated using the H method:

  $$s_{02i} - s_{2'} = -R \, \ln \frac{p_{02i}}{p_{2'}} \qquad \text{where} \quad p_{2'} = 1 \text{ bar},$$

  which yields

  $$s_{2'} = s_{02i} + R \, \ln p_{02i} = 6.6536 + 0.28716 \ln 36.468 = 7.6864 \text{ kJ/(kg K)}$$

  $$\overset{\text{table}}{\rightarrow} h_{02i} = 796.9 \text{ kJ/kg}$$

  so the ideal work in the compressor is

  $$w_{C_i} = h_{02i} - h_{01} = 796.9 - 287.3 = 509.6 \text{ kJ/kg}.$$

- **State 02**

  The pressure at state 02 is equal to the pressure at state 02i:

  $$p_{02} = p_{02i} = 36.468 \text{ bar.}$$

  The specific work in the compressor is calculated using the compressor efficiency

  $$\eta_C = \frac{w_{C_i}}{w_C} \rightarrow w_C = \frac{509.6}{0.85} = 599.5 \text{ kJ/kg}$$

  so that the enthalpy is

  $$h_{02} = h_{01} + w_C = 287.3 + 599.5 = 886.8 \text{ kJ/kg.}$$

  Having determined the pressure and enthalpy, the entropy is calculated using the S method:

  $$s_{02} - s_{2'} = -R \ln \frac{p_{02}}{p_{2'}} \qquad \text{where} \quad p_{2'} = 1 \text{ bar}$$

  $$s_{02} = s_{2'}(h_{02}) - R \ln p_{02} \overset{\text{table}}{=} 7.7963 - 0.28716 \ln 36.468 = 6.7635 \text{ kJ/(kg K).}$$

- **State 03**

  Having the turbine inlet temperature specified, the enthalpy is obtained from the table:

  $$T_{03} = 1160 \text{ K} \overset{\text{table}}{\rightarrow} h_{03} = 1230.8 \text{ kJ/kg.}$$

  Given the stagnation pressure drop in the combustor, the pressure at inlet in the turbine is

  $$p_{03} = p_{02} (1 - 4\%) = 36.468 \times 0.96 = 35.009 \text{ bar.}$$

  Having determined the pressure and enthalpy, we will use the S method to calculate the entropy:

  $$s_{03} - s_{3'} = -R \ln \frac{p_{03}}{p_{3'}}, \text{ where } p_{3'} = 1 \text{ bar}, \ s_{3'}(h_{03}) \overset{\text{table}}{=} 8.1403 \text{ kJ/(kg K)}$$

  $$s_{03} = 8.1403 - 0.28716 \ln 35.009 = 7.1193 \text{ kJ/(kg K).}$$

- **State 04i**

  At steady state, when the spool is neither accelerating nor decelerating, the power provided by the turbine is equal to the power required by the compressor, if we ignore the power needed to drive the oil and fuel pumps, the generator, and other consumers on the engine:

  $$\mathcal{P}_T = \mathcal{P}_C$$

  or

  $$\dot{m}_g w_T = \dot{m}_a w_C$$

  where $\dot{m}_g = \dot{m}_a + \dot{m}_f$. Since it was assumed that $\dot{m}_f = 0$, then

  $$w_T = w_C.$$

The ideal work in the turbine is calculated using the turbine efficiency:

$$\eta_T = \frac{w_T}{w_{T_i}} \to w_{T_i} = \frac{599.5}{0.88} = 681.3 \text{ kJ/kg}.$$

Using the ideal work in the turbine, the enthalpy is

$$h_{04i} = h_{03} - w_{T_i} = 1230.8 - 681.3 = 549.5 \text{ kJ/kg}.$$

The process between states 03 and 04i is isentropic

$$s_{04i} = s_{03} = 7.1193 \text{ kJ/(kg K)}.$$

Having determined the enthalpy and entropy, the pressure is calculated using the P method:

$$s_{04i} - s_{4'} = -R \, \ln \frac{p_{04i}}{p_{4'}}, \text{ where } p_{4'} = 1 \text{ bar}, \ s_{4'}(h_{04i}) \overset{\text{table}}{=} 7.3086 \text{ kJ/(kg K)}$$

so that

$$p_{04i} = \exp\left(-\frac{s_{04i} - s_{4'}}{R}\right) = \exp\left(-\frac{7.1193 - 7.3086}{0.28716}\right) = 1.9331 \text{ bar}.$$

- **State 04**

  The pressure at state 04 is equal to the pressure at state 04i

  $$p_{04} = p_{04i} = 1.9331 \text{ bar}.$$

  Using the specific work in the turbine, the enthalpy is

  $$h_{04} = h_{03} - w_T = 1230.8 - 599.5 = 631.3 \text{ kJ/kg}$$

  so that

  $$T_{04}(h_{04}) \overset{\text{table}}{=} 623.2 \text{ K}.$$

  Having determined the pressure and enthalpy, the entropy is calculated using the S method:

  $$s_{04} - s_{4''} = -R \, \ln \frac{p_{04}}{p_{4''}}, \text{ where } p_{4''} = 1 \text{ bar}, \ s_{4''}(h_{04}) \overset{\text{table}}{=} 7.4488 \text{ kJ/(kg K)}$$

  $$s_{04} = s_{4''} - R \, \ln p_{04} = 7.4488 - 0.28716 \ln 1.9331 = 7.2596 \text{ kJ/(kg K)}.$$

- **State 5i**

  No work is being done in the exit nozzle if we assume it to be fixed, that is, it has no moving parts. In addition, since the heat transfer per unit mass of fluid is much smaller than the enthalpy variation over the nozzle, we can assume that the process is adiabatic. Therefore, we conclude from (3.10) that $h_{04} = h_{05}$.

  Next we need to calculate the critical pressure in the nozzle. To calculate critical pressure, we need the value of the ratio of specific heat capacities, γ. From the tables we find the ratio of specific heat capacities at temperature $T_{04} = 623.3$ K, $\gamma_{04} = 1.37505$. Since we do not know the temperature T_{cr}, let us start an iterative process by using $\gamma^{(1)} = 1.37505$ to calculate

$$T_{cr}^{(1)} = 2T_{04}/(\gamma^{(1)} + 1) = 2 \times 623.2/2.37505 = 524.79 \text{ K,}$$

where $T_{cr}^{(1)}$ is the critical temperature after the first iteration. The mean temperature between states "04" and "cr" is

$$\bar{T}^{(1)} = (T_{04} + T_{cr}^{(1)})/2 = 573.99 \text{ K.}$$

From the tables we find that at temperature 573.99 K, the ratio of specific heat capacities is $\gamma^{(2)} = 1.37889$. The critical temperature after the second iteration becomes

$$T_{cr}^{(2)} = 2T_{04}/(\gamma^{(2)} + 1) = 2 \times 623.2/2.37889 = 523.94 \text{ K.}$$

The mean temperature after the second iteration becomes

$$\bar{T}^{(2)} = (T_{04} + T_{cr}^{(2)})/2 = 573.57 \text{ K.}$$

At this mean temperature we read from the tables $\gamma^{(3)} = 1.37895$ so that

$$T_{cr}^{(3)} = 2T_{04}/(\gamma^{(3)} + 1) = 2 \times 623.2/2.37895 = 523.93 \text{ K}$$

and the mean temperature is

$$\bar{T}^{(3)} = (T_{04} + T_{cr}^{(3)})/2 = 573.56 \text{ K}$$

for which the final value of the ratio of specific heat capacities is $\gamma = 1.37895$.

Let us calculate the critical pressure:

$$p_{cr} = p_{04} \left(\frac{2}{\gamma + 1} \right)^{\frac{\gamma}{\gamma-1}} = 1.9331 \left(\frac{2}{2.37895} \right)^{1.37895/0.37895} = 1.02813 \text{ bar.}$$

The problem states that the working fluid could expand in the exit nozzle down to $p_5 = 1.015 p_a = 1.015 \times 1.013 = 1.0282$ bar. Since $p_{cr} < p_5$, the working fluid will not reach critical conditions in the exit nozzle and will be able to expand down to $p_5 = 1.0282$ bar. Hence

$$s_{5i} = s_{04} = 7.2596 \text{ kJ/(kg K)}$$

$$p_{5i} = p_5 = 1.0282 \text{ bar.}$$

Having determined the entropy and pressure, we will use the H method to calculate the enthalpy

$$s_{5i} - s_{5'} = -R \ln \frac{p_{5i}}{p_{5'}}, \text{ where } p_{5'} = 1 \text{ bar}$$

$$s_{5'} = s_{5i} + R \ln p_{5i} = 7.2596 + 0.28716 \ln 1.0282 = 7.2676 \text{ kJ/(kg K)}$$

$$h_{5i}(s_{5'}) \overset{\text{table}}{=} 527.6 \text{ kJ/kg.}$$

- **State 5**

  The pressure at state 5 is equal to the pressure at state 5i

  $$p_5 = p_{5i} = 1.0282 \text{ bar}.$$

  The exit nozzle efficiency, φ_5, is defined as the ratio of the velocities c_5 and c_{5i}

  $$\varphi_5 = c_5/c_{5i}.$$

  The velocity c_{5i} is calculated from

  $$h_{05} = h_{5i} + c_{5i}^2/2$$

  so that using $h_{05} = h_{04}$ yields

  $$c_{5i} = \sqrt{2(h_{04} - h_{5i})} = \sqrt{2(631.3 - 527.6)10^3} = 455.4 \text{ m/s}$$

  and

  $$c_5 = c_{5i}\varphi_5 = 455.4 \times 0.97 = 441.7 \text{ m/s}.$$

  The enthalpy is

  $$h_5 = h_{05} - c_5^2/2 = h_{04} - c_5^2/2 = 631.3 \times 10^3 - 441.7^2/2 = 533{,}750 \text{ J/kg} = $$
  $$= 533.75 \text{ kJ/kg}.$$

  Having determined the pressure and enthalpy, we will use the S method to calculate the entropy:

  $$s_5 - s_{5'} = -R \ln \frac{p_5}{p_{5'}}, \text{ where } p_{5'} = 1 \text{ bar}, \ s_{5'}(h_5) \overset{\text{table}}{=} 7.2792 \text{ kJ/(kg K)}$$

  $$s_5 = s_{5'} - R \ln p_5 = 7.2792 - 0.28716 \ln 1.0282 = 7.2712 \text{ kJ/(kg K)}.$$

  The net work (6.39) can be written as

  $$w_{\text{net}} = h_{03} - h_5 - (h_{02} - h_{01}) = \underbrace{h_{03} - h_{04}}_{w_{\text{T}}} + h_{04} - h_5 - \underbrace{(h_{02} - h_{01})}_{w_{\text{C}}} = $$

  $$\overset{w_{\text{T}}=w_{\text{C}}}{=} h_{04} - h_5 = c_5^2/2 = 441.7^2/2 = 97.5 \times 10^3 \text{ J/kg} = 97.5 \text{ kJ/kg}.$$

  The thermal efficiency is

  $$\eta_{\text{th}} = \frac{w_{\text{net}}}{q_{\text{in}}} = \frac{c_5^2/2}{h_{03} - h_{02}} = \frac{441.7^2/2}{(1230.8 - 886.8) \times 10^3} = 0.284$$

  $$\mathcal{P} = w_{\text{net}} \, \dot{m} = 97.5 \times 1 = 97.5 \text{ kW}.$$

  Using the notation $u_{\text{e}} = c_5$ and $p_{\text{e}} = p_5$, the equivalent exit velocity $c_{5_{\text{eq}}}$ of the specific thrust equation

  $$T_{\text{sp}} = (1 + f)c_{5_{\text{eq}}} - u \tag{6.33}$$

Figure 6.16 $h - s$ diagram of turbojet real cycle.

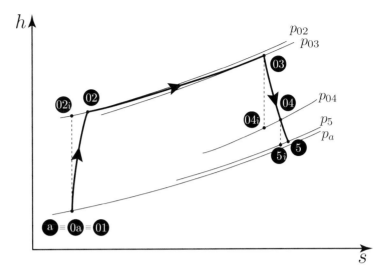

becomes

$$c_{5_{eq}} = c_5 \left[1 + \frac{1}{\gamma M_5^2} \left(1 - \frac{p_a}{p_5} \right) \right]. \tag{6.59}$$

To calculate M_5 we find from the tables $\gamma(h_5) = 1.38427$ so that

$$M_5 = c_5 / \sqrt{(\gamma - 1)h_5} = 441.7 / \sqrt{0.38427 \times 533,750} = 0.975.$$

Since $f = 0$ and $u = 0$, the specific thrust becomes

$$\mathcal{T}_{sp} = c_{5_{eq}} = c_5 \left[1 + \frac{1}{1.38427 \times 0.975^2} \left(1 - \frac{1.013}{1.0282} \right) \right] = 1.0112 \, c_5 = 446.7 \text{ m/s}.$$

The $h - s$ diagram of the turbojet cycle is shown in Fig. 6.16.

6.6.3.2 Turbojet Takeoff Real Cycle with Air Standard, Variable Properties

Takeoff is the phase of the flight when the aircraft goes from rest on the ground to flying in the air. The performance of the turbojet engine at takeoff is important since it determines whether the vehicle takes off or not, and if it does take off, how quickly this happens.

To calculate the turbojet takeoff real cycle we need to determine the processes that take place in the inlet nozzle. Once we understand how the air evolves in the inlet nozzle, we can calculate the pressure and temperature at the exit from the inlet nozzle, that is, at state 01, which is also the inlet into the compressor. The methodology for calculating the states downstream from 01 is identical to that presented in Section 6.6.3.1.

Let us assume that the engine is at a fixed point, or that the speed of the vehicle in which the engine is installed is negligible. As will be shown in Section 7.3, the velocity of the air

Figure 6.17 Variation of streamlines, velocity, pressure, and enthalpy in the inlet diffuser at takeoff.

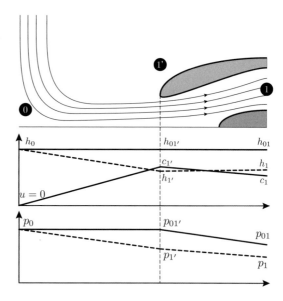

entering the compressor typically varies between 120 and 220 m/s. Therefore, the air in front of the engine will accelerate as it approaches the inlet nozzle. The streamlines and velocity variation upstream of the nozzle and in the nozzle are shown in Fig. 6.17.

The losses in the inlet diffuser at takeoff conditions are defined as

$$\sigma_{0_{\text{diff TO}}} := \frac{p_{01}}{p_0}$$

where p_{01} is the stagnation pressure at inlet in the compressor and p_0 is the stagnation pressure upstream of the engine.

Example 6.6.2 A jet engine is operated in a test cell while the atmospheric pressure and temperature are $p_a = 1.013$ bar and $T_a = 288$ K. The velocity at station $1'$ of the inlet diffuser is $c_{1'} = 200$ m/s and the velocity at station 1 is $c_1 = 180$ m/s. Determine the static pressures $p_{1'}$ and p_1 knowing that $\sigma_{0\text{diff TO}} = 0.985$.

Solution
The table of thermodynamic properties of air given in appendix B will be used to calculate the flow parameters in the inlet diffuser. Given the temperature T_a, we read from the air tables $h_a = 288.15$ kJ/(kg K). Using the S method of Section 6.5.3 we calculate $s_a = 6.6565$ kJ/(kg K).

Since the engine is at takeoff conditions, static state "a" is identical with stagnation states "0a" and "01'''". We then calculate the enthalpy at state $1'$ as

$$h_{1'} = h_{01'} - c_{1'}^2/2 = 288,150 - 200^2/2 = 268,150 \text{ J/kg}.$$

Figure 6.18 $h - s$ diagram of real process in inlet diffuser at takeoff.

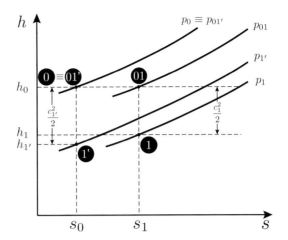

Knowing $h_{1'}$ and $s_{1'} = s_a$, we use the P method to calculate $p_{1'} = 0.792$ bar. We then calculate the stagnation pressure $p_{01} = \sigma_{0\mathrm{diff\ TO}} \times p_{0a} = 0.9978$ bar. Since $T_{01} = T_{01'}$, we calculate the entropy using the S method as $s_{01} = 6.6609$ kJ/(kgK).

For state "1" we know $s_1 = s_{01}$ and $h_1 = h_{01} - c_1^2/2 = 288,150 - 180^2/2 = 271,950$ J/(kg K). Using the P method we calculate $p_1 = 0.819$ bar (Fig. 6.18).

6.6.3.3 Turbojet In-Flight Real Cycle with Air Standard, Variable Properties

The in-flight real cycle is needed to predict engine performance at different speeds and altitudes. The methodology used to predict the processes occurring between the face of the compressor and the end of the exit nozzle is unchanged. The novel element is the prediction of the parameter variations in the inlet nozzle at in-flight conditions.

The losses in the inlet are defined similarly to the losses at takeoff, except that p_0 is replaced by p_{0a}, the stagnation pressure at the flight altitude:

$$\sigma_{0_{\mathrm{diff\ IF}}} := \frac{p_{01}}{p_{0a}}. \tag{6.60}$$

Depending on the relative values of the airplane speed, u, and the velocity at inlet in the diffuser, $c_{1'}$, two cases are possible: (1) $u > c_{1'}$ and (2) $u \leq c_{1'}$. Figures 6.19 and 6.20 show the variation of the streamlines, velocity, pressure, and enthalpy for the two cases.

Example 6.6.3 An airplane flies at 10,000 m with Mach $M = 0.8$. Knowing the pressure ratio in the inlet $\sigma_{0_{\mathrm{diff\ IF}}} = p_{01}/p_{0a} = 0.98$, determine (i) the stagnation pressure in front of the inlet diffuser, and (ii) the stagnation pressure downstream of the inlet diffuser.

Solution
Given the altitude of the airplane, we read from the ICAO Standard Atmosphere of Appendix A the pressure $p_a = 2.65 \times 10^4$ Pa and temperature $T_a = 223.2$ K.

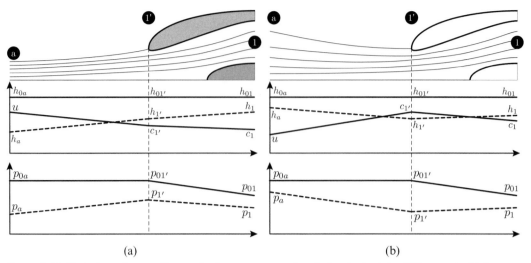

Figure 6.19 Variation of streamlines, velocity, pressure, and enthalpy in the inlet diffuser at in-flight conditions. (a) $u > c_{1'}$ and (b) $u < c_{1'}$

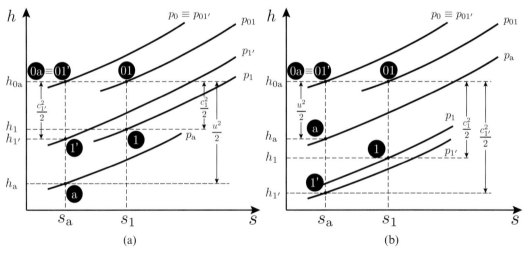

Figure 6.20 $h - s$ diagram of the real process in inlet diffuser at in-flight conditions. (a) $u > c_{1'}$ and (b) $u < c_{1'}$

The stagnation temperature and pressure in front of the inlet diffuser can be calculated in two ways: (1) assuming that the thermodynamic properties of air do not change from the static state to the stagnation state, and (2) taking into account the change of thermodynamic properties.

1. Assuming constant thermodynamic properties of air
   Having determined the pressure $p_a = 2.65 \times 10^4$ Pa from the ICAO Standard Atmosphere, we calculate the stagnation pressure

$$p_{0a} = p_a \left(1 + \frac{\gamma - 1}{2}M^2\right)^{\gamma/(\gamma-1)} = 2.65 \times 10^4 (1 + 0.2 \times 0.8^2)^{3.5} =$$
$$= 0.404 \times 10^5 \text{ Pa} = 0.404 \text{ bar}.$$

The stagnation pressure downstream from the inlet diffuser is

$$p_{01} = \sigma_{0\text{diff IF}} p_{0a} = 0.98 \times 0.404 = 0.396 \text{ bar}.$$

2. Accounting for variations of thermodynamic properties of air

Having determined the pressure p_a, temperature T_a and speed of sound a from the ICAO Standard Atmosphere, we calculate the velocity of the airplane

$$u = \text{M} \times a = 0.8 \times 299.5 = 239.6 \text{ m/s}.$$

We then read from the air tables the enthalpy $h_a = 223.12$ kJ/kg corresponding to temperature $T_a = 223.2$ K and calculate the stagnation enthalpy

$$h_{0a} = h_a + u^2/2 = 223,120 + 239.6^2/2 = 251,824 \text{ J/kg}.$$

Having the values of pressure p_a and temperature T_a, we calculate the entropy s_a using the S method, which gives $s_a = 6.7872$ kJ/(kgK). Since $s_{0a} = s_a$ and knowing the enthalpy h_{0a}, we use the P method to calculate $p_{0a} = 0.403$ bar. Then the stagnation pressure is

$$p_{01} = \sigma_{0\text{diff IF}} \times p_{0a} = 0.98 \times 0.403 = 0.395 \text{ bar}.$$

By comparing the results of the two methods, it turns out that assuming constant properties for air produces results that are very close to the results generated using the air tables that take into account the variation of thermodynamic properties of air. It is, however, safer to always use the second approach, especially when the velocity and pressure variations are large.

6.6.4 Turbojet Real Cycles with Actual Medium

The working fluids in a turbojet engine are air and combustion products. The working fluid in the inlet diffuser and the compressor is air. The working fluid in the combustor, turbine, and exit nozzle is the mixture of air and fuel that yields the combustion products. Consequently, the states of the real cycle with actual medium that define the processes in the inlet diffuser and in the compressor are identical with those of the real cycle with air standard, variable properties. The states that define the processes in the combustor, turbine, and exit nozzle use the thermodynamic properties of combustion products that are defined as a combination of the thermodynamic properties of air and stoichiometric combustion products, as presented in Chapter 5.

The steps needed to predict the real cycle of a turbojet engine with actual medium require that two of the independent parameters that define each state be determined. The third

Figure 6.21 $h - s$ diagram of the real cycle of a turbojet engine with actual medium.

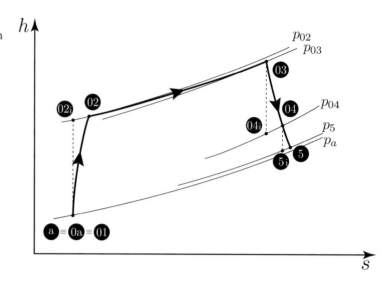

parameter is then calculated using one of the methods H, S, or P described in Section 6.5.3. The prediction of the real cycle requires that the following parameters be specified: pressure drop in inlet diffuser, efficiencies of the compressor and turbine, pressure drop in the combustor, exit nozzle efficiency.

Example 6.6.4 is presented to illustrate the methodology for calculating the real cycle of a turbojet engine with actual medium as the working fluid (Fig. 6.21).

Example 6.6.4 A turbojet operates in an engine test cell. The compressor has an inlet stagnation pressure of 101.3 kPa and an inlet stagnation temperature of 288.16 K. The mass flow rate of air entering the engine is $\dot{m} = 10$ kg/s. The compressor operates with a pressure ratio of 12 and an efficiency of 87%. The burner has an efficiency of 97% and a total pressure drop of 3%. The turbine inlet temperature is 1400 K. The turbine has an efficiency of 89%. The engine has a converging nozzle and the nozzle efficiency is $\varphi_5 = 97\%$. One can neglect the entropy correction $\Delta s'$. The engine uses standard fuel. Recall that standard fuel has a lower heating value $LHV = 43,500$ kJ/kg and $minL = 14.66$.

1. Determine the net power, cycle thermal efficiency, thrust, and thrust specific fuel consumption;
2. Draw the real cycle;
3. Determine the thrust and thrust specific fuel consumption if the convergent nozzle is replaced by a convergent-divergent nozzle.

Solution
The tables of thermodynamic properties of air and stoichiometric combustion products given in appendices B and C will be used to predict the states of the real cycle. Whenever these tables are used, it will be indicated by the word "table" as was done in example 6.6.1.

- **State 01**

  The temperature and pressure at state 01 are specified:

  $$T_{01} = 288.16 \text{ K}$$

  $$p_{01} = 101.3 \text{ kPa} = 1.013 \text{ bar.}$$

  The enthalpy is determined from the air table

  $$h_{01}(T_{01} = 288.16 \text{ K}) \overset{\text{table}}{=} 288.3 \text{ kJ/kg.}$$

  Knowing the temperature and pressure, the entropy is calculated using the S method as follows:

  $$s_{01_{p=1 \text{ bar}}}(T_{01} = 288.16 \text{ K}) = 6.6608 \text{ kJ/(kg K),}$$

  and using (2.93) yields

  $$s_{01} = s_{01_{p=1 \text{ bar}}} - R \ln p_{01} = 6.6608 - 0.28716 \ln 1.013 = 6.6571 \text{ kJ/(kg K).}$$

- **State 02i**

  The stagnation pressure at state 02i is equal to the stagnation pressure at state 02, which is calculated using the compressor pressure ratio (6.45)

  $$p_{02i} = p_{01} \, \pi_{0C} = 1.013 \times 12 = 12.156 \text{ bar.}$$

  The entropy at state 02i is equal to the entropy at state 01

  $$s_{02i} = s_{01} = 6.6571 \text{ kJ/(kg K).}$$

  Having determined the pressure and entropy, the enthalpy is calculated using the H method:

  $$s_{02i} - s_{2'} = -R \ln \frac{p_{02i}}{p_{2'}} \quad \text{where} \quad p_{2'} = 1 \text{ bar}$$

  $$s_{2'} = s_{02i} + R \ln p_{02i} = 6.6571 + 0.28716 \ln 12.156 = 7.3744 \text{ kJ/(kg K)} \overset{\text{table}}{\rightarrow}$$

  $$h_{02i} = 586.5 \text{ kJ/kg.}$$

  The ideal work in the compressor is

  $$w_{C_i} = h_{02i} - h_{01} = 586.45 - 288.31 = 298.14 \text{ kJ/kg.}$$

- **State 02**

  The pressure at state 02 is equal to the pressure at state 02i

  $$p_{02} = p_{02i} = 12.156 \text{ bar.}$$

  The work in the compressor is calculated using the efficiency of the compressor

  $$w_C = \frac{w_{C_i}}{\eta_C} = \frac{298.14}{0.87} = 342.69 \text{ kJ/kg}$$

so that the enthalpy is

$$h_{02} = h_{01} + w_C = 288.31 + 342.69 = 631.00 \text{ kJ/kg}.$$

Having determined the pressure and enthalpy, the entropy is calculated using the S method:

$$h_{02} \xrightarrow{\text{air table}} s_{02_{p=1\,\text{bar}}} = 7.4488 \text{ kJ/(kg K)}$$

and using (2.93) yields

$$s_{02} = s_{02_{p=1\,\text{bar}}} - R \ln p_{02} = 6.7312 \text{ kJ/(kg K)}.$$

- **State 03**
  The pressure drop in the combustor is 3%:

  $$p_{03} = p_{02} \, (1 - 3\%) = 12.156 \times 0.97 = 11.791 \text{ bar}$$

  and the turbine inlet temperature is specified to be

  $$T_{03} = 1400 \text{ K}.$$

  The enthalpies are then determined from the air and stoichiometric combustion tables:

  $$T_{03} \xrightarrow{\text{air table}} h_{03_{\text{air}}} = 1515.2 \text{kJ/kg}$$

  $$T_{03} \xrightarrow{\lambda=1 \text{ table}} h_{03_{\lambda=1}} = 1642.8 \text{kJ/kg}.$$

  The excess air, λ, is calculated using

  $$\lambda = \frac{h_{03\lambda=1}(1 + minL) - \xi_{\text{comb}} \, LHV - h_{03\text{air}} \, minL}{minL(h_{02} - h_{03\text{air}})}, \tag{5.39}$$

  which yields

  $$\lambda = 2.984.$$

  The enthalpy of the mixture for the excess air, λ, is calculated using

  $$h_{03_\lambda} = r \, h_{03_{\lambda=1}} + q \, h_{03_{\text{air}}} \tag{5.32}$$

  where the weighting functions are

  $$r = \frac{1 + minL}{1 + \lambda \, minL} = 0.35$$

  and

  $$q = \frac{(\lambda - 1) \, minL}{1 + \lambda \, minL} = 0.65,$$

  which yields

  $$h_{03_\lambda} = 1559.87 \text{ kJ/kg}.$$

To calculate the entropy of the mixture, the entropy corresponding to temperature T_{03} is obtained from the tables:

$$T_{03} \xrightarrow{\lambda=1 \text{ table}} s_{03_{\lambda=1, p=1 \text{ bar}}} = 8.7187 \text{ kJ/(kg K)}$$

$$T_{03} \xrightarrow{\text{air table}} s_{03_{\text{air}, p=1 \text{ bar}}} = 8.3620 \text{ kJ/(kg K)}.$$

The entropy of the mixture at $p = 1$ bar is then calculated using equation (5.41):

$$s_{03_{\lambda, p=1 \text{ bar}}} = r \, s_{03_{\lambda=1, p=1}} + q \, s_{03_{\text{air}, p=1}} + \underbrace{\Delta s'}_{\text{neglected}} =$$

$$= 0.35 \times 8.7187 + 0.65 \times 8.3620 = 8.4868 \text{ kJ/(kg K)}.$$

Using (2.93) yields

$$s_{03_{\lambda, p=11.791}} = s_{03_{\lambda, p=1 \text{ bar}}} - R \ln 11.791 = 8.4868 - 0.28716 \ln 11.791 =$$
$$= 7.7783 \text{ kJ/(kg K)}.$$

- **State 04i**

  The entropy at state 04i is equal to the entropy at state 03

  $$s_{04i} = s_{03} = 7.7783 \text{ kJ/(kg K)}.$$

The power of the turbine is equal to the power of the compressor if (1) the spool is rotating with constant angular velocity, (2) there are no mechanical losses, and (3) the power needed to drive the fuel and oil pumps, the generator, and other auxiliary devices on the engine is neglected:

$$\mathcal{P}_C = \mathcal{P}_T$$

so that

$$\dot{m}_g w_T = \dot{m}_a w_C \rightarrow w_T = w_C \frac{1}{1 + 1/(\lambda minL)} =$$

$$= 342.69 \frac{1}{1 + 1/(2.984 \times 14.66)} = 335.03 \text{ kJ/kg}.$$

Using turbine efficiency definition, (6.44) yields

$$w_{T_i} = \frac{w_T}{\eta_T} = 335.03/0.89 = 376.44 \text{ kJ/kg}.$$

Using the ideal work in the turbine (6.44) yields the enthalpy

$$h_{04i} = h_{03} - w_{T_i} = 1559.87 - 376.44 = 1183.43 \text{ kJ/kg}.$$

Table 6.2 Variation of enthalpy and function $f(T)$ (from Example 5.3.2) with temperature.

Iteration	Temperature K	Enthalpy $h_{\lambda=1}$ kJ/kg	Enthalpy h_a kJ/kg	Function $f(T)$ kJ/kg
1	1110.2	1260.9	1172.8	−20.23
2	1090.0	1234.9	1149.3	4.14
3	1093.4	1239.3	1153.3	0.00

The temperature corresponding to h_{04i} and λ is calculated using the iterative method described in Example 5.3.2. An initial guess of the temperature is obtained using the air and stoichiometric temperatures corresponding to $h = 1183.43$ kJ/kg:

$$T = rT_{\lambda=1} + qT_a = 0.350 \times 1049.6 + 0.650 \times 1118.6 = 1110.2 \text{ K}.$$

The temperature for the second iteration is chosen so that it brackets the solution of $f(T)$. The temperature of the third iteration is calculated using a linear approximation of $f(T)$ between iterations 1 and 2. The iterations shown in Table 6.2 yield $T_{04i} = 1093.4$ K.

Having determined the entropy and temperature, the pressure is calculated using the P method applied to combustion products. The air and stoichiometric combustion products tables give the entropy values corresponding to T_{04i}

$$s_{04i_{p=1, \lambda=1}} = 8.3935 \text{ kJ/(kg K)}; \quad s_{04i_{p=1, a}} = 8.0704 \text{ kJ/(kg K)}.$$

The entropy of the combustion products at the pressure of 1 bar is

$$s_{04i_{p=1, \lambda}} = r\, s_{04i_{p=1, \lambda=1}} + q\, s_{04i_{p=1, a}} + \underbrace{\Delta s'}_{\text{neglected}}$$

$$= 0.350 \times 8.3935 + 0.650 \times 8.0704 = 8.1835 \text{ kJ/(kg K)}.$$

Finally the pressure is obtained from the entropy variation equation (2.93)

$$s_{04i_{p04i, \lambda}} = s_{04i_{p=1, \lambda}} - R \ln p_{04i}$$

so that

$$p_{04i} = \exp\left(-\frac{7.7783 - 8.1835}{0.28716}\right) = 4.1005 \text{ bar}.$$

- **State 04**

  The pressure at state 04 is equal to that at state 04i

  $$p_{04} = p_{04i} = 4.1005 \text{ bar}.$$

  Knowing the work in the turbine and the enthalpy at state 03, the enthalpy at state 04 is

  $$h_{04} = h_{03} - w_T = 1559.87 - 335.03 = 1224.84 \text{ kJ/kg}.$$

The temperature T_{04} is calculated as shown in Example 5.3.2 and the result is $T_{04} = 1127.8$ K. Knowing the temperature T_{04}, the entropy values are obtained from the air and stoichiometric gases tables:

$$s04_{p=1,\,a} = 8.1060 \text{ kJ/(kg K)}; \quad s04_{p=1,\,\lambda=1} = 8.4333 \text{ kJ/(kg K)}.$$

The entropy of the mixture at the pressure of 1 bar is

$$s04_{p=1,\,\lambda} = r\, s04_{p=1,\,\lambda=1} + q\, s04_{p=1,\,a} =$$
$$= 0.350 \times 8.4333 + 0.650 \times 8.1060 = 8.2206 \text{ kJ/(kg K)}$$

and the entropy at the pressure p_{04} is

$$s04 = s04_{p=1,\,\lambda} - R\, \ln 4.093 = 8.2206 - 0.28716 \ln 4.1005 = 7.8154 \text{ kJ/(kg K)}.$$

• **State 5i**

The first step in determining the processes in the exit nozzle is to calculate the critical pressure based on state 04:

$$p_{cr} = p_{04} \left(\frac{2}{\gamma_{04-cr} + 1} \right)^{\frac{\gamma_{04-cr}}{\gamma_{04-cr} - 1}} \tag{6.61}$$

where γ_{04-cr} must be calculated as an average between the γ values at state 04 and the critical state corresponding the 04. Since the critical temperature

$$T_{cr} = T_{04} \left(\frac{2}{\gamma_{04-cr} + 1} \right)$$

is a function of γ_{04-cr}, an iterative process must be used [Flack, 2005]. Table 6.3 shows the variation of T_{cr}, $T_{avg} = (T_{04} + T_{cr})/2$ and the ratios of specific heat capacities during the iterative process. Note that γ_a is the ratio of specific heat capacities of air, $\gamma_{\lambda=1}$ is the ratio of specific heat capacities of stoichiometric combustion products, and $\gamma_{04-cr} = q\gamma_a + r\gamma_{\lambda=1}$ is the ratio of specific heat capacities of the combustion products.

Having determined γ_{04-cr}, the critical pressure (6.61) is $p_{cr} = 2.2237$ bar, which is larger than the atmospheric pressure $p_a = p_{01} = 1.013$ bar. Consequently, state 5i will be

Table 6.3 Iterative process for calculating the critical temperature T_{cr} corresponding to $T_{04} = 1127.8$ K and $\lambda = 2.984$.

Iteration	T_{cr} [K]	T_{avg} [K]	γ_a [-]	$\gamma_{\lambda=1}$ [-]	γ_{04-cr} [-]
1	1127.80	1127.80	1.32831	1.28263	1.31232
2	975.557	1051.73	1.33361	1.29242	1.31920
3	972.665	1050.28	1.33368	1.29227	1.31919
4	972.669	1050.28	1.33368	1.29227	1.31919

determined by $p_{5i} = p_{cr}$ and the entropy:

$$s_{5i} = s_{04} = 7.8154 \text{ kJ/(kg K)}.$$

Having determined the pressure and entropy, the enthalpy is calculated using the H method, which yields

$$h_{5i} = 1040.79 \text{ kJ/kg}.$$

The velocity c_{5i} is calculated as

$$c_{5i} = \sqrt{2(h_{04} - h_{5i})} = \sqrt{2(1224.84 - 1040.79) \times 1000} = 606.7 \text{ m/s}.$$

- **State 5**
  The pressure is

$$p_5 = p_{5i} = 2.2237 \text{ bar}.$$

Using the nozzle efficiency, the velocity $c_5 = c_{5i}\varphi_5 = 606.7 \times 0.97 = 588.5$ m/s. The enthalpy is then calculated as

$$h_5 = h_{04} - c_5^2/2 = 1224.84 \times 10^3 - 588.5^2/2 = 1051.67 \times 10^3 \text{ J/kg}.$$

Having determined the pressure and enthalpy, the entropy is calculated using the S method, which yields $s_5 = 7.8269$ kJ/(kg K).

Once the parameters of all states have been determined, the net work, thermal efficiency, thrust, specific thrust, and thrust specific fuel consumption are calculated as follows:

$$w_{net} = h_{03} - h_5 - (h_{02} - h_{01}) = 1559.87 - 1051.67 - (631.00 - 288.31) = 165.51 \text{ kJ/kg}.$$

The thermal efficiency is

$$\eta_{th} = \frac{w_{net}}{q_{in}} = \frac{w_{net}}{h_{03} - h_{02}} = \frac{165.51}{1559.87 - 631.00} = 0.1782.$$

The specific thrust is

$$T_{sp} = \frac{T}{\dot{m}_a} = (1 + f)c_{5eq} - u \quad \text{where} \quad c_{5eq} = c_5\left[1 + \frac{1}{\gamma_5 M_5^2}\left(1 - \frac{p_a}{p_5}\right)\right]$$

so that

$$T_{sp} = \left(1 + \frac{1}{\lambda \times minL}\right)c_5\left[1 + \frac{1}{\gamma_5 M_5^2}\left(1 - \frac{p_a}{p_5}\right)\right] - u.$$

Substituting $u = 0$, $\gamma_5 = \gamma_{cr} = 1.3236$, $c_5 = c_{cr} = 588.5$ m/s, $p_5 = p_{cr} = 2.2237$ bar and $M_5 = 1$ yields

$$T_{sp} = \left(1 + \frac{1}{2.984 \times 14.66}\right)588.5\left[1 + \frac{1}{1.3236}\left(1 - \frac{1.013}{2.2237}\right)\right] = 849.55 \text{ m/s}.$$

The thrust of the engine is

$$T = T_{sp}\dot{m}_a = 849.55 \times 10 = 8495.5 \text{ N}.$$

The momentum thrust is

$$\dot{m}_a(1 + f)c_5 = 6019.5 \text{ N}.$$

and the thrust due to the area-pressure term is

$$\dot{m}_a(1 + f)(c_{5_{eq}} - c_5) = 2476 \text{ N}.$$

The thrust specific fuel consumption is

$$\text{TSFC} = \frac{3600}{\lambda \times minL} \times \frac{1}{T_{sp}} = \frac{3600}{2.984 \times 14.66} \times \frac{1}{849.55} = 0.0969 \text{ kg/(N h)}.$$

If the convergent exit nozzle is replaced by a convergent-divergent (de Laval) nozzle, then the gases will expand all the way down to atmospheric pressure. In this case, the states 01 to 04 are unchanged.

- **State 5i**

  The entropy at state 5i is equal to that of state 04

  $$s_{5i} = s_{04} = 7.8154 \text{ kJ/(kg K)}.$$

  The pressure at state 5i is equal to the atmospheric pressure

  $$p_{5i} = p_a = 1.013 \text{ bar}.$$

  Having determined entropy and pressure, the enthalpy is calculated using the H method, which yields $h_{5i} = 840.78$ kJ/kg.

- **State 5**

  The pressure at state 5 is equal to the atmospheric pressure

  $$p_5 = p_a = 1.013 \text{ bar}.$$

  Using the nozzle efficiency

  $$\varphi_5 = c_5/c_{5i},$$

  the velocity c_5 is calculated as a function of c_{5i}:

  $$c_{5i} = \sqrt{2(h_{04} - h_{5i})} = \sqrt{2(1225.0 - 840.78) \times 1000} = 876.4 \text{ m/s}$$

  so that

  $$c_5 = c_{5i}\varphi_5 = 876.4 \times 0.97 = 850.1 \text{ m/s}.$$

  The enthalpy is calculated using h_{04} and c_5:

  $$h_5 = h_{04} - c_5^2/2 = 1224.83 \times 1000 - 850.1^2/2 = 863.48 \times 1000 \text{ J/kg} = 863.48 \text{ kJ/kg}.$$

Having determined the pressure and enthalpy, the entropy is determined using the S method, which yields $s_5 = 7.8437$ kJ/(kg K).

The net work of the turbojet with the convergent-divergent nozzle is

$$w_{net} = h_{03} - h_5 - (h_{02} - h_{01}) = 1559.87 - 863.48 - (631.00 - 288.31) = 353.7 \text{ kJ/kg.}$$

The thermal efficiency

$$\eta_{th} = \frac{w_{net}}{q_{in}} = \frac{w_{net}}{h_{03} - h_{02}} = \frac{353.7}{1559.87 - 631.00} = 0.3808$$

is approximately twice as large as that of the turbojet with convergent nozzle.

The specific thrust is

$$T_{sp} = \frac{T}{\dot{m}_a} = (1+f)c_{5eq} - u \quad \text{where} \quad c_{5eq} = c_5 \left[1 + \frac{1}{\gamma_5 M_5^2} \left(1 - \frac{p_a}{p_5} \right) \right].$$

Since $p_a = p_e$, the specific thrust reduces to

$$T_{sp} = \left(1 + \frac{1}{\lambda \times minL} \right) c_5 - u.$$

Substituting $u = 0$ and $c_5 = 850.1$ m/s yields

$$T_{sp} = \left(1 + \frac{1}{2.984 \times 14.66} \right) 850.1 = 869.55 \text{ m/s.}$$

The thrust of the engine is

$$T = T_{sp}\dot{m}_a = 869.55 \times 10 = 8695.5 \text{ N.}$$

For the engine with a convergent-divergent nozzle, the ram drag and the thrust due to the area-pressure term are both 0. The entire thrust is due to the momentum thrust. The thrust of the turbojet with convergent-divergent nozzle is approximately 1% smaller that the thrust of the turbojet with convergent nozzle.

The thrust specific fuel consumption

$$\text{TSFC} = \frac{3600}{\lambda \times minL} \times \frac{1}{T_{sp}} = \frac{3600}{2.984 \times 14.66} \times \frac{1}{869.55} = 0.0946 \text{ kg/(N h)}$$

is approximately 1% larger than that of the turbojet with convergent nozzle.

6.7 Turbofan

6.7.1 Introduction

As the turbine design improved, more power could be extracted from the expanding gases. This additional turbine power could be used to either operate a larger compressor or a fan. In the first case, since the mass flow rate of air through the compressor increases, the mass flow rate of fuel would also increase. In the second case, instead of providing the additional turbine power to a larger compressor, this power is used to operate a fan that compresses an additional mass flow rate of air. This mass flow rate of air is diverted around the gas generator and does not enter the combustion chamber. The latter approach led to the development of turbofan engines.

To quantify the motivation for developing turbofan engines, let us explore ways to increase propulsion efficiency, η_p. Assuming that $A_e(p_e - p_a) \ll \dot{m}_a [(1 + f)u_e - u]$ and taking into account that $f \ll 1$, the thrust can be approximated as

$$\mathcal{T} \simeq \dot{m}_a(u_e - u). \tag{6.62}$$

The propulsion efficiency can then be approximated as

$$\eta_p := \frac{\mathcal{T} u}{\dot{m}_a \left[(1+f)\frac{u_e^2}{2} - \frac{u^2}{2}\right]} \simeq \frac{\dot{m}_a(u_e - u)u}{\dot{m}_a(\frac{u_e^2}{2} - \frac{u^2}{2})} = \frac{2u}{u_e + u}. \tag{6.63}$$

From the expression of approximated thrust (6.62) one obtains the velocity $u_e = \mathcal{T}/\dot{m}_a + u$. Substituting velocity u_e in (6.63) yields

$$\eta_p = \frac{1}{1 + \dfrac{\mathcal{T}}{2\dot{m}_a u}}. \tag{6.64}$$

Therefore, for a given airplane speed u, propulsion efficiency η_p increases if specific thrust $\mathcal{T}/\dot{m}_a$ decreases. Consequently, the propulsion efficiency η_p increases if the velocity u_e of the gases at the exit from the engine decreases. This conclusion is also supported by the results shown in Fig. 6.2.

The decrease of the velocity u_e can be accomplished by transferring more of the kinetic energy of the gases to the turbine blades, that is, extracting more power in the turbine. If this additional power is supplied to a fan that compresses additional air and diverts it around the gas generator, then the mass flow rate of fuel will not increase. The thrust, however, will increase because of the thrust generated by the fan.

The mass flow rate that bypasses the gas generator is called the *cold mass flow rate*, $\dot{m}_{aC}$. The mass flow rate that enters the gas generator is called the *hot mass flow rate*, $\dot{m}_{aH}$. The ratio between the cold mass flow rate of air and the hot mass flow rate of air is defined as the *bypass ratio*, *BPR*:

$$BPR = \frac{\dot{m}_{aC}}{\dot{m}_{aH}}.$$

The value of the bypass ratio for current turbofans ranges between 0.2 and 12. Larger values of the bypass ratio, between 35 and 56, have been obtained on unducted fan engines. Based on the bypass ratio, turbofans are divided into low-bypass and high-bypass turbofans. As shown in Fig. 6.22, high-bypass turbofans have a higher propulsion efficiency than low-bypass turbofans. High-bypass turbofans are used on almost all jet airliners and most military transport aircraft. Low-bypass turbofans are used on fighter airplanes.

Figure 6.22 Propulsion efficiencies of jet engines [after Anonymous [2005]].

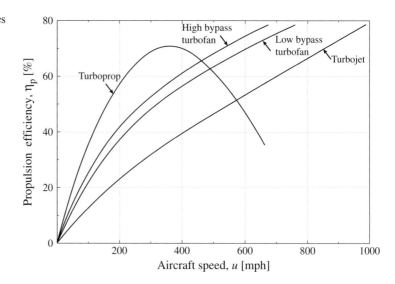

6.7.2   Turbofan Configurations

Several turbofan configurations are currently being used. The choice of turbofan configuration depends on the type of aircraft the engine is used on and the missions the aircraft must achieve. These configurations can be classified based on several criteria. The most common classifications are based on either the number of spools or on how the hot and cold fluxes mix.

Based on the number of spools, the turbofans can be single-spool, basic two-spool, boosted two-spool, or three-spool, as shown in Fig. 6.23. A variant of a two-spool turbofan is the aft-fan.

The *single-spool turbofan* engine has the simplest configuration, which consists of a fan and high-pressure compressor driven by a single turbine, as shown in Fig. 6.23a. A limited number of turbofans have this configuration, most notable being the SNECMA M53 engine used by the Dassault Mirage 2000 fighter aircraft.

The *basic two-spool turbofan* engine has a low-pressure (LP) spool and a high-pressure (HP) spool, as shown in Fig. 6.23b. The LP spool consists of a fan and an LP turbine, connected by the LP shaft. The HP spool consists of an HP compressor and an HP turbine, connected by the HP shaft. The HP shaft runs concentrically with the LP shaft. The LP spool runs at a lower angular velocity than the HP spool. Examples of basic two-spool turbofans include the Pratt & Whitney Canada PW600, the General Electric–Garrett CFE738 and the BMW–Rolls-Royce BR710 shown in Fig. 6.24.

Boosted two-spool turbofan engines typically produce larger thrust values than the basic two-spool. This is achieved by either increasing the pressure ratio of the HP compressor or by adding an intermediate-pressure (IP) compressor in front of the HP compressor. The IP compressor is mounted on the LP shaft with the fan and driven by the LP turbine. The overall pressure ratio is typically higher than 40:1 on more modern engines. Examples of boosted two-spool turbofan engines include the General Electric CF6, GE90, and GEnx, shown in Fig. 6.25, the Pratt & Whitney JT9D, PW4000, and PW6000, and the Engine Alliance GP7200.

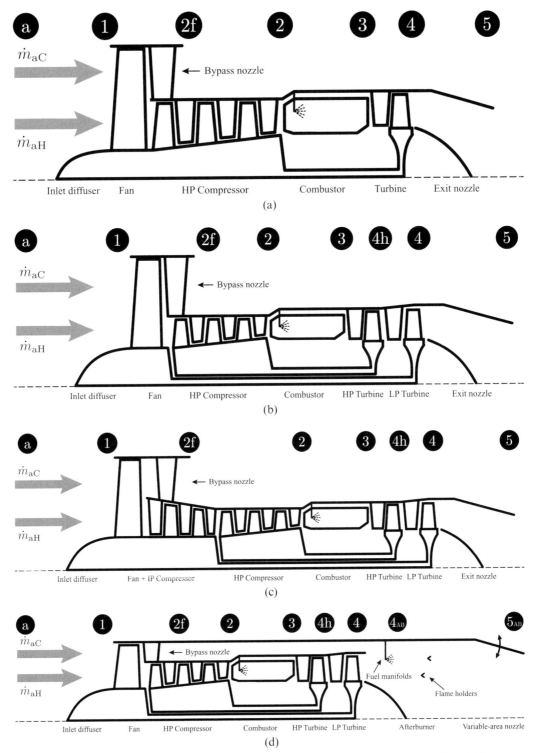

Figure 6.23 Turbofan layouts: (a) single-spool, (b) basic two-spool, (c) boosted two-spool, and (d) augmented two-spool.

Figure 6.24 Basic two-spool turbofan engine: BMW-Rolls-Royce BR710 (courtesy of Rolls-Royce).

Figure 6.25 Boosted two-spool turbofan with fan exhausted: General Electric GEnx (courtesy of GE Aviation).

Three-spool turbofan engines have an intermediate-pressure (IP) shaft that connects an IP compressor to an IP turbine. Typically the IP shaft runs concentrically with the LP and HP shafts. The three-spool turbofans were introduced to allow flexibility in setting the angular velocity of the IP compressor and turbine. One of the first three-spool turbofans was the Garrett ATF3 engine that is used on the Dassault Falcon 20 business jet. This engine, however, had an unusual configuration, where the third spool did not run concentrically with the other two. The classical configuration of a three-spool turbofan was introduced by Rolls-Royce

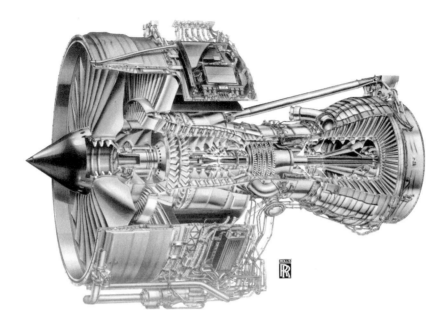

Figure 6.26 Three-spool turbofan: Rolls-Royce Trent 800 (courtesy of Rolls-Royce).

with the RB211 engine that powered the Lockheed L-1011 TriStar. This configuration was later used for the family of Trent engines, with the Trent 800 shown in Fig. 6.26.

Unducted fan or *aft-fan turbofan* engines were developed from turbojet or turbofan engines by adding a power turbine with a fan in the exhaust pipe of original turbojet/turbofan engine. Figure 6.27 shows the General Electric GE36 that was developed based on the F404 engine. The gases of the turbojet/turbofan expand in the power turbine which drives the fan. Typically a pair of counter-rotating fans is used. The number of blades varies between 8 and 12. The fan does not need to be geared to reduce the angular velocity, which is usually less than 2000 rpm. The bypass ratio is much larger than that of a classical turbofan, ranging between 35 on the GE36 and 56 on the Pratt & Whitney/Allison 578-DX. The thrust specific fuel consumption of an aft-fan turbofan engine can be 20 to 30% less than that of a classical turbofan.

The fan and turbine are spinning with the same angular velocity since they are connected by a shaft without a gearbox in between. The optimal angular velocities of the fan and the turbine, however, are not the same. As it will be shown in Chapter 7, the turbine produces more power if it spins faster while the fan requires less power if it spins slower. To have both the fan and the turbine operate closer to their optimal angular speeds, a planetary reduction gearbox was introduced between the turbine and the fan. This resulted in the *geared turbofan* configuration. The geared turbofan configuration was used in the past by Garrett, now Honeywell, on the TFE731 and ALF 502/507. More recently, the geared fan configuration was adopted by Pratt & Whitney for a larger size turbofan, the PW1000G. The PW1000G engine is 16% more fuel efficient that alternative turbofans in its thrust class.

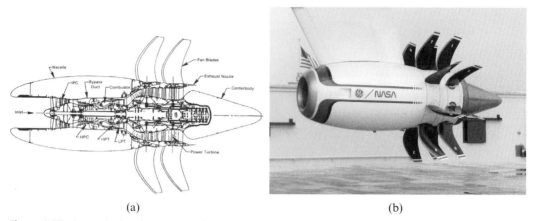

(a)   (b)

Figure 6.27 General Electric GE36 aft-fan turbofan, aka unducted fan (courtesy of GE Aviation). (a) Layout and (b) Photo.

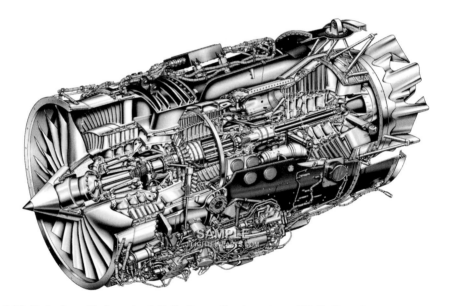

Figure 6.28 Turbofan with fan mixed: Rolls-Royce Tay (courtesy of Rolls-Royce).

Turbofan engines can also be classified as turbofan with (1) fan exhausted and (2) fan mixed. The *turbofan with fan exhausted*, shown in Figs. 6.25 and 6.26, has a short fan cowling so that the cold mass flow rate is not contained once it leaves the fan. Many of the turbofans used by airliners are turbofans with fan exhausted, such as the General Electric CF6, GE90, GEnx, the Pratt & Whitney PW4000, PW1000G, PW6000, and the Rolls-Royce Trent. The *turbofan with fan mixed*, shown in Fig. 6.28, mixes the cold mass flow rate with the hot mass flow rate coming out of the turbine. The mixed gases are accelerated in the exit nozzle to produce thrust. Examples of turbofan with fan mixed include the Pratt & Whitney JT8D, the Rolls-Royce Tay, and the General Electric GE F118.

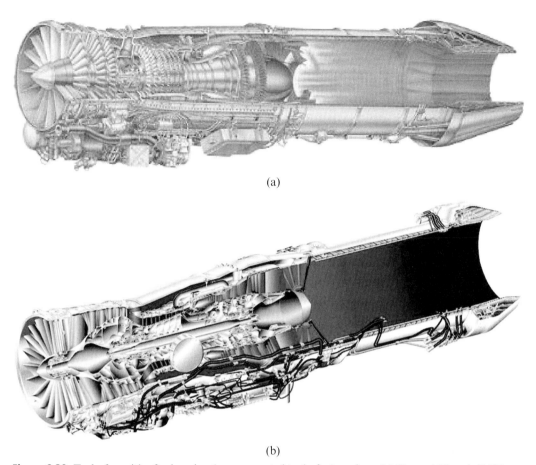

(a)

(b)

Figure 6.29 Turbofan with afterburning (or augmented turbofan) engines. (a) General Electric F404 (courtesy of GE Aviation) and (b) Pratt & Whitney F135 Conventional Takeoff & Landing (reproduced with permission from Pratt & Whitney).

To increase the thrust for a short duration, afterburners were added to turbofans. Most turbofans with afterburners were developed using turbofans with fan mixed, by adding afterburners downstream of the mixer. Examples of turbofans with afterburner include the Pratt & Whitney F100, shown in Fig. 6.29, and the General Electric F110, both used on the F-15 and F-16; the General Electric F404, shown in Fig. 6.29a, used on the F-18; the Pratt & Whitney F119 used on the F-22; and the F135, shown in Fig. 6.29b, used on F-35. Turbofan engines with afterburners are discussed in detail in Chapter 8.

6.7.3 Real Cycle Analysis

To illustrate the methodology, several cycle analyses will be presented. The first two examples present the real cycle of a single-spool turbofan: (1) with air as the working fluid, and (2) with actual medium as the working fluid. The third example presents the real cycle of a two-spool turbofan with actual medium as the working fluid.

Example 6.7.1 A single-spool turbofan engine operates in a test cell with a compressor inlet temperature of 288.16 K, inlet pressure 101.3 kPa, a fan efficiency of 85%, and a compressor efficiency of 87%. The overall compressor pressure ratio is 12. The turbine inlet temperature is 1400 K and the turbine efficiency is 89%. The air enters the turbofan at a rate of $\dot{m} = 50$ kg/s. The pressure drop in the combustor is 3%. The gas generator nozzle efficiency is 97% and the fan nozzle efficiency is 100%.

The bypass ratio is BPR = 7.5. The fan pressure ratio is $\pi_{0F} = 1.35$.

Calculate the thrust assuming the working fluid in the engine is air with variable properties.

Solution
The table of thermodynamic properties of air given in appendix B will be used to predict the states of the real cycle. Whenever this table is used, it will be indicated by the word "table" as it was done in example 6.6.1.

- **State 01**

$$T_{01} = 288.16 \text{ K} \xrightarrow{\text{table}} h_{01} = 288.3 \text{ kJ/kg}$$

$$p_{01} = 101.3 \text{ kPa} = 1.013 \text{ bar}.$$

Having determined the pressure and enthalpy, the entropy is calculated using the S method, where

$$s01_{p=1 \text{ bar}}(T_{01}) \overset{\text{table}}{=} 6.6608 \text{ kJ/(kg K)}$$

and

$$s_{01} = s01_{p=1 \text{ bar}} - R \ln p_{01} = 6.6608 - 0.28716 \ln 1.013 = 6.6571 \text{ kJ/(kg K)}.$$

- **State 02i**
  The pressure at state 02i is equal to the pressure at state 02. The latter is related by the compressor pressure ratio, π_{0C}, to the pressure at inlet in the compressor

$$p_{02i} = p_{01} \times \pi_{0C} = 1.013 \times 12 = 12.156 \text{ bar}.$$

Since the process between states 01 and 02i is isentropic,

$$s_{02i} = s_{01} = 6.6571 \text{ kJ/(kg K)}.$$

Having determined the pressure and the entropy, the enthalpy is calculated using the H method:

$$s2_{p=1 \text{ bar}} = s_{02i} + R \ln p_{02i} = 6.6571 + 0.28716 \ln 12.156 = 7.3744 \text{ kJ/(kg K)}$$

$$h_{02i}(s2_{p=1 \text{ bar}}) \overset{\text{table}}{=} 586.5 \text{ kJ/kg}.$$

The ideal work in the compressor is

$$w_{C_i} = h_{02i} - h_{01} = 586.5 - 288.3 = 298.2 \text{ kJ/kg}.$$

- **State 02**

  As mentioned at state 02i, the pressure at state 02 is equal to the pressure at state 02i

  $$p_{02} = p_{02i} = 12.156 \text{ bar}.$$

  The compressor efficiency is used to determine enthalpy

  $$w_C = \frac{w_{C_i}}{\eta_C} = \frac{298.2}{0.87} = 342.7 \text{ kJ/kg}$$

  so that

  $$h_{02} = h_{01} + w_c = 288.3 + 342.7 = 631.0 \text{ kJ/kg}.$$

  Having determined pressure and enthalpy, the entropy is calculated using the S method where

  $$s_{2_{p=1 \text{ bar}}}(h_{02}) \overset{\text{table}}{=} 7.4485 \text{ kJ/(kg K)}$$

  and

  $$s_{02} = s_{2_{p=1 \text{ bar}}} - R \ln p_{02} = 7.4485 - 0.28716 \ln 12.156 = 6.7312 \text{ kJ/(kg K)}.$$

- **State 03**

  The stagnation temperature at inlet in the turbine, T_{03}, is specified to be

  $$T_{03} = 1400 \text{ K};$$

  therefore using the table the enthalpy is determined to be

  $$h_{03} = 1515.2 \text{ kJ/kg}.$$

  Since the pressure drop in the combustor is 3%,

  $$p_{03} = p_{02} \times (1 - 3\%) = 12.156 \times 0.97 = 11.791 \text{ bar}.$$

  Having determined pressure and enthalpy, the entropy is calculated using the S method where

  $$s_{03_{p=1 \text{ bar}}}(T_{03}) \overset{\text{table}}{=} 8.3620 \text{ kJ/(kg K)}$$

  and

  $$s_{03} = s_{03_{p=1 \text{ bar}}} - R \ln p_{03} = 8.3620 - 0.28716 \ln 11.791 = 7.6535 \text{ kJ/(kg K)}.$$

Up to this point, the states were determined using the same steps as for a turbojet engine. To proceed further, the power of the turbine needs to be calculated. This requires that the fan power be predicted.

- **State 02fi**

  The entropy at state 02fi is equal to that at state 01:

  $$s_{02_{fi}} = s_{01} = 6.6571 \text{ kJ/(kg K)}.$$

  The stagnation pressure at exit from the fan is

  $$p_{02_{fi}} = p_{01} \times \pi_{0F} = 1.013 \times 1.35 = 1.368 \text{ bar}.$$

  Having determined the entropy and pressure, the enthalpy is calculated using the H method:

  $$s2_{f_{p=1 \text{ bar}}} = s_{02_{fi}} + R \ln p_{02_{fi}} = 6.6571 + 0.28716 \ln 1.368 = 6.7471 \text{ kJ/(kg K)}$$

  so that

  $$h_{02_{fi}}(s2_{f_{p=1 \text{ bar}}}) \overset{\text{table}}{=} 314.5 \text{ kJ/kg}.$$

  The ideal work in the fan is

  $$w_{F_i} = h_{02_{fi}} - h_{01} = 314.5 - 288.3 = 26.2 \text{ kJ/kg}.$$

- **State 02f**

  The pressure at state 02f is equal to the pressure at state 02fi

  $$p_{02_f} = p_{02_{fi}} = 1.368 \text{ bar}.$$

  The fan efficiency is used to calculate the enthalpy at state 02f

  $$w_F = \frac{w_{F_i}}{\eta_F} = \frac{26.2}{0.85} = 30.8 \text{ kJ/kg}$$

  and the enthalpy at state 02f is

  $$h_{02_f} = h_{01} + w_F = 319.1 \text{ kJ/kg}.$$

  Having determined pressure and enthalpy, the entropy is calculated using the S method where

  $$s2_{f_{p=1 \text{ bar}}}(h_{02_f}) \overset{\text{table}}{=} 6.7619 \text{ kJ/(kg K)}$$

  $$s_{02_f} = s2_{f_{p=1 \text{ bar}}} - R \ln p_{02_f} = 6.7619 - 0.28716 \ln 1.368 = 6.6719 \text{ kJ/(kg K)}.$$

- **State 5fi**

  The process between states 02f and 5fi is isentropic:

  $$s_{5_{fi}} = s_{02_f} = 6.6719 \text{ kJ/(kg K)}.$$

  The pressure at exit from the fan is equal to the atmospheric pressure, which is the same as the pressure at inlet in the compressor

  $$p_{5_{fi}} = p_{01} = 1.013 \text{ bar}.$$

Having determined entropy and pressure, the enthalpy is calculated using the H method:

$$s_{5_{fi}} - s_{5_{fi_{p=1\,bar}}} = -R\ln p_{5_{fi}}$$

so that

$$s_{5_{fi_{p=1\,bar}}} = s_{5_{fi}} + R\ln p_{5_{fi}} = 6.6719 + 0.28716\ln 1.013 = 6.6756\ \text{kJ/(kg K)}$$

and

$$h_{5_{fi}}(s_{5_{fi_{p=1\,bar}}}) \overset{\text{table}}{=} 292.5\ \text{kJ/kg}.$$

- **State 5f**

  Since the efficiency of the fan nozzle is 100%, the enthalpy of state 5f is equal to the enthalpy of state 5fi. In addition, the pressures at states 5f and 5fi are equal. As a result, state 5f is identical to state 5fi. If the fan nozzle efficiency were less than 100%, then the enthalpy of state 5f would be higher than that of state 5fi and the resulting entropy of state 5f would also be higher than that of state 5fi, as shown in Fig. 6.30. In this case, the entropy would be calculated using the S method.

  To calculate the velocity of the air leaving the fan, the stagnation enthalpy h_{05_f} is needed. Assuming that the fan exit nozzle is fixed, the work done in the nozzle is zero. In addition, assuming the heat transfer in the fan nozzle is much smaller than the enthalpy variation in the fan nozzle, then from (3.10) one obtains

$$h_{02_f} = h_{05_f}$$

so that

$$h_{02_f} - h_{5_f} = \frac{c_{5_f}^2}{2}$$

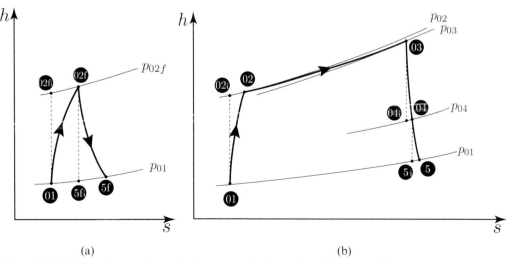

Figure 6.30   Mollier diagrams for a single-spool turbofan engine. (a) Fan and (b) gas generator.

which yields

$$c_{5_f} = \sqrt{2(h_{02_f} - h_{5_f})} = \sqrt{2(319.1 - 292.5) \times 10^3} = 230.6 \text{ m/s}.$$

- **State 04i**

  At steady state, when the spool is neither accelerating nor decelerating, the power provided by the turbine is equal to the power required by the compressor plus the fan, if we neglect the power needed to drive the oil and fuel pumps, the generator, and other consumers on the engine

  $$\mathcal{P}_F + \mathcal{P}_C = \mathcal{P}_T.$$

  Since the mass flow rate of fuel is neglected, then $\dot{m}_g = \dot{m}_{aH}$ so that

  $$w_F \times \dot{m}_{aC} + w_C \times \dot{m}_{aH} = w_T \times \dot{m}_{aH}.$$

  Dividing by $\dot{m}_{aH}$ and using the definition of BPR yields

  $$w_F \times BPR + w_C = w_T$$

  so that

  $$w_T = 30.8 \times 7.5 + 342.7 = 573.8 \text{ kJ/kg}.$$

  To find the ideal work of the turbine, let us use the turbine efficiency

  $$w_{T_i} = \frac{w_T}{\eta_T} = \frac{573.8}{0.89} = 644.7 \text{ kJ/kg}.$$

  The enthalpy is obtained from

  $$h_{03} - h_{04i} = w_{T_i}$$

  so that

  $$h_{04i} = h_{03} - w_{T_i} = 1515.2 - 644.7 = 870.5 \text{ kJ/kg}.$$

  The process from state 03 to state 04i is isentropic, therefore

  $$s_{04i} = s_{03} = 7.6535 \text{ kJ/(kg K)}.$$

  Having determined the enthalpy and the entropy, the pressure is calculated using the P method

  $$s_{04i} - s_{4_{p=1 \text{ bar}}} = -R \ln p_{04i}$$

  where $s_{4_{p=1 \text{ bar}}}(h_{04i}) \overset{\text{table}}{=} 7.7771 \text{ kJ/(kg K)}$ so that

  $$p_{04i} = \exp \frac{-(s_{04i} - s_{4_{p=1 \text{ bar}}})}{R} = \exp \frac{-(7.6535 - 7.7771)}{0.28716} = 1.538 \text{ bar}.$$

- **State 04**

  The pressure at state 04 is equal to the pressure at state 04i:

  $$p_{04} = p_{04i} = 1.538 \text{ bar.}$$

  Using the work in the turbine, the enthalpy h_{04} is

  $$h_{04} = h_{03} - w_{\text{T}} = 1515.2 - 573.8 = 941.4 \text{ kJ/kg}$$

  so that

  $$T_{04}(h_{04} = 941.4 \text{ kJ/kg}) \stackrel{\text{table}}{=} 907.7 \text{ K.}$$

  Having determined the pressure and enthalpy, the entropy is calculated using the S method, where

  $$s_{4_{p=1\,\text{bar}}}(h_{04}) \stackrel{\text{table}}{=} 7.8582 \text{ kJ/(kg K)}$$

  so that

  $$s_{04} = s_{4_{p=1\,\text{bar}}} - R \ln p_{04} = 7.8582 - 0.28716 \ln 1.538 = 7.7346 \text{ kJ/(kg K).}$$

- **State 5i**

  The process between states 04 and 5i is isentropic:

  $$s_{5i} = s_{04} = 7.7346 \text{ kJ/(kg K).}$$

  The pressure at exit from the core engine is equal to atmospheric pressure

  $$p_{5i} = p_{01} = 1.013 \text{ bar.}$$

  Having determined the entropy and pressure, the enthalpy is calculated using the H method:

  $$s_{5_{i_{p=1\,\text{bar}}}} = s_{5i} + R \ln p_{5i} = 7.7346 + 0.28716 \ln 1.013 = 7.7383 \text{ kJ/(kg K)}$$

  $$h_{5i}(s_{5_{i_{p=1\,\text{bar}}}} = 7.7383 \text{ kJ/(kg K)}) \stackrel{\text{table}}{=} 838.4 \text{ kJ/kg.}$$

  Using a similar argument to that used for the fan nozzle, the stagnation enthalpy at state 04 is equal to the enthalpy at state 05:

  $$h_{04} = h_{05}$$

  so that

  $$h_{04} - h_{5i} = \frac{c_{5i}^2}{2}$$

  which yields

  $$c_{5i} = \sqrt{2(h_{04} - h_{5i})} = \sqrt{2(941.4 - 838.4) \times 10^3} = 454.0 \text{ m/s.}$$

The critical pressure in the nozzle needs to be calculated next. Since the expansion process starts at state 04,

$$p_{cr} = p_{04} \left(\frac{2}{\gamma_{04-cr} + 1} \right)^{\gamma_{04-cr}/(\gamma_{04-cr}-1)}$$

where γ_{04-cr} is based on the averaged temperature between states 04 and "cr." The value of the ratio of specific heat capacities γ_{04-cr} is determined iteratively, as shown on page 224. In this case, $\gamma_{04-cr} = 1.3491$ so that

$$p_{cr} = 1.538 \left(\frac{2}{2.3491} \right)^{1.3491/0.3491} = 0.826 \text{ bar.}$$

Since the critical pressure is less than the atmospheric pressure, the working fluid will expand all the way down to the atmospheric pressure.

- **State 5**
  The pressure at state 5 is equal to the atmospheric pressure

$$p_5 = 1.013 \text{ bar.}$$

Using the gas generator nozzle efficiency, φ_5, the velocity of the working fluid at state 5 is

$$c_5 = \varphi_5 \times c_{5i} = 0.97 \times 454.0 = 440.4 \text{ m/s.}$$

The enthalpy at state 5 is

$$h_5 = h_{04} - c_5^2/2 = 941.4 \times 10^3 - 440.4^2/2 = 844.4 \times 10^3 \text{ J/kg} = 844.4 \text{ kJ/kg.}$$

Having determined the pressure and the enthalpy, the entropy is calculated using the S method where $s_{5_{p=1 \text{ bar}}}(h_5) = 7.7458 \text{ kJ/(kg K)}$ so that

$$s_5 = s_{5_{p=1 \text{ bar}}} - R \ln p_5 = 7.7458 - 0.28716 \ln 1.013 = 7.7421 \text{ kJ/(kg K).}$$

The thrust equation

$$T = (\dot{m}_{aH} + \dot{m}_f)u_{eH} - \dot{m}_{aH}u + \dot{m}_{aC}(u_{eC} - u) \tag{6.10}$$

can be rewritten using the notation specific for the engine, where

$$u_{eH} = c_5 \quad \text{and} \quad u_{eC} = c_{5_f}.$$

Since the mass flow rate of fuel is neglected and the engine is stationary,

$$T \overset{\dot{m}_f=0,\ u=0}{=} \dot{m}_{aH}c_5 + \dot{m}_{aC}c_{5_f} = \dot{m}_{aH}(c_5 + BPR \times c_{5_f}) =$$

$$= \frac{\dot{m}_a}{(1 + BPR)}(c_5 + BPR \times c_{5_f}) = \frac{50}{1 + 7.5}(440.4 + 7.5 \times 230.8) = 12,772 \text{ N.}$$

The thrust generated by the fan is

$$T_{fan} = \frac{\dot{m}_a}{(1 + BPR)}(BPR \times c_{5_f}) = \frac{50}{1 + 7.5}(7.5 \times 230.8) = 10,182 \text{ N}$$

and the thrust generated by the core engine is

$$T_{core} = \frac{\dot{m}_a}{(1 + BPR)} c_5 = \frac{50}{1 + 7.5} 440.4 = 2590 \text{ N}.$$

The fan produces $10,182/12,772 = 79.7\%$ of the total thrust of the engine.

Problem 6.7.1   Calculate the thrust of the single-spool turbofan engine defined in Example 6.7.1 assuming the working fluid in the engine is actual medium, that is, air up to the exit from the compressor, and combustion products downstream from the compressor.

Solution
States 01, 02i, 02, 02fi, 02f, 5fi, and 5f are identical to those of Example 6.7.1. The rest of the states are calculated following the same procedure as that of Example 6.7.1 except that the properties of the working fluid are estimated using the excess air, λ. Tables 6.4 and 6.5 compare the states and parameters that are different between the two predictions.

Table 6.4 Comparison of state parameters calculated assuming air and actual medium working fluids.

State	Air h [kJ/kg]	s [K]	p [bar]	Actual medium h [kJ/kg]	s [K]	p [bar]
03	1515.2	7.6535	11.791	1558.4	7.7743	11.791
5ifan	292.5	6.6719	1.013	292.5	6.6719	1.013
04i	870.5	7.6535	1.538	927.7	7.7743	1.664
04	941.4	7.7346	1.538	997.0	7.8507	1.664
5i	838.4	7.7346	1.013	871.6	7.8507	1.013
5	844.4	7.7421	1.013	879.1	7.8600	1.013

Table 6.5 Comparison of engine parameters calculated assuming air and actual medium working fluids.

Parameter	Units	Air	Actual medium
Excess air, λ	[-]	∞	3.084
Total thrust, T	[N]	12771.7	13101.7
Core thrust, T_{core}	[N]	2590.67	2920.67
Fan thrust/core thrust	[-]	79.7%	77.7%
TSFC	[kg/(N hr)]	0	3.57E-02
Turbine work, w_T	[kJ/kg]	573.798	561.385
Exit nozzle velocity, c_5	[m/s]	440.415	485.772

Let us now consider the case of a two-spool turbofan engine where the working fluid is actual medium, that is, air up to the exit from the compressor and combustion products downstream from the compressor.

Example 6.7.2 The General Electric CF6-6D engine is a two-spool turbofan engine that has a bypass ratio $BPR = 5.92$, a compressor overall pressure ratio $OPR = 25.2$, and a mass flow rate $\dot{m} = 590$ kg/s. Let us assume that the engine inlet diffuser efficiency is 0.98, the fan pressure ratio is $\pi_{0F} = 1.8$, the fan efficiency is $\eta_F = 0.85$, and the exit nozzle fan efficiency is 0.99. The compressor efficiency is $\eta_C = 0.87$. Let us assume the turbine inlet temperature is $T_{03} = 1430$ K, the pressure drop in the combustor is 3%, and the combustion efficiency is $\xi_{comb} = 0.9$. The high-pressure turbine efficiency is $\eta_{HPT} = 0.89$ and the low-pressure turbine efficiency is $\eta_{LPT} = 0.89$. The exit nozzle efficiency is 0.98. The engine uses standard fuel, that is, $minL = 14.66$ and $LHV = 43,500$ kJ/kg. Calculate the thrust and the thrust specific fuel consumption at takeoff, sea level conditions and compare them to the values reported in the literature.

Solution
The tables of thermodynamic properties of air (appendix B) and stoichiometric combustion products (appendix C) will be used to predict the states of the real cycle. Whenever these tables are used, it will be indicated by the word "table" as was done in previous examples.

- **State a**
  Assuming takeoff, sea level conditions leads to

  $$T_a = 288.16 \text{ K} \overset{\text{table}}{\rightarrow} h_a = 288.3 \text{ kJ/kg}$$

  and

  $$p_a = 101.3 \text{ kPa} = 1.013 \text{ bar.}$$

  Having determined the pressure and enthalpy, the entropy is calculated using the S method where

  $$s_a = 6.6571 \text{ kJ/(kg K)}.$$

- **State 0a**
  The stagnation state "0a" is identical to the static state "a" since the velocity of the engine is zero. Therefore

  $$T_{0a} = T_a = 288.16 \text{ K}$$
  $$h_{0a} = h_a = 288.3 \text{ kJ/kg}$$
  $$p_{0a} = p_a = 101.3 \text{ kPa} = 1.013 \text{ bar}$$
  $$s_{0a} = s_a = 6.6571 \text{ kJ/(kg K)}.$$

- **State 01**

  The enthalpy at inlet in the engine, that is, at the face of the fan is

  $$h_{01} = h_{0a} = 288.3 \text{ kJ/kg}$$

  and the pressure is

  $$p_{01} = p_{0a}\eta_{\text{diff}} = 1.013 \times 0.98 = 0.993 \text{ bar.}$$

  The entropy is calculated using the S method, which yields

  $$s_{01} = 6.6629 \text{ kJ/(kg K).}$$

- **State 02i fan**

  The entropy does not change from state "01" to "02i fan" since the process is isentropic

  $$s_{02fi} = s_{0a} = 6.6629 \text{ kJ/(kg K)}$$

  and the pressure is

  $$p_{02fi} = p_{01}\pi_{0F} = 0.993 \times 1.8 = 1.787 \text{ bar.}$$

  The enthalpy is calculated using the H method, which yields

  $$h_{02fi} = 341.1 \text{ kJ/kg.}$$

  The ideal work of the fan is

  $$h_{02fi} - h_{01} = 341.1 - 288.3 = 52.8 \text{ kJ/kg.}$$

- **State 02 fan**

  The pressure at state "02 fan" is the same as at state "02i fan"

  $$p_{02f} = p_{02fi} = 1.787 \text{ bar.}$$

  The enthalpy is calculated by determining the work of the fan

  $$w_F = w_{F_i}/\eta_F = 52.8/0.85 = 62.12 \text{ kJ/kg}$$

  so that

  $$h_{02f} = h_{01} + w_F = 288.31 + 62.12 = 350.43 \text{ kJ/kg.}$$

  The entropy is then calculated using the S method which gives

  $$s_{02f} = 6.6900 \text{ kJ/(kg K).}$$

- **State 5i fan**

  The air leaving the fan expands down to atmospheric pressure so that

  $$p_{5fi} = 1.013 \text{ bar.}$$

  Since the ideal process between states "02 fan" and "5i fan" is isentropic,

  $$s_{5fi} = 6.6900 \text{ kJ/(kg K).}$$

Knowing the pressure and entropy, the enthalpy is calculated using the H method, which yields

$$h_{5fi} = 297.87 \text{ kJ/kg}.$$

The ideal velocity at the exit of the fan nozzle is

$$c_{5fi} = \sqrt{2(h_{02f} - h_{5fi})} = \sqrt{2 \times (350.43 - 297.87) \times 10^3} = 324.2 \text{ m/s}.$$

- **State 5 fan**
  The pressure at state "5 fan" is equal to that of state "5i fan". The enthalpy is obtained using the efficiency of the fan nozzle

$$\eta_{\text{fan nozzle}} = \frac{c_{5f}}{c_{5fi}} = \sqrt{\frac{h_{02f} - h_{5f}}{h_{02f} - h_{5fi}}}$$

so that

$$h_{5f} = h_{02f} - \eta_{\text{fan nozzle}}^2 (h_{02f} - h_{5fi}) = 350.43 - 0.99^2 \times (350.43 - 297.87) =$$
$$= 298.92 \text{ kJ/kg}.$$

The velocity at the exit from the fan nozzle is

$$c_{5f} = c_{5fi} \, \eta_{\text{fan nozzle}} = 324.2 \times 0.99 = 321.0 \text{ m/s}.$$

Knowing the pressure and the enthalpy, the entropy is calculated using the S method which yields

$$s_{5f} = 6.6934 \text{ kJ/(kg K)}.$$

- **State 02i**
  The pressure ratio across the compressor is

$$\pi_{0C} = OPR/\pi_{0F} = 25.2/1.8 = 14.$$

The pressure at inlet in the compressor is equal to the pressure at the exit of the fan. Therefore, the pressure at exit from the compressor is

$$p_{02} = p_{02fi}\pi_{0C} = 1.787 \times 14 = 25.017 \text{ bar}$$

which is also equal to

$$p_{02i} = p_{02} = 25.017 \text{ bar}.$$

The entropy at inlet in the compressor is assumed to be equal to the entropy at exit of the fan

$$s_{02i} = s_{02f} = 6.6900 \text{ kJ/(kg K)}.$$

Knowing the pressure and entropy, the enthalpy is calculated using the H method, which yields

$$h_{02i} = 742.9 \text{ kJ/kg}.$$

The ideal work in the compressor is

$$w_{C_i} = h_{02i} - h_{02f} = 742.87 - 350.43 = 392.44 \text{ kJ/kg}.$$

- **State 02**

  The pressure at state 02 is $p_{02} = 25.017$ bar and the enthalpy can be calculated using the compressor work

  $$w_C = w_{C_i}/\eta_C = 392.44/0.87 = 451.08 \text{ kJ/kg}$$

  so that

  $$h_{02} = h_{02f} + w_C = 350.43 + 451.08 = 801.51 \text{ kJ/kg}.$$

  Knowing the pressure and enthalpy, the entropy is calculated using the S method, which yields

  $$s_{02} = 6.7676 \text{ kJ/(kg K)}.$$

- **State 03**

  The pressure at state 03 is

  $$p_{03} = p_{02}(1 - 3\%) = 25.017 \times 0.97 = 24.267.$$

  The enthalpies of air and stoichiometric combustion gases corresponding to a turbine inlet temperature $T_{03} = 1430$ K are read from tables B and C so that

  $$h_{03_{air}} = 1551.2 \text{ kJ/kg}$$

  and

  $$h_{03_{\lambda=1}} = 1683.1 \text{ kJ/kg}.$$

  The excess air, λ, is calculated using

  $$\lambda = \frac{h_{03\lambda=1}(1 + minL) - \xi_{comb} \, LHV - h_{03air} \, minL}{minL(h_{02} - h_{03air})} \qquad (5.39)$$

  $$= \frac{1683.1 \times 15.66 - 0.9 \times 43,500 - 1551.2 \times 14.66}{14.66 \times (801.5 - 1551.2)} = 3.23306.$$

  The enthalpy of the combustion products is calculated using

  $$h_{03_\lambda} = r \, h_{03_{\lambda=1}} + q \, h_{03_{air}} \qquad (5.32)$$

  where the weighting functions are

  $$r = \frac{1 + minL}{1 + \lambda \, minL} = 0.3236$$

and

$$q = \frac{(\lambda - 1)\, minL}{1 + \lambda\, minL} = 0.6764,$$

which yields

$$h_{03_\lambda} = 1593.89 \text{ kJ/kg}.$$

Knowing the pressure and enthalpy, the entropy is calculated using the S method, which yields

$$s_{03} = 7.5880 \text{ kJ/(kg K)}.$$

- **State 04Hi**
  State "04Hi" defines the ideal state at the exit from the high-pressure turbine. The power needed to drive the compressor must be equal to the power of the high-pressure turbine during steady state operation

$$\mathcal{P}_C = \mathcal{P}_{HPT}$$

which yields

$$w_C \dot{m}_{aH} = w_{HPT} \dot{m}_g = w_{HPT} \dot{m}_{aH}(1 + f) = w_{HPT} \dot{m}_{aH}\left(1 + \frac{1}{\lambda\, minL}\right)$$

so that

$$w_{HPT} = w_C / \left(1 + \frac{1}{\lambda\, minL}\right) = 451.08 / \left(1 + \frac{1}{3.23306 \times 14.66}\right) =$$
$$= 441.76 \text{ kJ/kg}.$$

The ratio between the mass flow rate of fuel and the mass flow rate of hot air is $f = 0.0211$. The ideal work in the high-pressure turbine is

$$w_{HPTi} = w_{HPT}/\eta_{HPT} = 441.76/0.89 = 496.36 \text{ kJ/kg}.$$

The enthalpy can now be calculated from

$$h_{04Hi} = h_{03} - w_{HPTi} = 1593.89 - 496.36 = 1097.53 \text{ kJ/kg}.$$

The entropy at state "04i" is equal to that at state "03"

$$s_{04Hi} = 7.5880 \text{ kJ/(kg K)},$$

and the pressure is calculated using the P method, which yields

$$p_{04Hi} = 5.863 \text{ bar}.$$

- **State 04H**

  State "04H" defines the state at the exit from the high-pressure turbine. The pressure is the same as at state "04Hi" and the enthalpy is calculated using the work in the high-pressure turbine

  $$h_{04H} = h_{03} - w_{HPT} = 1593.89 - 441.76 = 1152.13 \text{ kJ/kg.}$$

  Knowing the pressure and the enthalpy, the entropy is calculated using the S method, which yields

  $$s_{04H} = 7.6402 \text{ kJ/(kg K).}$$

- **State 04Li**

  State "04Li" defines the ideal state at the exit from the low-pressure turbine. The entropy is equal to that of state "04H." The low-pressure turbine work is calculated using the equality between the fan power and the low-pressure turbine; at steady state operation

  $$\mathcal{P}_F = \mathcal{P}_{LPT}$$

  which yields

  $$w_F \dot{m}_{aC} = w_{LPT} \dot{m}_g = w_{LPT} \dot{m}_{aH}(1 + f)$$

  so that

  $$w_{LPT} = w_F \frac{1}{1 + f} \frac{\dot{m}_{aC}}{\dot{m}_{aH}} = w_F \frac{1}{1 + f} BPR = 62.12 \frac{1}{1 + 0.0211} 5.92 =$$
  $$= 360.15 \text{ kJ/kg.}$$

  The ideal work in the low-pressure turbine is

  $$w_{LPTi} = w_{LPT}/\eta_{LPT} = 360.15/0.89 = 404.66 \text{ kJ/kg}$$

  so that the enthalpy is

  $$h_{04Li} = h_{04H} - w_{LPTi} = 1152.13 - 404.66 = 747.47 \text{ kJ/kg.}$$

  Knowing the entropy and the enthalpy, the pressure is calculated using the P method, which yields

  $$p_{04Li} = 1.193 \text{ bar.}$$

- **State 04L**

  The pressure at state "04L" is the same as that at state "04Li". The enthalpy is calculated using the low-pressure turbine work

  $$h_{04L} = h_{04H} - w_{LPT} = 1152.13 - 360.15 = 791.98 \text{ kJ/kg.}$$

  Knowing the pressure and the enthalpy, the entropy is calculated using the S method, which yields

  $$s_{04L} = 7.7005 \text{ kJ/(kg K).}$$

- **State 5i**
  State "5i" defines the ideal state at the exit from the core engine nozzle. The entropy is the same as that of state "04L." The pressure is atmospheric pressure, so that $p_{5i} = 1.013$ bar. Knowing the entropy and the pressure, the enthalpy is calculated using the H method which yields

  $$h_{5i} = 757.32 \text{ kJ/kg.}$$

  The ideal velocity of the gases at the exit of the core engine nozzle is

  $$c_{5i} = \sqrt{2(h_{04L} - h_{5i})} = \sqrt{2 \times (791.98 - 757.32) \times 10^3} = 263.27 \text{ m/s.}$$

- **State 5**
  State "5" defines the state at the exit from the core engine nozzle. The pressure is the same as that of state "5i." The velocity at the exit of the nozzle is

  $$c_5 = c_{5i} \, \eta_{\text{nozzle}} = 263.27 \times 0.98 = 258.01 \text{ m/s.}$$

  The enthalpy at nozzle exit is

  $$h_5 = h_{04L} - c_5^2/2 = 791.98 - 258.01^2/2 \times 10^{-3} = 758.69 \text{ kJ/kg.}$$

  Knowing the pressure and the enthalpy, the entropy is calculated using the S method, which yields

  $$s_5 = 7.7027 \text{ kJ/(kg K).}$$

  Having calculated all the states of the real cycle, the thrust results from

  $$T = (\dot{m}_{aH} + \dot{m}_f)u_{e_H} - \dot{m}_{aH}u + \dot{m}_{aC}(u_{e_C} - u) \tag{6.10}$$

  where $u_{e_H} \equiv c_5$ and $u_{e_C} \equiv c_{5f}$. The total mass flow rate of air entering the engine is $\dot{m}_a = \dot{m}_{aC} + \dot{m}_{aH} = 520$ kg/s and the bypass ratio $BPR = \dot{m}_{aC}/\dot{m}_{aH} = 5.92$ so that

  $$\dot{m}_{aC} = \dot{m}_a \frac{BPR}{BPR + 1} = 590 \frac{5.92}{6.92} = 504.74 \text{ kg/s}$$

  and

  $$\dot{m}_{aH} = \dot{m}_a \frac{1}{BPR + 1} = 590 \frac{1}{6.92} = 85.26 \text{ kg/s.}$$

  The mass flow rate of fuel

  $$\dot{m}_f = \dot{m}_{aH} f = 85.26 \times 0.0211 = 1.8 \text{ kg/s.}$$

  The total thrust of the engine at takeoff, $i.e.$, $u = 0$, is

  $$T = (85.26 + 1.8) \times 258.01 + 504.74 \times 320.98 = 184,474 \text{ N} = 41,500 \text{ lbf.}$$

  The thrust generated by the fan at takeoff is

  $$T_{\text{fan}} = 504.74 \times 320.98 = 162,012 \text{ N}$$

and the thrust generated by the core engine at takeoff is

$$\mathcal{T}_{core} = (85.26 + 1.8) \times 258.01 = 22,462 \text{ N}.$$

At takeoff the fan generates 162,012/184,474 = 87.8% of the total thrust.
    The power produced by the low-pressure turbine in order to drive the fan is

$$\mathcal{P}_F = \mathcal{P}_{LPT} = (\dot{m}_{aH} + \dot{m}_f)\, w_{LPT} = (85.26 + 1.8) \times 360.15 = 31,354 \text{ kW}$$

and the power produced by the high-pressure turbine in order to drive the compressor is

$$\mathcal{P}_C = \mathcal{P}_{HPT} = (\dot{m}_{aH} + \dot{m}_f)\, w_{HPT} = (85.26 + 1.8) \times 441.76 = 38,459 \text{ kW}.$$

    The thrust specific fuel consumption is

$$\text{TSFC} = 1.8/184,474 = 9.75 \times 10^{-6} \text{ kg/(N s)} = 9.75 \text{ g/(kN s)} =$$
$$= 3.51 \times 10^{-2} \text{ kg/(N hr)}.$$

The thrust reported in the literature for the CF6-6D engine is 41,500 lbf, which is identical to the thrust calculated above. The thrust specific fuel consumption reported in the literature is 9.9 g/(kN s), which is close to the value calculated, 9.75 g/(kN s).

Figure 6.31 shows the Mollier diagram of the two-spool turbofan engine drawn to scale. While the entropy increase in the fan is small compared to the entropy increase in the core engine, the mass flow rate of air through the fan is 5.92 larger than that through the core engine. As a result, the thrust produced by the fan is 7.2 times larger than the thrust produced by the core engine.

Figure 6.31 Mollier diagram of a two-spool turbofan engine.

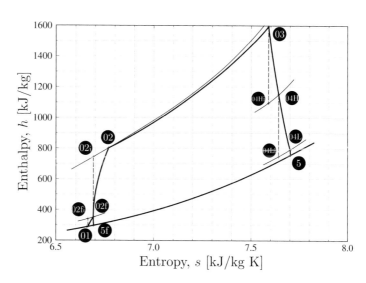

6.8   Turboprop and Turboshaft

6.8.1   Introduction

Turboprop and turboshaft engines are gas turbine engines that transmit all the useful power via a shaft. Turboprop engines are used to provide power for airplanes. For a turboprop, propulsion is achieved by the combined action of a propeller and the thrust produced by the exhaust gases. In the case of a turboshaft, the gases in the turbine are expanded to a lower pressure than in the turboprop, therefore having a small exhaust velocity but a large shaft power. Turboshaft engines are used to provide power for helicopters, ships, tanks, and stationary equipment.

Turboprop and turboshaft engines were developed from turbojet engines by adding a power turbine and a gearbox to reduce the angular velocity of the power shaft. The turbine of a turbojet engine extracts only the power needed to drive the compressor and the accessories, leaving the energy of the exhaust gases to produce thrust. The turbines of a turboprop/turboshaft, however, not only extract power to drive the compressor and the accessories, but also provide the power to the propeller shaft. Because of the propeller, gearbox, and power turbine, the weight of a turboprop engine could be 50% larger than that of a turbojet with the same gas generator.

Typically 90% of the turboprop thrust is due to the propeller, while the rest comes from the exhaust gases. This split between the thrust produced by the propeller and the exhaust gases varies with altitude and air speed, as well as with the engine operating parameters. At higher speed, the thrust produced by the jet increases. At lower speed, the thrust produced by the propeller increases. The thrust produced by the propeller is directly proportional to the propeller efficiency. Propeller efficiency is typically flat up to an airplane Mach number of 0.5, after which it falls rapidly. A good estimate of propeller efficiency for Mach numbers less than 0.5 is $\eta_{pr} = 80\%$.

The thrust produced by the propeller is the result of accelerating a large mass flow rate by a small amount. This is more efficient than accelerating a small mass flow rate by a large amount, as happens with the exhaust gases. By comparison, the turbojet engine achieves all its thrust by accelerating a small mass flow rate by a large amount. Consequently, at takeoff and low flight speeds, the turboprop is more efficient than a turbojet and can compensate for the additional weight due to the propeller, gearbox, and power turbine.

The turbojet and turbofan are rated in thrust while the turboprop is rated in shaft horsepower and thrust. To be able to compare the turboprop against the other gas turbine engines, a conversion between shaft horsepower (SHP or $\mathcal{P}_s$) and thrust is needed. This conversion is also needed to account for the fact that part of the thrust of the turboprop/turboshaft is due to the propeller and part to the exhaust gases.

For static conditions,

1 shaft horsepower (or 736 W) $\equiv$ 2.5 lbf of thrust (or 11.1 N).

As a result, one can define the *static equivalent shaft horsepower* (ESHP$_{\text{static}}$ or $\mathcal{P}_{\text{es}}$) as:

$$\boxed{\text{ESHP}_{\text{static}} = \text{SHP}_{\text{propeller}} + \frac{T_{\text{jet}}}{2.5}},$$

where T_{jet} is in lbf. The *in-flight equivalent shaft horsepower* (ESHP$_{\text{in-flight}}$) is

$$\boxed{\text{ESHP}_{\text{in-flight}} = \text{SHP}_{\text{propeller}} + \frac{T_{\text{jet (net)}}}{2.5}},$$

where T_{jet} is in lbf.

When the thrust is given in N, the equivalent shaft horsepower (in hp) is defined as

$$\boxed{\text{ESHP}_{\text{static}} = \text{SHP}_{\text{propeller}} + \frac{T_{\text{jet}}}{11.1}}.$$

Note that the power at propeller is

$$\text{SHP}_{\text{propeller}} = \text{SHP}\,\eta_{\text{pr}}.$$

Recall that the equivalent brake specific fuel consumption (EBSFC) is

$$\text{EBSFC} = \frac{\dot{m}_{\text{f}}}{\text{ESHP}}.$$

6.8.2 Turboprop and Turboshaft Configurations

There are minor differences between the turboshaft and the turboprop versions of an engine. Usually the turboprop version is reinforced more than the turboshaft version because it supports the gearbox and the propeller. For the turboshaft, the gearbox is separate. Because of the similarities between turboprop and turboshaft engines, the rest of this section will discuss only turboprop engines.

Two configurations of turboprop engines are most common: (1) split shaft or free power turbine, and (2) fixed shaft. The main difference between these configurations is whether the power turbine is free or not. Figure 6.32 shows a turboprop with free power turbine. In this case, the shaft of the gas generator is not mechanically connected to the power turbine. This has the advantage that the free power turbine can rotate at a different angular velocity than the gas generator.

Figure 6.33 shows the Allison 501-D13 turboprop engine. This is a fixed-shaft turboprop engine where the shaft of the propeller is mechanically connected to the gas generator. The large dimensions of the gearbox used in this engine are noteworthy.

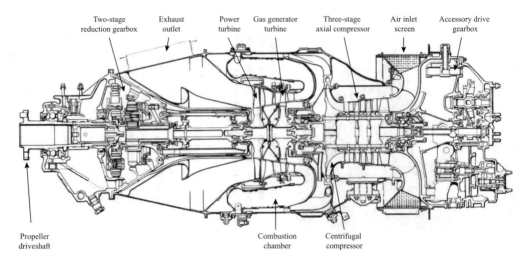

Figure 6.32 Turboprop with free power turbine: Pratt & Whitney PT6 (reproduced with permission from Pratt & Whitney).

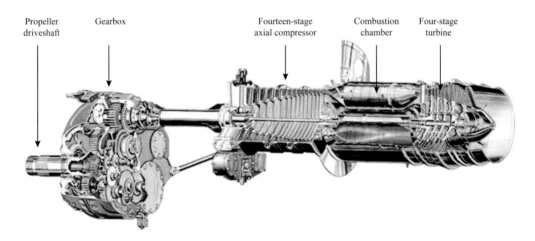

Figure 6.33 Fixed-shaft turboprop: Allison 501-D13 (courtesy of Rolls-Royce).

6.8.3 Real Cycle Analysis

For the real cycle analysis we will consider a turboprop whose gas generator is identical to that of the turbojet example 6.6.4.

Example 6.8.1 A turboprop engine operates in an engine test cell. The mass flow rate of air entering the engine is 20 kg/s. The compressor has an inlet stagnation temperature of 288.16 K and an inlet stagnation pressure of 101.3 kPa. The compressor has a pressure ratio of 12 and an efficiency of 87%. The combustor efficiency is 97%. The pressure drop in the combustor is 3%. The turbine inlet temperature is 1400 K. The core engine turbine efficiency of 89%.

The power turbine efficiency is 88%. The nozzle efficiency is 97%. The propeller efficiency is 80%. The shaft power is 6,000 kW. The correction of the entropy $\Delta s'$ can be neglected. The engine uses standard fuel, which has a lower heating value of 43,500 kJ/kg, and $minL = 14.66$.

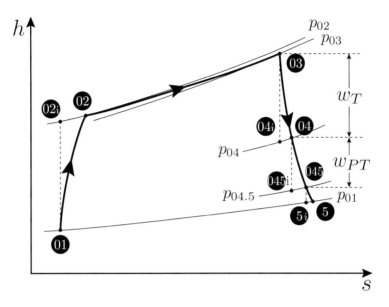

Figure 6.34 Turboprop real cycle.

1. What is the equivalent power?
2. What is the equivalent fuel consumption?
3. What is the ratio propeller thrust/exhaust gases thrust?

Solution

The states 01, 02i, 02, 03, 04, given in Table 6.6 are identical to those calculated in the turbojet example 6.6.4. In addition, the value of excess air, $\lambda = 2.984$, and the weighting functions $r = 0.35$ and $q = 0.65$, are also the same as those of example 6.6.4.

- **State 04.5i**

  The process between states 04 and 04.5i is isentropic

  $s_{04.5i} = s_{04} = 7.8160$ kJ/(kg K).

  The specific work of the power turbine is calculated using the shaft horse power of the power turbine

  $$w_{PT} = \frac{SHP}{\dot{m}_g} = \frac{SHP}{\dot{m}_a(1+f)} \overset{(5.13)}{=} \frac{SHP}{\dot{m}_a \times \left(1 + \frac{1}{\lambda \times minL}\right)} =$$

  $$= \frac{6000}{20\left(1 + \frac{1}{2.984 \times 14.66}\right)} = 293.3 \text{ kJ/kg}.$$

Table 6.6 Turboprop states corresponding to gas generator.

State	h [kJ/kg]	s [kJ/(kg K)]	p [bar]
01	288.3	6.6571	1.013
02i	586.5	6.6571	12.156
02	631.1	6.7315	12.156
03	1559.8	7.7783	11.791
04i	1183.5	7.7783	4.093
04	1225.0	7.8160	4.093

The ideal specific work of the power turbine is obtained using the given efficiency of the power turbine

$$w_{PTi} = \frac{w_{PT}}{\eta_{PT}} = \frac{293.3}{0.88} = 333.3 \text{ kJ/kg}.$$

The enthalpy of state 04.5i is then calculated using the ideal specific work of the power turbine and the enthalpy at state 04

$$h_{04.5i} = h_{04} - w_{PTi} = 1224.9 - 333.3 = 891.6 \text{ kJ/kg}.$$

Using the excess air, λ, and the enthalpy, $h_{04.5i}$, the temperature of the mixture is calculated following an iterative process described in example 5.3.2, which yields $T_{04.5_i} = 845.2$ K.

Having determined the entropy and temperature, the pressure is calculated using the P method. Note that for combustion products, which are modeled as a mixture of air and stoichiometric combustion products, the P method must use temperature instead of enthalpy. This is because the components of the mixture have a common temperature as opposed to a common enthalpy.

Given the temperature $T_{04.5i}$, the entropies of air and stoichiometric combustion products are determined from the tables:

$$T_{04.5i} \xrightarrow{\text{table}} \begin{cases} s_{04.5_{i_{p=1,\,a}}} = & 7.7786 \text{ kJ/(kg K)} \\ s_{04.5_{i_{p=1,\,\lambda=1}}} = & 8.0717 \text{ kJ/(kg K)} \end{cases}$$

and then the entropy of the mixture at pressure of 1 bar is obtained:

$$s_{04.5i_{p=1,\,\lambda}} = q\, s_{04.5i_{p=1,a}} + r\, s_{04.5i_{p=1,\lambda=1}} =$$
$$= 0.65 \times 7.7786 + 0.35 \times 8.0717 = 7.8812 \text{ kJ/(kg K)}.$$

Using the entropy variation equation (2.93) yields the pressure

$$p_{04.5i} = \exp[-(s_{04.5i} - s_{04.5_{i_{p=1\,\text{bar}}}})/R] = \exp[-(7.8160 - 7.8812)/0.28716] =$$
$$= 1.255 \text{ bar}.$$

- **State 04.5**

The pressure at state 04.5 is equal to that at state 04.5i

$$p_{04.5} = p_{04.5i} = 1.255 \text{ bar.}$$

The enthalpy at state 04.5 is calculated using the specific work in the power turbine

$$h_{04.5} = h_{04} - w_{PT} = 1224.9 - 293.3 = 931.6 \text{ kJ/(kg K).}$$

Using the excess air, λ, and the enthalpy, $h_{04.5}$, the temperature of the mixture is calculated using the iterative process described in example 5.3.2, which yields $T_{04.5} = 880.0$ K.

Having determined the pressure and temperature, the entropy is calculated using the S method:

$$T_{04.5} \xrightarrow{\text{table}} \begin{cases} s_{04.5_{p=1, \, a}} = & 7.8235 \text{ kJ/(kg K)} \\ s_{04.5_{p=1 \text{ bar}, \, \lambda=1}} = & 8.1207 \text{ kJ/(kg K).} \end{cases}$$

The entropy of the mixture at pressure of 1 bar is calculated as

$$s_{04.5_{p=1 \text{ bar}, \, \lambda}} = q \, s_{04.5_{p=1 \text{ bar}, a}} + r \, s_{04.5_{p=1 \text{ bar}, \lambda=1}} =$$
$$= 0.65 \times 7.8235 + 0.35 \times 8.1207 = 7.9275 \text{ kJ/(kg K)}$$

and the entropy at pressure $p = 1.258$ bar is calculated using (2.93)

$$s_{04.5} = s_{04.5_{p=1 \text{ bar}, \lambda}} - R \ln p_{04.5} = 7.9275 - 0.28716 \ln 1.255 = 7.8622 \text{ kJ/(kg K).}$$

- **State 5i**

The process from state 04.5 to 5i is isentropic

$$s_{5i} = s_{04.5} = 7.8622 \text{ kJ/(kg K)}$$

and the pressure is equal to atmospheric pressure

$$p_{5i} = p_a = p_{01} = 1.013 \text{ bar.}$$

Having determined the entropy and pressure, the temperature is calculated next. The temperature, T, must satisfy the equation

$$f(T) \equiv s_{5i} - q s_{5i_a}(T) - r s_{5i_{\lambda=1}}(T) = 0 \qquad (6.65)$$

where s_{5i_a} and $s_{5i_{\lambda=1}}$ are the entropies at pressure p_{5i} for air and stoichiometric products. These entropies are then written as a function of the entropies at the pressure of 1 bar

$$s_{5i_a} = s_{5i_{a,p=1 \text{ bar}}} - R \ln p_{5i}$$
$$s_{5i_{\lambda=1}} = s_{5i_{\lambda=1,p=1 \text{ bar}}} - R \ln p_{5i}$$

and the function $f(T)$ of (6.65) becomes

$$f(T) \equiv s_{5i} - q s_{5i_{a,p=1 \text{ bar}}}(T) - r s_{5i_{\lambda=1,p=1 \text{ bar}}}(T) - R \ln p_{5i} = 0 \qquad (6.66)$$

since $r + q = 1$. For a given temperature T, the values of $s5_{i_{a,p=1 \text{ bar}}}$ and $s5_{i_{\lambda=1,p=1 \text{ bar}}}$ are obtained from the tables. The temperature that satisfies (6.66) is determined using an iterative process.

An initial guess of temperature T is obtained by calculating the entropy at pressure of 1 bar using (2.93)

$$s5_{i_{p=1 \text{ bar}}} = s5_i + R \ln p_{5i} = 7.8622 + 0.28716 \ln 1.013 = 7.8659 \text{ kJ/(kg K)},$$

which yields the temperatures for air and stoichiometric combustion products

$$s5_{i_{p=1 \text{ bar}}} \xrightarrow{\text{table}} \begin{cases} T5_{i_a} = & 913.9 \text{ K} \\ T5_{i_{\lambda=1}} = & 711.6 \text{ K.} \end{cases}$$

An initial guess for T_{5i} is approximated as

$$T_{5i} = q\, T5_{i_a} + r\, T5_{i_{\lambda=1}} =$$
$$= 0.65 \times 913.9 + 0.35 \times 711.6 = 843.1 \text{ K.}$$

The result of the iteration process is $T_{5i} = 834.4$ K. The enthalpy is then calculated as

$$h_{5i} = q\, h5_{i_a}(T_{5i}) + r\, h5_{i_{\lambda=1}}(T_{5i}) =$$
$$= 0.65 \times 859.8 + 0.35 \times 915.2 = 879.2 \text{ kJ/kg.}$$

Assuming a fixed exit nozzle where the heat transfer is small compared to the enthalpy variation across the nozzle allows us to calculate the velocity of the fluid at state 5i from

$$h_{04.5} - h_{5i} = \frac{c_{5i}^2}{2}$$

so that

$$c_{5i} = \sqrt{2(h_{04.5} - h_{5i})} = \sqrt{2(931.6 - 879.2) \times 10^3} = 323.7 \text{ m/s.}$$

Using the nozzle efficiency

$$\varphi_{\text{nozzle}} = \frac{c_5}{c_{5i}},$$

the velocity of the gases at exit from the nozzle is

$$c_5 = c_{5i} \times \varphi_{\text{nozzle}} = 323.7 \times 0.97 = 314.0 \text{ m/s.}$$

The equivalent shaft power is

$$P_{\text{es}} = \eta_{\text{propeller}} \times P_{\text{s}} + \frac{\mathcal{T}}{11.1},$$

where

$$\mathcal{T} = \dot{m}_g \times c_5 = \dot{m}_a(1 + f)c_5 = \dot{m}_a \left(1 + \frac{1}{\lambda \times minL}\right) c_5$$

$$= 20 \times \left(1 + \frac{1}{2.984 \times 14.66}\right) \times 314.0 = 6423.9 \text{ N}$$

so that

$$\mathcal{P}_{es} = 0.8 \times 6000/0.736 + \frac{6423.9}{11.1} = 7100 \text{ hp} = 5226 \text{ kW}.$$

The mass flow rate of fuel is

$$\dot{m}_f = \dot{m}_a \times \frac{1}{\lambda \times minL} = 20 \times \frac{1}{2.984 \times 14.66} = 0.457 \text{ kg/s} = 1646 \text{ kg/h}$$

so that the equivalent brake specific fuel consumption is

$$\text{EBSFC} = \frac{\dot{m}_f}{\mathcal{P}_{es}} = \frac{1646}{5226} = 0.315 \text{ kg/kWh}.$$

Note that the typical range for EBSFC is (0.27,0.36) kg/kWh. The ratio between the thrust generated by the propeller and the thrust generated by the jet is

$$\frac{\eta_{pr}\mathcal{P}_s}{\mathcal{T}/11.1} = \frac{6521.7}{578.7} = 11.27,$$

that is, 8.2% of the thrust is produced by the exhaust gas.

6.9 Ramjet

6.9.1 Introduction

The ramjet, invented by René Lorin in 1913, is the simplest air-breathing engine. The ramjet uses the dynamic pressure of the incoming air and therefore does not need a compressor to raise the pressure at inlet in the combustor. Since a compressor is not needed, a turbine to power the compressor is also not needed. As a result, the ramjet has no rotating parts and consists only of an inlet diffuser, a combustion chamber, and an exhaust nozzle, as shown in Fig. 6.35.

Air is compressed in the inlet diffuser and then mixed with fuel and burned in the combustion chamber. The combustion gases are then accelerated in the exit nozzle by the pressure difference between the burner and the ambient surroundings.

The ramjet relies on the dynamic pressure at the inlet diffuser in order to operate. Therefore, since the pressure ratio is strictly dependent on the flight speed and inlet diffuser performance, the ramjet cannot develop static thrust. As a result, the ramjet cannot accelerate a vehicle that is stationary.

The ramjet can operate at subsonic flight speeds; however, it is more advantageous if it flies at supersonic speeds because a higher pressure can be achieved in the combustor. The flow at inlet in the combustor must be decelerated to a Mach number of approximately 0.2 to 0.3. If the ramjet operates in supersonic flight, this deceleration is done through a system of shock waves. As a result of the deceleration, the pressure in the combustor increases substantially. For example, assuming that the flow is decelerated from M = 3.5 to M = 0.25 through an

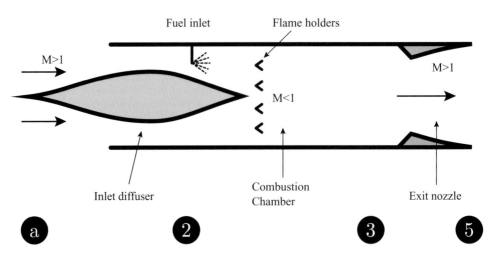

Figure 6.35 Conventional ramjet configuration.

isentropic process, the static pressure in the combustor would be 73 times higher than the ambient pressure. In the real compression process, however, the pressure increase is only a fraction of that of the isentropic process because of the losses in the shock wave system.

Fuel is added to the high-pressure air in the combustor, and the mixture burns with the flames anchored on the flame holders. The temperature of the mixture increases and the combustion products expand in the exit nozzle. The thrust (6.7) generated by the momentum of the working fluid is applied by the pressure and shear forces acting on the internal surfaces of the ramjet.

Having a much simpler configuration than that of a gas turbine, the ramjet can be better cooled than the turbine blades of the gas turbine engine. The cooling of the ramjet is done by injecting fuel along the walls in order to create a shielding layer. It is much easier to cool the fixed walls of the ramjet than the gas turbine rotating blades subjected to large centrifugal forces. As a result, the maximum temperature in the ramjet is higher than the turbine inlet temperature of a gas turbine engine. Having a higher maximum temperature than that of the gas turbine engine is beneficial for the ramjet performance. However, if the temperature exceeds 2500 K, dissociation of the combustion products becomes significant and can penalize performance.

6.9.2 Ramjet Configurations

The ramjet engine can have two configurations: (i) subsonic-combustion ramjet, aka conventional ramjet, and (ii) supersonic-combustion ramjet, aka scramjet. At flight speeds exceeding Mach 3, the conventional ramjet engine is more efficient than a jet engine using rotating machinery. As the flight speed increases, however, the losses due the terminal shock needed for subsonic combustion increase significantly. In addition, the temperature increases to the point that the completion of recombination reactions is inhibited, which leads to large energy

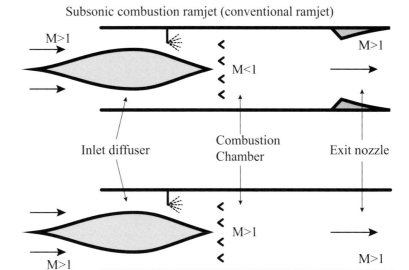

Figure 6.36 Schematic configurations of subsonic and supersonic combustion ramjet engines.

losses. To avoid these issues, supersonic flow should be maintained throughout the engine so that the heat addition is done at supersonic speed.

The two configurations, the conventional ramjet and the scramjet, are shown in Fig. 6.36. The conventional ramjet has a physical throat that is required by the subsonic conditions in the burner in order to maintain the inlet operational conditions. The scramjet, on the other hand, needs an area increase as combustion occurs. For this reason, on the scramjet the mechanical throat is substituted by a thermal throat that decelerates the flow through tailored heat release. In spite of the benefits of the scramjet over the conventional ramjet at high Mach numbers, the scramjet presents considerable technological difficulties [Curran, 2001; Segal, 2009].

6.9.3 Real Cycle Analysis

Figure 6.37 shows the $h - s$ diagram of the real process in the ramjet. The air is compressed in the inlet diffuser from the static state "a" to stagnation state 02. The air–fuel mixture is combusted in the burner, a process that takes place between stagnation states 02 and 03. The gases then expand in the exit nozzle between stagnation state 03 and static state 5. Note that an alternative notation for state 5 is "e," for exit. The losses in the diffuser, combustor, and nozzle are measured by the ratio of stagnation states:

Diffuser: $\sigma_{0d} = p_{02}/p_{0a}$
Combustor: $\sigma_{0c} = p_{03}/p_{02}$
Nozzle: $\sigma_{0n} = p_{05}/p_{03}.$

Figure 6.37 Ramjet real cycle.

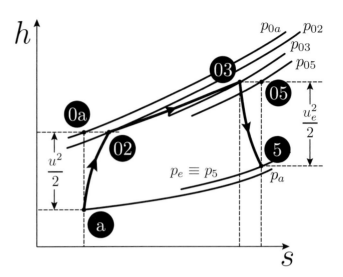

Assuming that the ratio of specific heat capacities, γ, is constant in the ramjet and using the equation for isentropic processes (3.33) yields

$$\frac{p_{0a}}{p_a} = \left(1 + \frac{\gamma - 1}{2} M^2\right)^{\frac{\gamma}{\gamma - 1}} \tag{6.67}$$

$$\frac{p_{05}}{p_5} = \left(1 + \frac{\gamma - 1}{2} M_e^2\right)^{\frac{\gamma}{\gamma - 1}}. \tag{6.68}$$

Combining (6.67) and (6.68) yields

$$M_e^2 = \frac{2}{\gamma - 1}\left[\left(1 + \frac{\gamma - 1}{2} M^2\right)\left(\frac{p_{05}\, p_a}{p_{0a}\, p_5}\right)^{\frac{\gamma - 1}{\gamma}} - 1\right]. \tag{6.69}$$

Using $p_{05}/p_{0a} = \sigma_{0d}\sigma_{0c}\sigma_{0n}$ in (6.69) yields

$$M_e = \sqrt{\frac{2}{\gamma - 1}\left[\left(1 + \frac{\gamma - 1}{2} M^2\right)\left(\sigma_{0d}\sigma_{0c}\sigma_{0n}\frac{p_a}{p_5}\right)^{\frac{\gamma - 1}{\gamma}} - 1\right]}. \tag{6.70}$$

Using the notation

$$\beta = \left(1 + \frac{\gamma - 1}{2} M^2\right)\left(\sigma_{0d}\sigma_{0c}\sigma_{0n}\frac{p_a}{p_5}\right)^{\frac{\gamma - 1}{\gamma}},$$

(6.70) becomes

$$M_e = \sqrt{\frac{2}{\gamma - 1}(\beta - 1)}. \tag{6.71}$$

Assuming the heat transfer from the engine can be neglected, the exit velocity, that is, the velocity at state 5 is

$$u_e = M_e \sqrt{\gamma R T_5} \overset{T_{03}=T_{05}}{=} M_e \sqrt{\gamma R \frac{T_{03}}{1 + \frac{\gamma-1}{2}M_e^2}} \overset{(6.71)}{=} \sqrt{\frac{2\gamma}{\gamma-1} \frac{\beta-1}{\beta} R T_{03}}. \tag{6.72}$$

To find the fuel to air ratio, f, let us use the energy conservation equation in the combustor

$$\dot{m}_a h_{02} + \dot{m}_f (\xi_{comb} LHV + h_f) = (\dot{m}_a + \dot{m}_f) h_{03}. \tag{5.30}$$

Since the fuel enthalpy h_{fuel} is much smaller than the fuel lower heating value, (5.30) yields

$$h_{02} + f \xi_{comb} LHV = (1 + f) h_{03} \tag{6.73}$$

and therefore

$$f = \frac{h_{03} - h_{02}}{\xi_{comb} LHV - h_{03}} \overset{h_{02}=h_{0a}}{=} \frac{h_{03}/h_{0a} - 1}{\xi_{comb} LHV/h_{0a} - h_{03}/h_{0a}} =$$

$$= \frac{T_{03}/T_{0a} - 1}{\xi_{comb} LHV/c_p T_{0a} - T_{03}/T_{0a}}. \tag{6.74}$$

The specific thrust

$$T_{sp} = u_e(1 + f) - u + \frac{A_e}{\dot{m}_a}(p_e - p_a), \tag{6.30}$$

with the third term written as in (6.31) becomes

$$T_{sp} = (1 + f) \sqrt{\frac{2\gamma}{\gamma-1} \frac{\beta-1}{\beta} R T_{03}} \left[1 + \frac{\gamma-1}{2\gamma} \frac{1}{\beta-1} \left(1 - \frac{p_a}{p_5} \right) \right] - M \sqrt{\gamma R T_{0a}}. \tag{6.75}$$

Figure 6.38 shows the variation with Mach number of the three terms of the specific thrust equation: $u_e(1 + f)$, u, and $A_e(p_e - p_a)/\dot{m}_a$. It is apparent that $u_e(1 + f)$ reaches an asymptotic value as Mach number increases, while u varies linearly with Mach number. The value of $A_e(p_e - p_a)/\dot{m}_a$ is much smaller than the other two terms, and for this reason a different scale was used to show its variation. The $A_e(p_e - p_a)/\dot{m}_a$ term decreases with Mach number.

Figure 6.39 shows the variation of specific thrust and thrust specific fuel consumption vs. Mach number for a ramjet operating at three T_{03} temperatures: 2000 K, 2400 K, and 2800 K. It was assumed that the inlet temperature was $T_a = 216$ K, the pressure losses were $\sigma_{0d} = 0.7$, $\sigma_{0c} = 0.95$, $\sigma_{0n} = 0.98$ and the combustion efficiency was $\xi_{comb} = 0.96$. The pressure ratio at the exit nozzle was $p_a/p_5 = 0.99$ and the lower heating value of the fuel was $LHV = 43,500$ kJ/kg. The specific thrust increased with T_{03}, reaching its maximum value at M near 3. As the temperature T_{03} increased, the Mach number corresponding to the maximum specific thrust also increased.

The thrust specific fuel consumption decreased with increasing Mach number. The fuel consumption reached an asymptotic value at Mach numbers larger than 3. There was a small variation of the thrust specific fuel consumption with temperature T_{03}.

Figure 6.38 Variation of specific thrust terms with Mach number for a ramjet operating with $T_a = 216$ K, $T_{03} = 2400$ K, $\sigma_{0d} = 0.7$, $\sigma_{0c} = 0.95$, $\sigma_{0n} = 0.98$, $\xi_{comb} = 0.96$, $p_a/p_5 = 0.99$, $\gamma = 1.4$ and $LHV = 43,500$ kJ/kg.

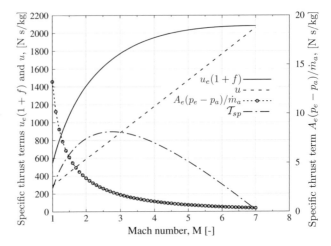

Figure 6.39 Specific thrust and thrust specific fuel consumption vs. Mach number for a ramjet operating with $T_a = 216$ K, $\sigma_{0d} = 0.7$, $\sigma_{0c} = 0.95$, $\sigma_{0n} = 0.98$, $\xi_{comb} = 0.96$, $p_a/p_5 = 0.99$, $\gamma = 1.4$ and $LHV = 43,500$ kJ/kg.

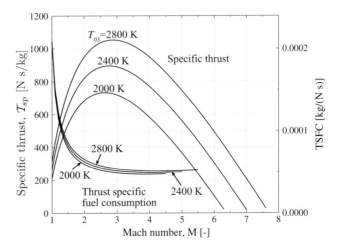

Figure 6.40 shows the variation of propulsion, thermal, and overall efficiencies, η_p, η_{th}, and η_o with Mach number. The figure also shows the variation of fuel to air ratio, f, with Mach number. The propulsion efficiency has a minimum value at approximately $M = 1.5$ after which it increases with Mach number. The thermal and overall efficiencies increase with Mach number over the entire Mach number range. The fuel to air ratio is much smaller than 1 and decreases with Mach number.

It is important to mention that the results of the cycle analysis presented above assume that the pressure loss in the diffuser, σ_{0d}, is not affected by the Mach number. It is expected, however, that these losses be strongly dependent on the Mach number because the shock system changes when Mach number changes. For this reason, the diffuser should have a variable geometry to reduce the pressure losses as the vehicle operates over a range of Mach numbers.

Figure 6.40 Efficiencies and fuel to air ratio variation with Mach number for a ramjet operating with $T_a = 216$ K, $\sigma_{0d} = 0.7$, $\sigma_{0c} = 0.95$, $\sigma_{0n} = 0.98$, $\xi_{comb} = 0.96$, $p_a/p_5 = 0.99$, $\gamma = 1.4$ and $LHV = 43,500$ kJ/kg.

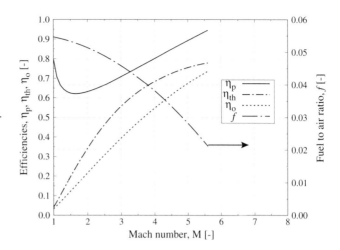

Problems

1. A turbojet engine operates on an airplane that flies at altitude $H = 6000$ m and Mach number, M $= 0.6$. The compressor pressure ratio is $\pi_{0C} = 15$. The efficiency of the compressor is 85%. Air enters in the compressor at a rate of 50 kg/s. The pressure drop in the combustor is 3%. The turbine inlet temperature is 1400 K and the turbine efficiency is 91%. The nozzle exit gas pressure is 1% above atmospheric pressure and the exit nozzle efficiency is 97%.

   (a) What are the net power, cycle thermal efficiency, and specific thrust if one assumes air standard with variable properties for the working fluid in the engine;
   (b) What are the net power, cycle thermal efficiency, and specific thrust if one assumes an ideal (Brayton) cycle;
   (c) Draw to scale, on the same plot, the Mollier ($h - s$) diagrams of the two cycles.

2. The work in the compressor of a turbojet engine is 342.66 kJ/kg. The pressure at the exit of the compressor is 12.156 bar. The pressure drop in the combustor is 3%. The turbine inlet temperature is 1400 K. The turbine has an efficiency of 89%. What are the temperature and pressure at the exit of the turbine assuming the working fluid in the engine is air with variable properties?

3. A turbojet engine operates on an airplane that flies at altitude $H = 8000$ m and Mach number, M $= 0.7$. The compressor pressure ratio is $\pi_{0C} = 9$. The efficiency of the compressor is 85%. Air enters in the compressor at a rate of 50 kg/s. The pressure drop in the combustor is 4% and the combustor efficiency is 96%. The turbine inlet temperature is 1400 K and the turbine efficiency is 89%. The nozzle exit gas pressure is 1% above the compressor inlet pressure and the exit nozzle efficiency is 97%.

   (a) What are the net power, cycle thermal efficiency, specific thrust, and thrust specific fuel consumption if one assumes actual medium for the working fluid in the engine.
   (b) Draw to scale the Mollier ($h - s$) diagram of the cycle.

4. The stagnation temperature and stagnation pressure at the exit of the compressor are $T_{02} = 700$ K and $p_{02} = 8$ bar, respectively. The pressure drop in the burner is 3%. The excess air in the burner is $\lambda = 3$. The burner efficiency is 90%. The entropy correction $\Delta s'$ is neglected. Standard fuel is used in the engine.
   (a) What is the enthalpy of the gases at the exit from the combustor?
   (b) What is the turbine inlet temperature?
   (c) How much heat has been added in the combustor?
   (d) What is the entropy increase in the combustor?

5. A turbojet engine operates on an airplane that is ready to take off. The compressor pressure ratio is $\pi_{0C} = 10$. The efficiency of the compressor is 89%. Air enters in the compressor at a rate of 20 kg/s. The pressure drop in the combustor is 4% and the combustor efficiency is 97%. The turbine inlet temperature is 1400 K and the turbine efficiency is 92%. The exit nozzle efficiency, φ, is 97%.
   (a) What are the net power, cycle thermal efficiency, specific thrust, and thrust specific fuel consumption if one assumes actual medium for the working fluid in the engine.
   (b) Draw to scale the Mollier ($h - s$) diagram of the cycle.

6. The mass flow rate of gases through the turbine of a turbojet is $\dot{m}_g$. The excess air in the combustor is λ. The power of the turbine is $\mathcal{P}_T$. What is the efficiency of the compressor, if the ideal work in the compressor is w_{C_i}?

7. What is the bypass ratio of a turbofan engine that uses standard fuel and has
   (a) fan pressure ratio of 1.6
   (b) fan efficiency of 85%
   (c) compressor power, $\mathcal{P}_C = 2000$ kW
   (d) turbine power, $\mathcal{P}_T = 5000$ kW
   (e) mass flow rate of gases, $\dot{m}_g = 20$ kg/s
   (f) excess air, $\lambda = 3$
   (g) inlet pressure, $p_{01} = 1$ bar
   (h) inlet temperature $T_{01} = 288$ K.
   You should know that the air constant is $R = 287.16$ J/(kg K) and that for standard fuel $minL = 14.66$.

8. A turbofan has a bypass ratio BPR, a specific work in the compressor w_C, a specific work in the fan w_F, a turbine-inlet temperature T_{03}, a turbine-inlet pressure p_{03}, a turbine efficiency η_T, the excess air is λ, and the minimum air for stoichiometric combustion is $minL$. The values of the specific heat capacity at constant pressure, c_p, and the ratio of specific heat capacities, γ, are also known. What is the expression of the expansion ratio in the turbine, p_{03}/p_{04}, as a function of the above known parameters?

9. A turbofan engine has a bypass ratio of 4, a specific work in the compressor $w_C = 200$ kJ/kg, a specific work in the fan $w_F = 50$ kJ/kg, a turbine inlet temperature $T_{03} = 1400$ K, a turbine inlet pressure $p_{03} = 12$ bar, a turbine efficiency $\eta_T = 0.9$, and the excess air is $\lambda = 3.41$. The engine uses standard fuel. What is the stagnation pressure at exit from the turbine?

10. A single-spool turbofan has a bypass ratio BPR, a compressor with an overall pressure ratio (that is, the ratio between the pressure at the exit from the compressor and the

pressure at inlet in the engine) π_{0C} and efficiency η_C, and a fan with a pressure ratio π_{0F} and efficiency η_F. The stagnation temperature at inlet in the engine is T_{01}, the excess air in the combustor is λ, and one can assume that heat capacity at constant pressure is approximately constant. What is the expression of the specific work in the turbine as a function of the parameters given above (assuming standard fuel)?

11. The Williams International F107 (WR19) is the turbofan engine used on the Tomahawk missile. The turbofan has a bypass ratio $BPR = 1$. The maximum thrust of the engine is 3.1 kN and the TSFC = 0.0695 kg/(N h). What is the mass flow rate of the engine needed to achieve the performance given above?

    Hints:
    (a) one could assume that the turbine inlet temperature is approximately 1200 K;
    (b) one needs to estimate the pressure ratio over the fan and over the compressor;
    (c) one needs to estimate the efficiencies of fan, compressor, and turbine;
    (d) one needs to estimate the pressure drop in the combustor.

12. A turbofan has a bypass ratio BPR, a total air mass flow rate $\dot{m}_a$, a fan efficiency η_F, and a stagnation enthalpy at inlet in the engine h_{01}. The power needed by the compressor is $\mathcal{P}_C$ and the power generated by the turbine is $\mathcal{P}_T$. Assume that γ and c_p do not vary with temperature.
    (a) Derive the analytical expression of the pressure ratio in the fan as a function of the parameters given above except c_p, i.e., $\pi_{0F}(h_{01}, BPR, \dot{m}_a, \eta_F, \mathcal{P}_C, \mathcal{P}_T, \gamma)$.
    (b) Assuming $\gamma = 1.4$, $\dot{m}_a = 100$ kg/s, $BPR = 9$, and $h_{01} = 288$ kJ/kg, what is the typical range of power difference between the turbine and the compressor? (One needs to select a typical range for the fan efficiency and fan pressure ratio.)

13. An airplane flies at $H = 8000$ m with a velocity of 250 m/s. The airplane uses a turbofan engine. The inlet diffuser efficiency is $\sigma_{0d} = 0.98$. The fan efficiency is 84% and the compressor efficiency is 87%. The fan pressure ratio is 1.7 and the compressor pressure ratio is 9. The bypass ratio is 5.5. The air enters the engine at a rate of 50 kg/s. The pressure drop in the combustor is 3%. The turbine inlet temperature is 1300 K. The turbine efficiency is 92%. The gas generator nozzle efficiency is 97% and the fan nozzle efficiency is 99%.
    (a) What are the cycle thermal efficiency, thrust, specific thrust, and TSFC of the engine, assuming actual medium for the working fluid in the engine?
    (b) What are the cycle thermal efficiency, thrust, and specific thrust assuming an ideal cycle?
    (c) Draw to scale, on the same plot, the Mollier ($h - s$) diagrams of the two cycles.

14. A turboprop engine operates on an airplane that flies at $H = 3000$ m with a velocity of 140 m/s. The inlet diffuser efficiency is $\sigma_{0\mathrm{diffuser}} = 0.98$. The compressor efficiency is 87%. The compressor pressure ratio is 8. The air enters the engine at a rate of 25 kg/s. The pressure drop in the combustor is 3%. The turbine inlet temperature is 1400 K. The turbine efficiency is 90% and the efficiency of the free turbine is 88%. The propeller efficiency is 80% and the free turbine shaft power is 7200 kW. The exit nozzle efficiency is 98%.
    (a) Draw on the Mollier diagram the real cycle of the turboprop (assume actual medium for the working fluid in the engine).

(b)   What are the equivalent power and *EBSFC* of the engine?

(c)   What is the ratio between the propeller thrust and the gas exhaust thrust?

(d)   How does the ratio between the propeller thrust and the gas exhaust vary if the airplane is at takeoff?

15.   A turboprop engine has a power turbine that produces 6600 kW using standard fuel. The air mass flow rate of the engine is $\dot{m}_a = 20$ kg/s. The excess air is $\lambda = 3$ and the enthalpy at inlet in the power turbine is $h_{04} = 1200$ kJ/kg. Assuming the efficiency of the power turbine is 100% (to make your life easier), what is the temperature of the gases at the exit of the power turbine?

16.   Derive an analytical expression for the shaft power $\mathcal{P}_s$ of a turboprop, as a function of the following parameters:

(a)   power turbine expansion pressure rate, $\pi_{0PT} = p_{04}/p_{04.5}$

(b)   air mass flow rate, $\dot{m}_a$

(c)   power turbine inlet total temperature, T_{04}

(d)   power turbine efficiency, η_{PT}

(e)   excess air, λ, and minimum air for stoichiometric combustion, $minL$

(f)   specific heat at constant pressure, c_p

(g)   ratio of specific heats, γ.

To simplify the expression of shaft power, one can neglect the variation of c_p and γ with temperature.

17.   A turboprop engine has a power turbine that produces 4000 kW using standard fuel. The air mass flow rate of the engine is $\dot{m}_a = 16$ kg/s. The specific work of the compressor is 250 kJ/kg. The excess air is $\lambda = 2.5$ and the enthalpy at the exit from the power turbine is $h_{04.5} = 800$ kJ/kg. What is the turbine inlet temperature?

18.   A turboprop engine has a power turbine (aka free turbine) that produces 370 kW (using standard fuel). The air mass flow rate of the engine is $\dot{m}_a = 1.56$ kg/s. The excess air is $\lambda = 2.984$, the pressure at the inlet in the power turbine is $p_{04} = 3.2$ bar, and the entropy is $s_{04} = 7.9112$ kJ/(kg K). What is the temperature at the exit from the power turbine if the efficiency of the power turbine is $\eta_{PT} = 0.89$?

Bibliography

Anonymous, editor. *The Jet Engine*. Rolls-Royce plc, 2005. 242

R. D. Archer and M. Saarlas. *An Introduction to Aerospace Propulsion*. Prentice Hall, 1996. 204, 206

W. W. Bathie. *Fundamentals of Gas Turbines*. John Wiley & Sons, Inc., second, edition, 1996. 198, 206

E. T. Curran. Scramjet engines: The first forty years. *Journal of Propulsion and Power*, 17(6):1138–1148, Nov.-Dec. 2001. 273

R. D. Flack. *Fundamentals of Jet Propulsion with Applications*. Cambridge University Press, 2005. 206, 237

P. Hill and C. Peterson. *Mechanics and Thermodynamics of Propulsion*. Addison-Wesley, second edition, 1992. 199, 206

C. Segal. *The Scramjet Engine*. Cambridge University Press, 2009. 273

7   Jet Engine Components

7.1   Introduction

This chapter presents the main jet engine components: inlet diffuser, compressor, combustor, turbine, and exit nozzle. Typical configurations are presented for each component, followed by a description of the main processes and parameters. The performance of each component is then related to the engine real cycle, which establishes a tight connection between this chapter and Chapter 6. The section describing the combustors is also connected to Chapters 5 and 8.

7.2   Inlet Diffusers

The inlet (or diffuser) of any jet engine has the function of bringing the air from the ambient conditions to the conditions required at the inlet in the compressor or fan. The air should be supplied with the smallest (i) loss in total pressure and (ii) drag force on the air vehicle as possible. Inlets designed for purely subsonic flight differ greatly from those designed for supersonic flight. The latter must operate at both subsonic and supersonic flight conditions.

7.2.1   Subsonic Inlets

The type of subsonic inlet varies depending on the type of jet engine and the location of the engine on the air vehicle. The design of inlets for turboprop and turboshaft engines is complicated by the presence of the gear box and the propeller or helicopter blades. Aircraft engines may be located in different places: (1) under the wing; (2) in the fuselage with inlet located above or under the fuselage, or in the root of the wing; (3) at the base of the vertical stabilizer. The inlets can also be classified as either single-entrance or divided-entrance. Most inlets are single-entrance while divided-entrance inlets are used on some fighter aircraft, where

(a)   (b)

(c)   (d)

Figure 7.1 Subsonic inlets. (a) Under-wing engine on Boeing 737 (courtesy of aviation-images.com/ Universal Images Group/Getty Images), (b) engine in vertical stabilizer on McDonnell Douglas MD-11 (courtesy of Joel Saget/AFP/Getty Images), (c) In-fuselage engine, under fuselage inlet on Vought A-7 Corsair II (courtesy of © Corbis/Getty Images), and (d) in-fuselage engine, divided-entrance inlet on Aero L-39 Albatros.

the inlets are located near the root of the wing. Figure 7.1 shows the different types of subsonic inlets.

The inlet must take the incoming air at Mach number M_0 and supply it to the compressor or fan at the axial Mach number M_1. M_1 is determined mainly by the rotational Mach number of the compressor or fan. Consequently M_1 depends on the rotational speed of the compressor or fan and the inlet air temperature, T_0. M_1 has the smallest value at low engine speed and high inlet temperature, that is, low altitude. M_1 has the largest values at full engine speed and low temperature, that is, high altitude.

Although the inlet must properly supply air to the compressor or fan over the entire flight envelope and engine regimes, the most important requirements are for (1) takeoff at full engine speed and high T_0, and (2) cruise at low T_0 and reduced engine speed.

As the altitude increases from takeoff to cruise, air temperature T_0 decreases. In addition, engine speed decreases from takeoff to cruise. When T_0 decreases, M_1 increases; when engine

Figure 7.2 Dividing streamline variation with aircraft Mach number for inlet $M_1 = 0.5$.

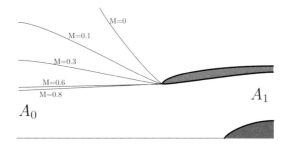

speed is reduced, M_1 decreases. Therefore the decrease of T_0 and increase of M_0 are offset by the reduction of engine speed, so that the variation of M_1 is not large. Typically, M_1 decreases by approximately 20% from takeoff to high subsonic cruise [Kerrebrock, 1992].

While the variation of M_1 is rather limited, M_0 can vary from zero at takeoff to approximately 0.8 at cruise. As a result, large changes of the streamlines entering the inlet occur, which can cause flow separation due to the large flow angle variation. Example 3.4.5 shows that for a $M_1 = 0.5$, the ratio A_0/A_1 varies between infinity at $M_0 = 0$ and 0.775 at $M_0 = 0.8$. Assuming areas A_0 and A_1 are circular, Fig. 7.2 shows the variation of dividing streamlines for different aircraft Mach numbers, M_0. It is apparent that at low M_0 numbers the flow turning on the inlet casing is large. To reduce the risk of flow separation, it is beneficial to increase the curvature radius of the inlet leading edge at low M_0. As the large curvature radius creates higher losses when M_0 increases, a compromise must be reached between the optimum geometries at low and high M_0.

7.2.2 Supersonic Inlets

The design of supersonic inlets is more challenging than that of the subsonic inlets, mainly due to the presence of shock waves. Shock waves introduce a new loss mechanism that can result in large decreases in stagnation pressure. Additionally, the shock waves can lead to bi-stable operation that creates large fluctuations of mass flow rates, losses, and drag between the two modes. Figure 7.3 shows the inlets of two classical airplanes: the McDonnel Douglas F-4 Phantom II and Mikoyan-Gurevich MiG-21.

A typical inlet consists of a supersonic diffuser and subsonic diffuser, as shown in Fig. 7.4. The difference between the capture area and the throat (minimum) area is the supersonic area change. The area variation in front of the cowl lip is called the *external compression*. In the external compression, the streamtube is only bounded on one side by solid surfaces. The variation of supersonic area between the cowl lip and the throat is called the *internal compression*. In internal compression, the flow is bounded on both sides by solid surfaces.

The supersonic inlet must provide good performance over a large range of operating conditions, from takeoff to supersonic flight. As a result of the wide range of flight conditions, there are large variations of the capture streamtube area. For an inlet that operates at both $M_0 = 1$ and $M_0 = 3$, the capture area varies by a factor of 4.2.

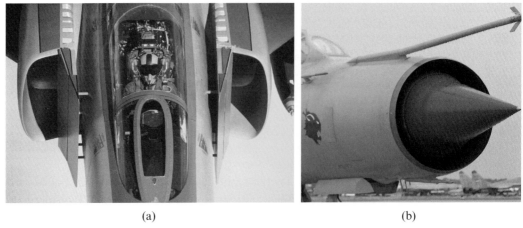

(a)   (b)

Figure 7.3 Supersonic inlets (courtesy of Getty Images). (a) McDonnel Douglas F-4 Phantom II (courtesy of Aviation-images.com/Universal Images Group/Getty Images) and (b) Mikoyan-Gurevich MiG-21 (courtesy of Andrej Isakovic/AFP/Getty Images).

Figure 7.4 Supersonic inlet.

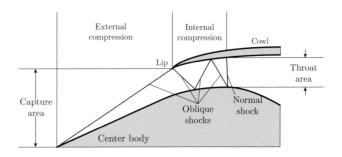

The supersonic inlet must also provide a uniform pressure distribution at the inlet in the compressor. In addition, the drop in stagnation pressure in the inlet should be as small as possible. Assuming a small stagnation pressure drop in the inlet, that is, a large recovery pressure, as the velocity of the vehicle increases, the inlet compression becomes a larger fraction of the cycle overall compression pressure ratio. Therefore, the specific thrust depends more on the inlet pressure recovery.

The geometry of a supersonic inlet is typically either axisymmetric or two-dimensional. Some inlets have a variable geometry center body to optimize the shock wave locations. Some inlets have boundary layer bleeds to reduce boundary layer ingestion and flow disturbances in the compressor.

The main performance characteristics of a supersonic inlet are: (1) total pressure recovery, (2) cowl drag, (3) boundary layer bleed flow, (4) ratio between capture area and throat area, and (5) inlet weight.

There are three types of supersonic inlets: (1) external compression inlets, (2) internal compression inlets, and (3) mixed compression inlets. The following sections will explore external and internal compression.

7.2.2.1 External Compression

The supersonic diffuser decelerates the flow by a combination of shock waves. As shown in Fig. 3.36, if the flow is decelerated by a series of weak shocks, the losses are much smaller and the total pressure recovery much higher than if fewer but stronger shocks are used. For an incoming flow at $M_0 = 3.5$, the total pressure recovery increases from 0.789 with two shocks to 0.960 with five shocks. In addition, the entropy rise decreases from 68.1 to 11.8 kJ/(kg K).

Let us consider the external compression on the supersonic inlet shown in Fig. 7.5 where the flow is decelerated by one oblique shock followed by a normal shock. Given an incoming Mach number M_0, the wedge angle δ can be optimized to minimize the losses and therefore maximize the total pressure recovery, π_d.

Table 7.1 shows, for an incoming Mach number $M_0 = 4$, the variation of the shock angle σ; Mach numbers after the oblique and normal shocks, M_1 and M_2, respectively; stagnation pressure ratios; and total pressure recovery for three wedge angles. These parameters were calculated using (3.103), (3.98), and (3.96). The maximum pressure recovery corresponds to a wedge angle of approximately 23.5°, as also shown in Fig. 7.6. The entropy variation is a minimum where the pressure recovery is a maximum.

Figure 7.7 shows the variation of pressure recovery with incoming Mach number M_0 and wedge angle δ. As M_0 increases, the optimum wedge angle increases slightly, from approximately 22.5° to 24°. The maximum pressure recovery, however, varies significantly, from 0.59 at $M_0 = 3$ to 0.17 at $M_0 = 5$.

If M_0 decreases from 4 to 3.5, then for a wedge angle $\delta = 24°$ the angle of the oblique shock increases from 37.2° to 39.4°, as shown in Fig. 7.8. In this case, the capture area decreases and

Table 7.1 Flow parameter variation for an oblique shock followed by a normal shock for an incoming Mach number $M_0 = 4$.

Wedge angle, δ	20°	24°	28°
Shock angle, σ	32.5°	37.2°	42.4°
M_1	2.569	2.281	1.994
p_{01}/p_{00}	0.652	0.532	0.424
M_2	0.507	0.537	0.578
p_{02}/p_{01}	0.472	0.592	0.724
π_d	0.308	0.315	0.307

Figure 7.5 Supersonic inlet with one oblique and one normal shock.

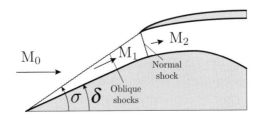

Figure 7.6 External-compression inlet: pressure recovery and entropy variation vs. wedge angle at upstream Mach 4.

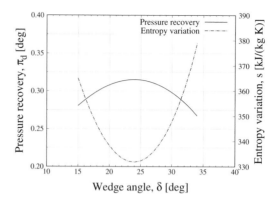

Figure 7.7 External-compression inlet: pressure recovery vs. wedge angle and upstream Mach number.

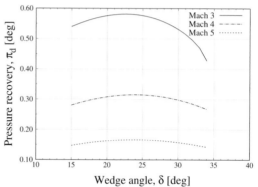

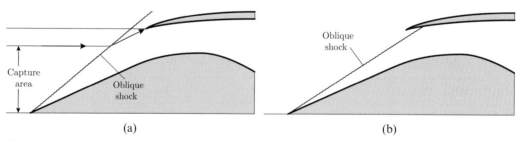

Figure 7.8 Supersonic inlet with (a) subcritical operation and (b) supercritical operation [Bathie, 1996].

the inlet is said to be at subcritical operation. If M_0 increases from 4 to 4.5, then the angle of the oblique shock decreases from $37.2°$ to $35.7°$ and the inlet is said to be at supercritical operation.

The pressure recovery can be increased by using multiple oblique shocks ahead of the normal shock. By further increasing the number of oblique shocks, one reaches an isentropic wedge diffuser [Kerrebrock, 1992, p. 129]. The isentropic wedge diffuser, however, is not practical for two reasons: (1) the boundary layer is prone to separate under the impact of the shock waves, and (2) the flow deflection angle is large, and therefore a large cowl drag results as the near-sonic flow is turned back to the axial direction. Therefore, a compromise must be reached between increasing pressure recovery and reducing inlet drag.

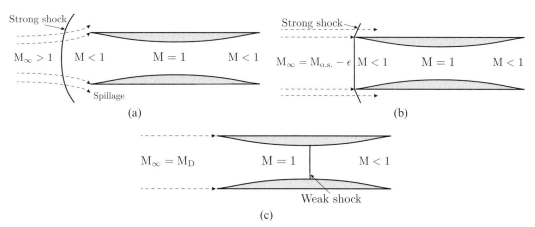

Figure 7.9 Internal compression supersonic inlet conditions: (a) subcritical, (b) critical, and (c) supercritical.

7.2.2.2 Internal Compression

External compression leads to significant wave drag due to outward flow on the inlet cowl, even when the shock waves are attached. This section will explore whether supersonic compression can be achieved by inward flow turning in an enclosed duct.

The internal compression inlet is a convergent-divergent channel that decelerates the supersonic flow by a series of weak compression waves to sonic flow, followed by a subsonic deceleration in the divergent part of the channel. Depending on the location of the shock, three conditions can be identified, as shown in Fig. 7.9: (1) *subcritical*, when the shock is detached and spillage occurs, (2) *critical*, when the shock is at the inlet lip, and (3) *supercritical*, when the shock is swallowed. M_∞ denotes the flight Mach number, M_D is the design flight Mach number, and $M_{o.s.}$ is the overspeed Mach number, which will be defined later in this section.

To understand the operation mode of the internal compression supersonic inlet, let us consider the variation of flow parameters in the inlet as the aircraft speed increases. It will be assumed that the component attached downstream from the inlet can ingest the entire inlet mass flow rate. Consequently, the mass flow rate in the inlet is limited only by the inlet throat area, A_t.

If the flight velocity is subsonic and the Mach number is sufficiently low, the inlet is not choked. As a result, the flow through the inlet, and therefore the upstream capture area A_∞, is set by the static pressure downstream of the inlet. This condition corresponds to regime (a) shown in Fig. 7.10.

As the subsonic flight velocity increases, the flow in the inlet reaches sonic velocity at the throat area, so that the throat area is equal to the critical area, $A_t = A_{cr}$. Consequently, the mass flow rate through the inlet is limited by the chocking condition at the throat. Since the flow is assumed isentropic, the upstream capture area, A_∞, is obtained using (3.44):

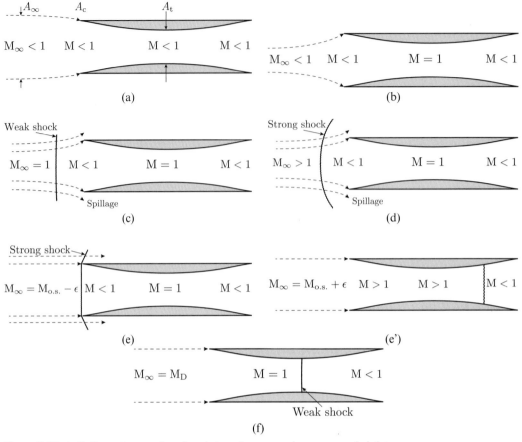

Figure 7.10 (a-f) Operation modes of an internal compression supersonic inlet.

$$\frac{A_\infty}{A_t} = \frac{A_\infty}{A_{cr}} = \frac{1}{M_\infty}\left[\frac{2}{\gamma+1}\left(1+\frac{\gamma-1}{2}M_\infty^2\right)\right]^{\frac{\gamma+1}{2(\gamma-1)}}. \tag{7.1}$$

For a sufficiently high subsonic flight velocity, the capture area A_∞ is smaller than the cowl area at the front of the inlet, A_c. Therefore, the flow will spill around the inlet, as shown for regime (b) in Fig. 7.10b.

As the flight velocity increases further and reaches sonic flight speeds, a shock develops just upstream of the inlet. This is due to the fact that as the flow enters the area A_c without deviation, the throat area limits the flow. Therefore there is an accumulation of mass, and the pressure increases in the inlet. As the pressure increases rapidly, a weak shock develops upstream of the inlet that generates the needed spillage to reduce the backlog, as shown in Fig. 7.10c.

When the flight speed increases past $M_\infty = 1$, the detached bow shock upstream of the inlet strengthens, as shown in Fig. 7.10d. Because of the presence of the shock wave, the flow is non-isentropic. As shown in (3.40), the mass flow rate is proportional to the stagnation pressure and inversely proportional to the square root of the stagnation temperature. Since the stagnation temperature is constant across the shock, the mass flow rate is only proportional to stagnation pressure. Due to the drop in stagnation pressure across the shock, the actual mass flow rate in the inlet is less than the mass flow rate expected based on the throat area and stagnation pressure corresponding to flight Mach number, M_∞. The actual mass flow rate of the air that enters the inlet is obtained using (3.40) where the stagnation pressure corresponding to the flight speed is substituted by the stagnation pressure downstream of the shock.

Let us determine the area $A_{\infty 2}$ of the flow that enters the inlet, where the subscript "$\infty 2$" denotes a point just downstream the shock. Assuming the flow is isentropic from point "$\infty 2$" to the throat, the area ratio is

$$\frac{A_{\infty 2}}{A_t} = \left(\frac{A}{A_{cr}}\right)_{\infty 2} = \frac{1}{M_{\infty 2}} \left[\frac{2}{\gamma + 1}\left(1 + \frac{\gamma - 1}{2}M_{\infty 2}^2\right)\right]^{\frac{\gamma+1}{2(\gamma-1)}} \tag{7.2}$$

where $M_{\infty 2}$ is the Mach number just downstream of the shock.

If we assume that the bow shock in front of the inlet can be approximated by a normal shock, then the Mach number $M_{\infty 2}$ downstream of the shock is obtained from (3.76):

$$M_{\infty 2} = \sqrt{\frac{M_\infty^2 + 2/(\gamma - 1)}{M_\infty^2 2\gamma/(\gamma - 1) - 1}}. \tag{7.3}$$

The variation of capture area to throat area, $A_{\infty 2}/A_t$ vs. M_∞, is shown in Fig. 7.11.

As the flight speed increases to the design Mach number, M_D, the inlet area A_c is still larger than the capture area $A_{\infty 2}$ and spillage continues. If the flight speed is increased past the design point to $M_{o.s.}$ where $A_c = A_{\infty 2}$, then the inlet can ingest the entire mass flow rate without spillage. As a result, the shock will be located at the lip of the inlet, as shown in Fig. 7.10e. A small increase in flight speed will cause the shock to enter the inlet, and since it cannot reach a stable position in the convergent part of the nozzle, it will quickly travel to the divergent part of the nozzle where it will attain a stable position determined by the conditions downstream of the inlet. Since the shock wave is downstream of the throat, as shown in Fig. 7.10e', the flow through the throat is isentropic. Therefore, as shown in Fig. 7.11, the capture area to critical area ratio, A_∞/A_{cr}, increases to the (e') value calculated using (7.1), as opposed to the ratio calculated by (7.2) for state (e).

The incoming supersonic flow is decelerated from A_c to A_t and then accelerated in the divergent part of the inlet until a shock forms downstream of the throat. Once an isentropic flow is reached over the throat, the flight speed can be reduced from $M_{o.s.}$ to M_D so that the flow is sonic at the throat, as shown in Fig. 7.10g. However, if there is a slight reduction in flight speed or increase in back pressure, the shock will move from throat to inlet, spillage will occur, and the operating point will move from (f) to (d). To avoid unstable operation

Figure 7.11 Internal-compression inlet: best pressure recovery, (π_d), throat Mach number, M_t, and capture-to-throat area ratio, $A_{\infty 2}/A_t$. The design Mach number, M_D, and the overspeed Mach number, $M_{o.s.}$, correspond to the cowl inlet to throat area ratio $A_c/A_t = 1.3$.

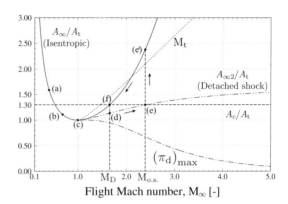

with the shock at the throat, it is safer to maintain a throat Mach number slightly larger than 1. If over-speeding is not possible, that is reaching $M_{o.s.}$, the shock wave can be swallowed by momentarily increasing the throat area and then returning to the original A_t value.

The operation modes summarized in Fig. 7.10 reveal two possible cases: (i) the shock reaches the entry plane before the flow is sonic at the throat, or (ii) the flow is sonic at the throat before the shock reaches the entry plane.

If the shock is attached and $M_t < 1$, there is no spillage and $A_\infty = A_c$. An increase of exit area will cause the shock to move inside the duct and reach a stable position in the divergent part of the nozzle, as shown in Fig. 7.10e′ [Hill and Peterson, 1992, p. 230]. The location of the shock is such that the exit area times stagnation pressure behind the shock is equal to the same product corresponding to the attached shock. An inlet operating at this condition is called "started" because supersonic compression is occurring between the entry plane and the shock. The inlet operates at supercritical conditions.

If the shock is detached and $M_t = 1$, as shown in Fig. 7.10d, increasing the inlet exit area will not affect the flow upstream of the throat but will generate a second shock downstream of the throat. In this case, the subsonic flow extends from the detached shock to the throat. An inlet in this operation mode is called "unstarted" and the inlet operates at subcritical conditions.

The condition needed to have a started inlet is that the freestream shock reaches the inlet entry at the same time the flow is sonic at throat. In this case, the inlet operates at critical conditions. This condition defines the limit of the contraction area A_t/A_c; any smaller contraction area, that is, any greater degree of compression, would lead to the inlet being unstartable.

Since the freestream shock is attached to the inlet entry, $A_c = A_{\infty 2}$ in (7.2),

$$\frac{A_t}{A_c} = \left(\frac{A_{cr}}{A}\right)_{\infty 2} = M_{\infty 2}\left[\frac{\gamma + 1}{(\gamma - 1)M_{\infty 2}^2 + 2}\right]^{\frac{\gamma + 1}{2(\gamma - 1)}}. \tag{7.4}$$

Figure 7.12 Limiting contraction ratio, χ, and sonic area relationship, A_{cr}/A, vs. flight Mach number.

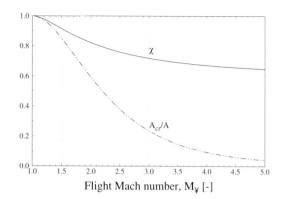

Flight Mach number, $M_\yen$ [-]

Substituting (7.3) in (7.4) yields

$$\chi = \left(\frac{A_t}{A_c}\right)_{\text{lim}} = \left[\frac{\gamma - 1}{\gamma + 1} + \frac{2}{(\gamma + 1)M_\infty^2}\right]^{\frac{1}{2}} \left[\frac{2\gamma}{\gamma + 1} - \frac{\gamma - 1}{(\gamma + 1)M_\infty^2}\right]^{\frac{1}{\gamma - 1}} \tag{7.5}$$

where χ is the *limiting contraction ratio* [Seddon and Goldsmith, 1999, p. 137]. Figure 7.12 compares the limiting contraction ratio to the sonic area relationship A_{cr}/A, where the latter represents the ideal contraction from freestream speed to sonic condition. The starting criterion imposes a significant limitation on the allowable contraction. The lowest value of χ is reached when M_∞ tends to infinity

$$\chi(M_\infty \to \infty) = \left(\frac{\gamma - 1}{\gamma + 1}\right)^{\frac{1}{2}} \left(\frac{2\gamma}{\gamma + 1}\right)^{\frac{1}{\gamma - 1}} \tag{7.6}$$

which for $\gamma = 1.4$ is $\chi(M_\infty \to \infty) = 0.6$. Note that the graph of $1/\chi$ is identical to that of $A_{\infty 2}/A_t$ in Fig. 7.11. If A_c/A_t is larger than $1/0.6 = 1.66$ then there is no $M_{\text{o.s.}}$ that will start the inlet.

Figure 7.11 also shows the variation of Mach number, M_t, ahead of the shock at the throat as a function of the flight Mach number, M_∞. The Mach number, M_t, is calculated from the sonic area relationship

$$\left(\frac{A}{A_{cr}}\right)_{M_t} = \left(\frac{A}{A_{cr}}\right)_{M_\infty} \left(\frac{A_t}{A_c}\right)_{\text{lim}}.$$

Using this Mach number at throat, M_t, in (3.34) yields the maximum pressure recovery $(\pi_d)_{\text{max}}$ the inlet can achieve [Kerrebrock, 1992, p. 127]. As shown in Fig. 7.11, the maximum pressure recovery drops off quickly for $M_\infty > 1.5$.

This study of supersonic internal compression ignored viscosity effects. If boundary layer presence is accounted for, the analysis becomes more complicated and the prediction of starting the inlet more difficult.

Figure 7.13 Mixed compression inlet.

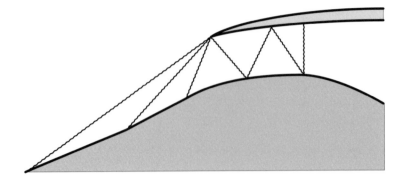

7.2.2.3 Mixed Compression

Mixed compression inlets use a combination of external and internal compression, as shown in Fig. 7.13. The flow is slowed down by a series of oblique shock waves before reaching a normal shock at the throat of the duct.

Most supersonic inlets use either external or mixed compression. The selection of the inlet type depends on the cruise Mach number and cruise altitude, as well as the fraction of the flying time spent at supersonic speeds. At off-design, operation the flow in the inlet can be controlled by variable geometry, boundary layer bleed, and bypass doors [Bathie, 1996]. Variable geometry solutions include: (1) translating center body, (2) variable throat area, (3) variable angle cowl lip, and (4) variable ramp (wedge or cone) angle. Boundary layer bleeding removes the low-momentum air, typically at the locations where the shock waves impinge on the duct walls. Bleeding delays or precludes flow separation in the inlet and therefore provides a more uniform flow to the compressor. Bypass doors are used to spill the excess air while the engine is shut-down or throttled.

7.3   Fans and Compressors

The compressor is a device that raises the pressure of the working fluid passing through it. The fan is a large, low-pressure compressor situated at the front of a turbofan engine. The working fluid of a jet engine fan or compressor is air.

The purpose of the compressor is to raise the air pressure at the inlet of the combustor. Depending on the direction of the flow at the exit of the compressor, the compressor can be either an axial-flow compressor or a radial-flow (or centrifugal) compressor. As the name suggests, the flow at the exit of an axial compressor is mainly in the axial direction, while the flow at the exit of a centrifugal compressor is in radial directions. Axial and centrifugal compressors can be used in combination to form an axi-centrifugal compressor. Figures 7.14 and 7.15 show typical layouts of axial and centrifugal compressors.

An axial compressor consists of one, two, or three rotor assemblies that carry rotor blades. The rotor assemblies are rotated by the turbine, through the shaft. The rotor assembly is

Figure 7.14 Layout of an axial compressor.

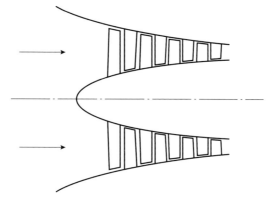

Figure 7.15 Layout of a centrifugal compressor.

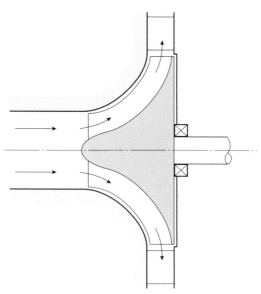

supported by bearings attached to the casing structure. The casing also holds the stator vanes. Both the rotor blades and the stator vanes have airfoil cross sections. A row of rotor blades and the neighboring row of stator vanes form a *stage*. The axial compressor shown in Fig. 7.14 has four stages. The number of airfoils on a row varies and is usually between 30 and 60.

The centrifugal compressor consists of an impeller and diffuser vanes. Some centrifugal compressors also have an inducer mounted on the front of the impeller. The impeller is rotated by the turbine. As the impeller rotates, air is continuously induced into the center of the impeller. The inducer provides an initial swirl to the flow, to increase compressor efficiency. As the air enters the impeller, the impeller blades force the air to rotate in the circumferential direction. Consequently, the centrifugal force acting on the air induced in the compressor increases. The shape of the impeller then forces the air to move in the radial direction. As the air moves in the radial direction and the radius increases, the centrifugal force increases proportionally. Therefore, the velocity and pressure increase as the air travels through the impeller. Once the air leaves the impeller, the air passes through the radial diffuser. The divergent nozzles of the diffuser reduce the air velocity. The kinetic energy of the air is converted into compressor work. Consequently, the stagnation enthalpy and stagnation pressure increase.

The pressure rise in the compressor results from the energy imparted to the air by the rotor blades. There is, however, a critical difference between how the pressure increase is achieved in the two types of compressors. In an axial compressor, Figs. 7.16 and 7.17, the pressure increase is obtained as a result of the pressure difference between the pressure and suction sides of the airfoils. Consequently, the pressure rise in the axial compressor is intimately related to the lift force on the airfoil. In the centrifugal compressor, Figs. 7.18 and 7.19, the pressure increase is driven by the centrifugal force that acts on the working fluid, that is, air. The difference between the two mechanisms used to increase air pressure, one based on the lift force and the other based on the centrifugal force, leads to significant differences between the two types of compressors.

The differences between axial and centrifugal compressors make them more or less suitable for certain applications. Axial compressors are used almost exclusively for large jet engines. Centrifugal compressors are typically used on engines with a mass flow rate smaller than approximately 25 kg/s. A succinct discussion of the advantages and disadvantages of the two types of compressors is included in the following paragraphs.

The axial compressor has a smaller frontal area for a given mass flow rate than a centrifugal compressor, because the variation of the outer diameter of the axial compressor is much smaller than that of a centrifugal compressor. A smaller frontal area is important for reducing the shape drag of the engine. Since the flow at discharge from a stage of an axial compressor does not turn in the radial direction as in a centrifugal compressor, the axial compressor is better suited for multistaging. This is important since typically the needed pressure rise cannot be achieved in only one stage. Axial compressors usually have a higher efficiency than centrifugal compressors. Consequently, a jet engine using axial compressors is more efficient.

Advantages of the centrifugal compressor vs. the axial compressor are: (1) higher stage pressure ratio, (2) simplicity and ruggedness of construction, (3) less reduction in performance due to "dirty" blades, (4) shorter length for the same overall pressure ratio, (5) wider range of stable operation between surging and choking, (6) flow direction at discharge convenient for the installation of an inter-cooler and/or heat exchanger, (7) lighter than axial with smaller inertia, and (8) less expensive to manufacture than axial compressors.

7.3.1 Compression Process

The compressor pressure ratio (total-to-total) is defined as

$$\pi_{0C} := \frac{p_{02}}{p_{01}}. \tag{7.7}$$

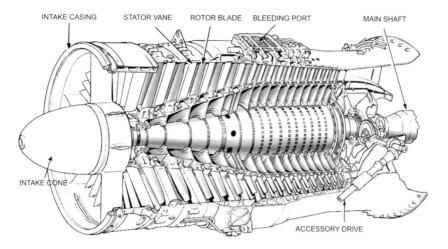

INTAKE CASING   STATOR VANE   ROTOR BLADE   BLEEDING PORT   MAIN SHAFT

INTAKE CONE

ACCESSORY DRIVE

Figure 7.16 Axial compressor (courtesy of Rolls-Royce).

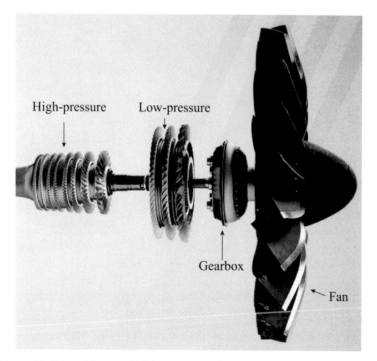

High-pressure   Low-pressure

Gearbox

Fan

Figure 7.17 Rotor with fan and low- and high-pressure axial compressors of Pratt & Whitney GTF geared turbofan engine (reproduced with permission from Pratt & Whitney).

Typical values of the pressure ratio per stage are 4 for radial compressors, 1.3 for subsonic axial compressors, and 2 for supersonic axial compressors. Recall that the stage is defined as one row of rotor blades plus one row of stator vanes.

Figure 7.18 Centrifugal compressor.

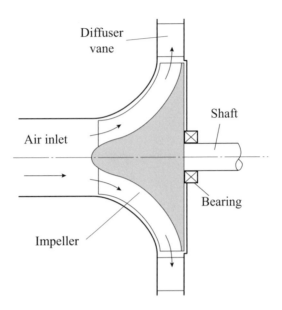

Figure 7.19 Impeller of centrifugal compressor.

The compression mechanism consists of the following processes:

1. the rotor blade transmits work from the shaft to the fluid (air);
2. the kinetic energy of the fluid increases;
3. the stator transforms the kinetic energy into compressor work, by increasing total enthalpy.

This mechanism is valid for both axial and centrifugal compressors. However, the work is transmitted from the rotor blade to the fluid in two different ways: (1) in the axial compressor due to the airfoil lift; (2) in the centrifugal compressor due to the centrifugal force.

7.3.2 Compressor Performance

The jet engine cycle analysis presented in Chapter 6 relied, among other parameters, on the stagnation pressure at outlet from the compressor, p_{02} (or the pressure ratio, π_{0C}), and compressor adiabatic efficiency, η_C. Both p_{02} and η_C, for either axial or centrifugal compressors, depend on the following parameters:

T_{01}	-	inlet stagnation temperature
p_{01}	-	inlet stagnation pressure
R	-	gas constant
D	-	characteristic dimension (usually diameter)
N	-	rotor rotational speed
$\dot{m}$	-	mass flow rate of air
γ	-	ratio of specific heat capacities
ν	-	kinematic gas viscosity
design	-	specification of the dimensionless geometric shape of the compressor, as a set of ratios of every important dimension to the characteristic dimension, D

so that

$$p_{02}, \eta_C = f(\dot{m}, p_{01}, T_{01}, N, \gamma, R, \nu, D, \text{design}).$$

The list of independent parameters that describe compressor performance is long. Therefore, it is cumbersome to independently assess the influence of all parameters on compressor operation. By using dimensional analysis, the number of parameters in this thermodynamic problem can be reduced by four, since there are four significant primary variables: mass, length, time, and temperature. Consequently, the list of nine parameters is reduced to five dimensionless terms, out of which two are γ and "design". The dimensionless parameters are shown in Table 7.2.

Table 7.2 Compressor dimensionless parameters.

Dimensionless term	Significance
$\dfrac{\dot{m}\sqrt{RT_{01}}}{p_{01}D^2}$	Mass flow rate
$\dfrac{ND}{\sqrt{RT_{01}}}$	Mach number at rotor tip, $M_{\text{tip rotor}}$
$\dfrac{ND^2}{\nu}$	Reynolds number
γ	Ratio of specific heat capacities
"design"	Compressor geometric shape

Using the dimensionless parameters, the pressure ratio and the compressor adiabatic efficiency can be written as

$$\pi_{0C}, \eta_C = f\left(\frac{\dot{m}\sqrt{RT_{01}}}{p_{01}D^2}, \frac{ND}{\sqrt{RT_{01}}}, \frac{ND^2}{\nu}, \gamma, \text{design}\right).$$

The list of dimensionless parameters can be simplified further. The variation of the ratio of specific heat capacities, γ, is small in the compressor. In addition, the Reynolds number is typically high but its variation is small. Consequently, for a given compressor design, the pressure ratio and adiabatic efficiency are functions of only two dimensionless parameters

$$\pi_{0C}, \eta_C = f\left(\frac{\dot{m}\sqrt{RT_{01}}}{p_{01}D^2}, \frac{ND}{\sqrt{RT_{01}}}\right).$$

For a compressor with a given diameter, D, and a working fluid with a given gas constant, R, the pressure ratio and adiabatic efficiency become

$$\pi_{0C}, \eta_C = f\left(\frac{\dot{m}\sqrt{T_{01}}}{p_{01}}, \frac{N}{\sqrt{T_{01}}}\right). \tag{7.8}$$

Using (7.8) the compressor performance can be described by measuring $\dot{m}$, N, p_{01}, and T_{01}. Since during the experiments the temperature and pressure might vary, it is advantageous to replace p_{01} and T_{01} by

$$\delta = p_{01}/(p_{01})_{\text{std day}}$$

and

$$\theta = T_{01}/(T_{01})_{\text{std day}},$$

where the standard day values are $(p_{01})_{\text{std day}} = 101,325$ Pa and $(T_{01})_{\text{std day}} = 288.15$ K. Using the dimensionless pressure and temperature, the compressor pressure ratio and adiabatic efficiency can be written as

$$\pi_{0C}, \eta_C = f\left(\frac{\dot{m}\sqrt{\theta}}{\delta}, \frac{N}{\sqrt{\theta}}\right) \tag{7.9}$$

where $\dot{m}\sqrt{\theta}/\delta$ is the inlet *corrected mass flow rate* and $N/\sqrt{\theta}$ is the *corrected speed*.

7.3.3 Compressor Map

The compressor performance is illustrated by the relations (7.9). The plot of compressor pressure ratio and adiabatic efficiency vs. the inlet corrected mass flow rate and the corrected speed is called the compressor map. The compressor map not only illustrates the performance of the compressor but also provides important information about the performance of the entire engine. The compressor map can be determined experimentally or by using numerical simulations.

The compressor map of Fig. 7.20 shows constant speed lines and constant efficiency lines. In addition, the surge line marks the left side limit of compressor operation. The compressor

Figure 7.20 Compressor map.

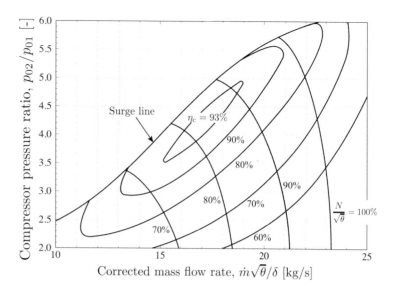

surge is the condition of violent unsteady flow caused by flow separation. A brief discussion of surge and rotating stall is given in Section 7.3.4.

Example 7.3.1 The mass flow in a compressor is 20 kg/s at $T_H = 288.15$ K and $p_H = 101,325$ Pa. The compressor pressure ratio is $\pi_{0C} = 4.5$. The compressor performance map is shown in Fig. 7.20. What should the mass flow and rotational speed be in order to keep the compressor operating at the same point on the performance map if the $T_{H_c} = 308.15$ K and $p_{H_c} = 100,100$ Pa?

Solution

The operating point is determined by the corrected mass flow rate and the corrected speed. Since the operating point should be kept constant as the pressure and temperature change, the corrected mass flow rate and the corrected speed should not change:

$$\dot{m}_a \frac{\sqrt{\theta_1}}{\delta_1} = \dot{m}_{a_c} \frac{\sqrt{\theta_{1_c}}}{\delta_{1_c}} \tag{7.10}$$

and

$$\frac{N}{\sqrt{\theta_1}} = \frac{N_c}{\sqrt{\theta_{1_c}}}. \tag{7.11}$$

Let us calculate the dimensionless pressures, δ, and temperatures, θ:

$$\theta_1 = \frac{T_H}{288.15} = 1$$

$$\delta_1 = \frac{p_H}{101,325} = 1$$

$$\theta_{1_c} = \frac{308.15}{288.15} = 1.0694$$

$$\delta_{1_c} = \frac{100,100}{101,325} = 0.9879.$$

Using (7.10) yields

$$\dot{m}_{a_c} = \dot{m}_a \frac{\sqrt{\theta_1}}{\delta_1} \frac{\delta_{1_c}}{\sqrt{\theta_{1_c}}} = 20 \frac{0.9879}{\sqrt{1.0694}} = 19.1 \text{ kg/s}.$$

Given the corrected mass flow rate of 20 kg/s and the pressure ratio of 4.5, one can read from the compressor map that the rotor speed is 93% of the design speed. Using (7.11) yields

$$N_c = N \frac{\sqrt{\theta_{1_c}}}{\sqrt{\theta_1}} = 93\% \times \sqrt{1.0694} = 96.2\%.$$

7.3.4   Rotating Stall and Surge

Highly maneuverable aircraft require that the propulsion system (jet engine) operates during sudden accelerations and rapid changes in inlet conditions. Consequently, the compressor of the jet engine must occasionally operate at low flow rates and high angles of attack. The high-angle-of-attack low-flow regime of compressor operation is often plagued by rotating stall and surge. A brief description of rotating stall and surge is given in the following paragraphs.

As the compressor mass flow rate is reduced, the pressure rise increases up to a point beyond which a further reduction of the mass flow rate results in a sudden change in the flow pattern. If the mass flow is reduced beyond this catastrophic point, the compressor enters into either stall or surge. Regardless of whether the compressor goes into stall or surge, the point of instability is called the *surge point* [Cumpsty, 1989, p. 359]. Depending on the geometry and load of the compressor, the stall could be progressive (a small drop in performance) or abrupt (a very large drop in the pressure ratio and mass flow) [McKenzie, 1997, pp. 112–3]. In both cases, the flow is no longer axisymmetric but has a circumferentially non-uniform pattern of regions of separated flow, which rotate around the annulus [Iura and Rannie, 1954].

The regions of separated flow are called stall cells. There may be one or more stall cells, and the cells may extend from hub to casing (full-span stall) or only over part of the span (part-span stall), as shown in Fig. 7.21 [Cumpsty, 1989, p. 360]. Figure 7.22 shows that the blockage generated by a stall cell leads to a reduction of angle of attack on one side of the cell and an increase on the other side. As a result, the stall cell, which could cover more than one blade passage [McDougall, 1988], moves along the circumference with a fraction of the wheel speed. This phenomenon, in which the total flow rate through the annulus does not vary with time, is called a rotating (or propagating) stall. The rotating stall was first reported by Cheshire [1945] for centrifugal compressors and by Iura and Rannie [1954] and Emmons et al. [1955] for axial compressors.

Figure 7.21 Part-span and full-span stall, where the gray regions indicate low axial velocity (stall cells).

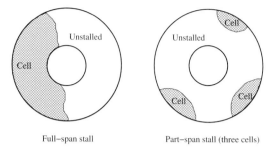

Full–span stall   Part–span stall (three cells)

Figure 7.22 Stall cell propagation.

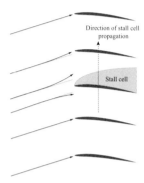

Surge, on the other hand, is defined as the phenomenon where the net flow through the entire annulus fluctuates with time [Iura and Rannie, 1954]. Violent flow instabilities during surge include symptoms such as audible thumping and honking at inlet and exit, at frequencies as low as 1 Hz, and severe mechanical vibrations [Emmons et al., 1955]. Certainly, violent surge cannot be tolerated in an aircraft jet engine because of the danger of mechanical failure or interruption of the combustion process.

Research into understanding compression system instabilities began with the low-order, hydrodynamic stability model of rotating stall and surge developed by Moore and Greitzer [1986]. This model was developed for an incompressible, undistorted flow field. The model consists of a set of three simultaneous nonlinear third-order partial differential equations for pressure rise, and average and disturbed values of flow coefficient, as functions of time and circumferential angle. This model provided the theoretical framework for understanding how to suppress rotating stall and surge using active control [Epstein et al., 1989]. More recently, a reduced-order model of the unsteady viscous flow in a compressor cascade has been developed, which is ideally suited for the design of active control strategies to suppress rotating stall [Florea et al., 1997, 1998].

Rotating stall has been shown to be a two-dimensional or three-dimensional form of instability [Day, 1993a]. Surge, on the other hand, has traditionally been regarded as a one-dimensional disturbance with axisymmetric velocity fluctuations affecting all parts of the system [Greitzer, 1978]. Rotating stall is limited to blade rows and depends only on the compressor, while surge involves the entire pumping system: compressor, throttles, ducting and plenums [Paduano et al., 1993]. Recent experimental results have revealed that rotating stall plays an important part in initiating a surge event, particularly for highly loaded axial compressors [Day, 1993a]. Consequently, delaying rotating stall also delays surge.

Day [1993b] used a system of twelve casing-mounted air valves to generate damping disturbances by injecting small puffs of air into the main stream near the tips of the first rotor in a four-stage axial compressor. The valves were opened and closed 120 times per second, that is, 12 times faster than the rotational frequency of the modal perturbations. The improvement in the stalling flow coefficient was 4%. A 23% reduction of the stalling mass flow of a low-speed, single-stage, axial research compressor was obtained by "wiggling" the inlet guide vanes to generate circumferentially traveling waves with appropriate phase and amplitude to control the first three spatial harmonics [Silkowski, 1990; Paduano et al., 1993].

The casing-mounted air valves provided a simpler actuation mechanism compared to the moving inlet guide vanes. It is somewhat difficult to judge how much more effective the moving inlet guide vanes were compared to the casing-mounted air valves because they were applied to different compressors, a single-stage and a four-stage, respectively. Certainly the effectiveness of the casing injection was reduced because it mainly affected the flow at the tip of the blades.

Steady mass injection upstream of the tip of a high-speed compressor rotor reduces the stalling flow coefficient at design speed by 6% using an injected mass flow equivalent to 2% of the annulus flow [Suder et al., 2001]. Larger reductions of the stalling flow coefficient were obtained at reduced percentage speeds. For compressors operating at part speed for extended periods of time, however, tip injection is not a viable option because the aerodynamic performance is reduced due to the efficiency penalty incurred using tip injection [Suder et al., 2001].

Unsteady injection on the suction side of a compressor stator airfoil in a linear cascade has been used to reduce the total pressure loss by 65%, when the injected mass flow rate was 1.6% of the freestream mass flow rate [Carter et al., 2001]. Steady and unsteady injection was also used on the suction side of a stator airfoil in a four-stage low-speed compressor. In this case, a 25% reduction in area-averaged loss was obtained when the injected mass flow rate was 1% of the compressor throughflow [Culley et al., 2004].

The methods currently used to suppress surge and stall are: (1) bleed off-take, (2) active control of variable inlet guide vane and variable stator vanes, and (3) casing air injection upstream of the blade tip.

The interstage bleeds remove a portion of the air at an intermediate stage and dump the bleed air into the bypass flow. As a result, energy is wasted through the work done to compress the (bleed) air that is then not used for combustion.

The variable inlet guide vanes and variable stator vanes are quite complex (see Fig. 7.23), require power to operate and add to the overall weight of the engine. In addition, the variable vanes are not fail-safe, and their malfunction can severely affect engine performance.

Air tip injection could be implemented in a jet engine, but the efficiency penalty incurred using tip injection makes it inapplicable to compressors that operate at part speed for extended duration. Furthermore, the mass flow rate of air that needs to be injected in order to prevent

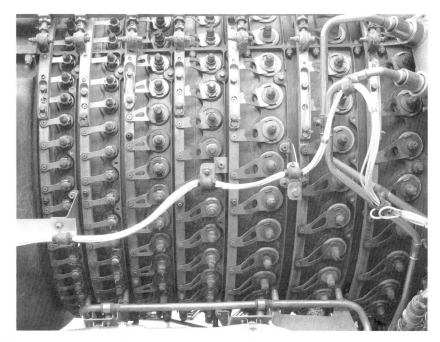

Figure 7.23 Details of the variable stator vanes of the CF6 turbofan engine of the Texas A&M Propulsion Laboratory.

flow separation in the hub region is significantly larger than the amount of air needed locally to prevent separation. An alternative approach is to block the advancement of the rotating stall cell by using a pulse modulated flow control actuator over the entire span of a compressor blade [Johnson et al., 2017].

7.3.5 Axial Compressors

7.3.5.1 Velocity Diagram for Axial Compressors

To understand how the geometry of the blades and vanes affects the performance of the compressor, let us start by establishing the compressor blade/vane notation. Figure 7.24 shows a constant-radius cross section through a row of blades. The notation is somewhat similar to that used for wings except for several new parameters that are specific to airfoils in cascades. The new parameters are: (1) pitch, s, which is the distance between two identical points on adjacent airfoils, (2) stagger angle, γ, which is the angle between the axial direction and the line that connects the leading and trailing edges, and (3) solidity, σ, which is the ratio between the chord and the pitch.

The velocity entering the row of blades is W_{inlet} and the angle, measured from the axial direction, is β_{inlet}. The flow leaves the blade with velocity W_{outlet} after the airfoil has turned the flow by an angle θ. The velocity W_{outlet} makes a β_{outlet} angle with the axial direction.

Figure 7.24 Compressor blade notation.

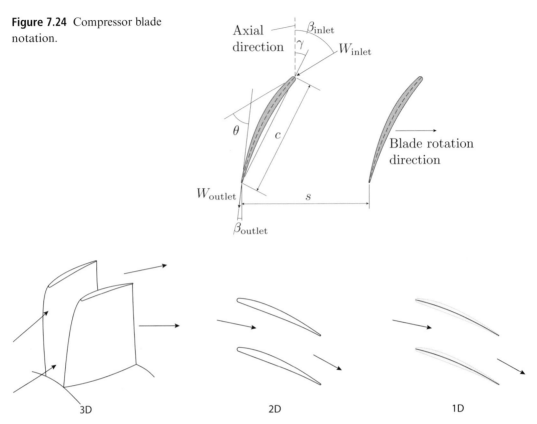

Figure 7.25 Compressor flow models: three-dimensional (3D), two-dimensional (2D) and one-dimensional (1D).

In Fig. 7.24 it was assumed that the velocities W_{inlet} and W_{outlet} are tangent to the camberline at the leading edge and trailing edge, respectively. The velocity W_{inlet} might or might not be tangent to the camber line at the leading edge, depending on the upstream flow conditions and shaft angular velocity. The velocity W_{outlet} always turns less than the airfoil camber line at the trailing edge would require, so that the actual β_{outlet}, called the flow angle, is higher than the β_{outlet} of the airfoil, called the metal angle. The lag in turning is a function of solidity: the smaller the solidity, the smaller the lag.

The flow in the compressor is three-dimensional and highly unsteady. There are, however, several levels of simplifications that can be made for modeling the flow in the compressor, as shown in Fig. 7.25. The three-dimensional flow can be modeled as two-dimensional if one neglects: (i) the flow in the radial direction due to centrifugal force, and (ii) the flow over the tip of the blade, from the pressure side to the suction side. The flow in a well designed compressor is as close to two dimensional as possible. The modeling of the two-dimensional flow can be further simplified if the details of the airfoil shape are neglected and the airfoil is approximated by its camber line. This simplification leads to one-dimensional modeling.

Figure 7.26 Compressor velocity diagram.

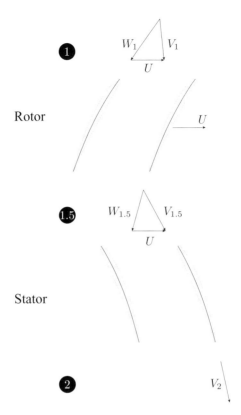

The modeling of the flow as one-dimensional is done by using the velocity diagram. The velocity diagram consists of the velocity vectors at the inlet and exit of each row of the compressor. The velocity vectors depict the absolute, relative, and transport velocities, as shown in Fig. 7.26. The one-dimensional flow model, although rather simple, provides relevant information on the compressor and therefore it is usually used for pre-design analysis. As will be shown in the following sections, once the velocity diagrams are determined, the specific work of the compressor and the pressure ratio can be estimated. The relationship between the absolute velocity, V, the relative velocity, W, and the transport velocity, U is

$$\vec{V} = \vec{U} + \vec{W}$$

where the transport velocity is $U = r\omega$.

Example 7.3.2 Air at 101.3 kPa and 288 K enters an axial compressor stage with a velocity of 170 m/s. There are no inlet guide vanes. The rotor blade has a tip diameter of 0.66 m, a hub diameter of 0.457 m, and rotates at 8000 rpm. The air enters the rotor blades and leaves the stator vanes in the axial direction with no change in velocity or radius. The air is turned through 15° as it passes through the rotor blades. The following assumptions can be made: (1) constant specific heats with $\gamma = 1.4$, (2) the air enters and leaves the blades at the blade angles, and (3) the axial component of the velocity remains constant over the compressor stage.

Figure 7.27 Velocity triangles for the entire stage.

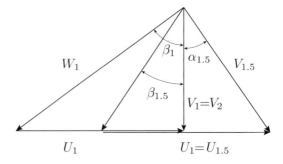

1. Construct the velocity diagrams at the mean blade height for this stage.
2. Determine the shape of the rotor blades and stator vanes at mean radius.

Solution

1. Velocity diagrams (see Fig. 7.27)

    <u>Location 1</u> (in front of the rotor blade)

$$r_{mean} = \frac{1}{2}(0.66 + 0.457)\frac{1}{2} = 0.2793 \text{ m}$$

$$U = r_{mean}\omega = r_{mean} \, 2\pi n/60 = 0.27932\pi \, 8000/60 = 233.9 \text{ m/s}$$

$$W_1 = \sqrt{V_1^2 + U^2} = \sqrt{170^2 + 233.9^2} = 289.2 \text{ m/s}$$

$$\beta_1 = \tan^{-1}\frac{U}{V_1} = \tan^{-1}\left(\frac{233.9}{170}\right) = 54°.$$

    <u>Location 1.5</u> (between the rotor blade and stator vane)

$$\beta_{1.5} = \beta_1 - 15 = 39°$$

$$W_{U_{1.5}} = V_1 \tan\beta_{1.5} = 137.7 \text{ m/s}$$

$$V_{U_{1.5}} = U - W_{U_{1.5}} = 233.9 - 137.7 = 96.2 \text{ m/s}$$

$$\tan\alpha_{1.5} = \frac{V_{U_{1.5}}}{V_1} \rightarrow \alpha_{1.5} = 29.5°.$$

    <u>Location 2</u> (downstream of the stator)

$$V_2 = V_1 = 170 \text{ m/s}.$$

2. To approximate the camber line, simply draw the velocity directions at the leading edge and trailing edge, as determined in Section 1, and then draw a curve that is tangent to these directions, as shown in Fig. 7.28. For the rotor blade, the tangent directions are given by the relative velocity, while for the stator vane, the tangent directions are given by the absolute velocity.

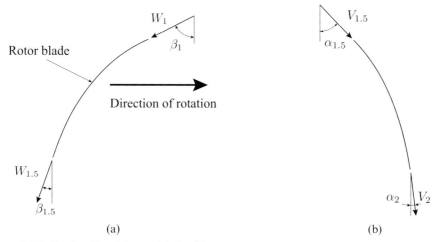

Figure 7.28 Camber line: (a) rotor blade, (b) stator vane.

In the velocity diagrams shown in Figs. 7.27 and 7.29a the velocities have a common apex. An alternative is to generate a diagram with a common transport velocity, as shown in Fig. 7.29b.

It is important to note that for the blade the relevant velocity is the relative velocity. For the vane, the relevant velocity is the absolute velocity.

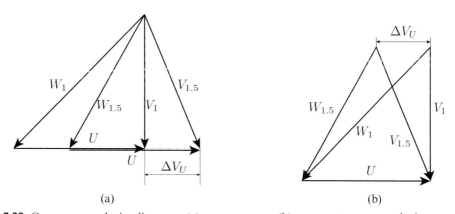

Figure 7.29 Compressor velocity diagrams: (a) common apex, (b) common transport velocity.

Figure 7.30 Compressor radius variation.

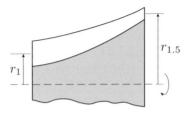

7.3.5.2 Energy Transfer

The torque on the shaft, T_{shaft}, for a compressor stage is obtained from the equation of moment of momentum (2.42)

$$T_{shaft} = \dot{m}\left(r_1 V_{1_u} - r_{1.5}V_{1.5_u}\right)$$

where station 1 is in front of the rotor blade and station 1.5 is between the rotor blade and the stator vane, as shown in Figs. 7.26 and 7.30.

Using the angular velocity, ω, the power of the compressor is

$$\mathcal{P} = \omega\, T_{shaft} = \dot{m}\left(\omega r_1 V_{1_u} - \omega r_{1.5}V_{1.5_u}\right)$$

or

$$\boxed{\mathcal{P} = \dot{m}\left(U_1 V_{1_u} - U_{1.5}V_{1.5_u}\right)}. \tag{7.12}$$

This is **Euler's equation for turbomachinery**. The specific work of the compressor, w, is

$$\boxed{w = \frac{\mathcal{P}}{\dot{m}} = U_1\, V_{1_u} - U_{1.5}V_{1.5_u}}. \tag{7.13}$$

Note that this expression of the specific work is valid for both axial and centrifugal compressors. A similar expression will be obtained for turbines. Note also that for compressors, the value of w is negative.

Once the velocity diagram is determined, using the Euler's equation for turbomachinery allows us to calculate the specific work of the compressor, the power needed to operate the compressor, and the pressure ratio of the compressor. The following example illustrates this process.

Example 7.3.3 For the compressor stage specified in Example 7.3.2, calculate the following:

1. mass flow rate;
2. power required to drive the compressor;
3. ideal total-to-total pressure ratio for this stage.

Solution

1. Mass flow rate

$$\dot{m} = \rho_1 A_1 V_{1a} = \frac{p_1}{RT_1} \frac{\pi}{4} \left(D_{\text{tip}}^2 - D_{\text{hub}}^2 \right) V_1 =$$

$$= \frac{101,300}{287.16 \times 288} \frac{\pi}{4} (0.66^2 - 0.457^2) \times 170 = 37.08 \text{ kg/s.}$$

2. Compressor power

   The torque is

$$T = \dot{m}(r_1 V_{1_u} - r_{1.5} V_{1.5_u}) = 37.08 \times (0.2793 \times 0 - 0.2793 \times 96.2) =$$
$$= -996.3 \text{ N m}$$

$$\mathcal{P}_c = T \omega = -996.3 \times 2\pi \frac{8000}{60} = -834,658 \text{ W} = -834.7 \text{ kW}$$

$$\mathcal{P}_c = \dot{m} w_c$$

   so that

$$w_c = \frac{\mathcal{P}_c}{\dot{m}} = \frac{-834.7}{37.08} = -22.5 \text{kJ/kg.}$$

3. Stage pressure ratio (ideal, total-to-total)

$$h_{01} = h_1 + \frac{V_1^2}{2}.$$

   Assuming that $c_p = \text{const}$ and using $\gamma = 1.4$, one obtains $c_p = 1,004 \text{ J/(kg K)}$.

$$c_p T_{01} = c_p T_1 + V_1^2/2 = 1,004\ 288 + 170^2/2 = 303,602 \text{ J/kg}$$
$$T_{01} = 303,602 / 1,004 = 302.4 \text{ K}$$
$$h_{02} = h_{01} + |w_C| = 303.6 + 22.5 = 326.1 \text{ kJ/kg}$$
$$T_{02} = h_{02}/c_p = 326.1/1.004 = 324.8 \text{ K.}$$

   Assuming that the compression process is isentropic, that is, $\Delta s_{12} = 0$, then

$$\frac{p_{02}}{p_{01}} = \left(\frac{T_{02}}{T_{01}} \right)^{\frac{\gamma}{\gamma-1}} = \left(\frac{324.8}{302.4} \right)^{\frac{1.4}{0.4}} = 1.284.$$

   Note 1 – If one doubles the velocity V_1 and U then the shape of the blades is the same but the mass rate of flow is doubled, the torque is four times larger, and the power is eight times larger.

   Note 2 – Increasing the turning angle of the rotor increases ΔV_U, increasing the torque and work per unit mass and therefore increasing the stage pressure ratio.

7.3.5.3 Compressor Parameters

This section will present the most important parameters that describe the compressor.

Flow Coefficient, ϕ

The flow coefficient is defined as the ratio between the axial component of the absolute velocity and the transport velocity, that is,

$$\phi := \frac{V_a}{U}.$$

The flow coefficient is a dimensionless parameter for the through flow, which is proportional to V_a. ϕ provides a measure of the axial component of the absolute velocity as a fraction of the transport velocity. Typical values of ϕ are in the neighborhood of 0.5.

Work Coefficient (or Blade Loading Coefficient), Ψ

The work coefficient is defined as the ratio between the stage (or more precisely rotor) specific work and the specific kinetic energy due to transport velocity, that is,

$$\Psi := \frac{-\Delta h_0}{U^2/2}$$

where $\Delta h_0 = h_{02} - h_{01}$ for the compressor. Note that the work on a stage is done only by the rotor because there are no moving parts in the stator. The work coefficient of a compressor has negative values because the compressor work is negative. Note that the work coefficient of a turbine has positive values because $\Delta h_0 = h_{04} - h_{03}$ and $h_{04} < h_{03}$.

For highly loaded stages $|\Psi| > 1.0$, while for lightly loaded stages $|\Psi| < 0.6$.

The work coefficient can be written as a function of the flow coefficient and the angles of the velocity triangle shown in Fig. 7.31:

$$\Psi = \frac{U(V_{1_u} - V_{1.5_u})}{\frac{U^2}{2}} = \frac{2(V_a \tan\alpha_1 - V_a \tan\alpha_{1.5})}{U}.$$

Figure 7.31 Compressor velocity diagram.

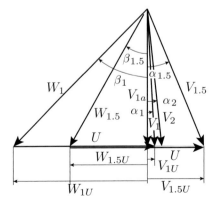

Figure 7.32 Hub and tip radii.

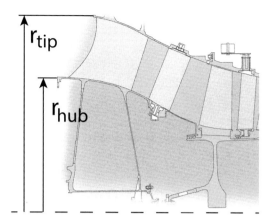

Note that angles α and β are positive if they turn the flow in the direction of the transport velocity, U. For example, in Fig. 7.31, angles $\alpha_{1.5}$ and α_2 are positive, while angles β_1 and $\beta_{1.5}$ are negative.

Using the definition of the flow coefficient ϕ, the work coefficient becomes

$$\Psi = 2\phi(\tan\alpha_1 - \tan\alpha_{1.5}).$$

Load Coefficient, $\overline{w}$

A parameter related to the work coefficient is the *load coefficient*, or *load factor*, defined as

$$\overline{w} = w/U^2 \tag{7.14}$$

where $\overline{w} = \Psi/2$.

Mach Number, M

The value of the M number at the tip of the blade

$$M = \frac{V}{\sqrt{\gamma RT}}$$

is used to determine whether the flow on the blade is subsonic or not. Usually if $M_{tip} < 0.8$ the flow in the compressor is subsonic. For larger values of M_{tip} the flow is transonic.

Hub-to-Tip Ratio, r_{hub}/r_{tip}

The ratio between the radius at hub and tip is an indication of the blade span size relative relative to the diameter of the compressor. Small values of the hub-to-tip ratio, *e.g.* 0.3, are typical for fan blades. Large values of the hub-to-tip ratio (*e.g.*, 0.7) are typical for the last stages of the compressor.

de Haller Number, $W_{1.5}/W_1$

The de Haller number shows how much the relative velocity slows down on the rotor blade. Therefore, the de Haller number is a measure of the compression ratio over the rotor blade. Typically, the de Haller number is larger than 0.72. A smaller de Haller number leads to larger turning and, therefore, the danger of flow separation.

Degree of Reaction, R'

The degree of reaction measures the ratio between the enthalpy variation over the rotor and the enthalpy variation over the stage:

$$\boxed{R' := \frac{h_{1.5} - h_1}{h_2 - h_1}.} \tag{7.15}$$

For incompressible flow, using the ideal gas law (2.66), the expression of the degree of reaction becomes

$$R' = \frac{p_{1.5} - p_1}{p_2 - p_1}.$$

Let us expand the numerator and denominator of (7.15) in order to obtain an expression of the degree of reaction as a function of velocity triangles parameters.

In the relative frame of reference, the stagnation enthalpy is

$$h_{0W} = h + W^2/2.$$

Because the blade does not move in the relative frame of reference, the work done on the blade is zero. Assuming the heat transfer is small, the energy conservation yields

$$h_{0W1.5} = h_{0W1}$$

or

$$h_{1.5} + W_{1.5}^2/2 = h_1 + W_1^2/2,$$

which yields

$$h_{1.5} - h_1 = \frac{W_1^2 - W_{1.5}^2}{2}.$$

In the absolute frame of reference, the stagnation enthalpy is

$$h_0 = h + V^2/2$$

and therefore the denominator of (7.15) can be written as

$$h_2 - h_1 = h_{02} - V_2^2/2 - (h_{01} - V_1^2/2) = h_{02} - h_{01} + (V_1^2 - V_2^2)/2.$$

Using (6.36), where $w_{1-2} < 0$, yields

$$h_2 - h_1 = -w_{1-2} + (V_1^2 - V_2^2)/2.$$

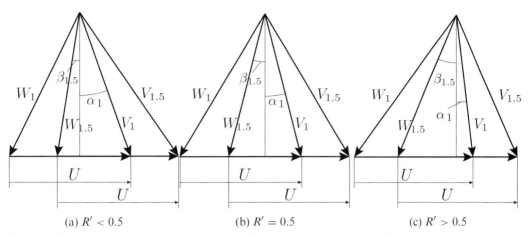

(a) $R' < 0.5$ (b) $R' = 0.5$ (c) $R' > 0.5$

Figure 7.33 Compressor velocity diagrams for different degrees of reaction R'.

Assuming $V_1 \approx V_2$ and using the Euler equation for turbomachinery (7.13) yields

$$h_2 - h_1 = -(U_1 V_{1_u} - U_{1.5} V_{1.5_u}).$$

If the flow radius does not change, then

$$h_2 - h_1 = -U(V_{1_u} - V_{1.5_u}).$$

Typically the variation of the axial velocity from one stage to another one is small. Consequently, the expression of the degree of reaction can be simplified by assuming that the axial velocity remains constant:

$$R' = \frac{W_1^2 - W_{1.5}^2}{-2U(V_{1_u} - V_{1.5_u})} = \frac{W_{1_u}^2 + V_{1_a}^2 - (W_{1.5_u}^2 + V_{1.5_a}^2)}{-2U(V_{1_u} - V_{1.5_u})} \overset{V_{1a} = V_{1.5a}}{=} \frac{W_{1_u}^2 - W_{1.5_u}^2}{-2U(V_{1_u} - V_{1.5_u})}$$

$$= \frac{-(W_{1_u} + W_{1.5_u})}{2U}.$$

Using the velocity triangle shown in Fig. 7.31, the velocity W_{1_u} is written as a function of U and V_{1_u} so that

$$R' = \frac{-(V_{1_u} - U + W_{1.5_u})}{2U} = \frac{1}{2} - \frac{V_{1_u} + W_{1.5_u}}{2U} = \frac{1}{2} - \frac{V_a}{2U}(\tan \alpha_1 + \tan \beta_{1.5}).$$

The value of the degree of reaction provides information about the velocity triangles, as shown in Fig. 7.33. The case $R' = 0.5$ is a particular case in which the enthalpy variation over the rotor is equal to the enthalpy variation over the stator. In this case, the velocity triangles display symmetry.

Example 7.3.4 What are the degree of reaction, flow coefficient, and work coefficient for the compressor given at the previous example?

Solution

$$R' = \frac{1}{2} - \frac{V_a}{2U}(\tan\beta_{1.5} + \tan\alpha_1) = \frac{1}{2} - \frac{170}{2 \times 233.9}(\tan(-39) + \tan 0) = 0.794$$

$$\phi = \frac{V_a}{U} = \frac{170}{233.9} = 0.727$$

$$\Psi = 2\phi(\tan\alpha_1 - \tan\alpha_{1.5}) = 2 \times 0.727(\tan 0 - \tan 29.5) = -0.823.$$

7.3.6 Centrifugal Compressors

Centrifugal compressors (or radial-flow compressors) achieve part of the compression by forcing the fluid to move in the radial direction. As a result, the centrifugal force produced by the rotation of the impeller increases the fluid pressure as the radius increases. The pressure rise in the centrifugal compressor results from the energy imparted to the fluid by the rotor blades, similarly to the axial compressor. There is, however, a major difference between how the pressure increase is achieved in centrifugal and axial compressors. In the axial compressor, the pressure rise is the result of exchanging kinetic for thermal energy in the diffusion process. In the centrifugal compressor, the pressure rise results from the change in the potential energy in the centrifugal force field of the impeller (or rotor).

In an axial compressor, the pressure rise is intimately related to the lift force on the airfoil. In a centrifugal compressor, the pressure rise is driven by the centrifugal force that acts on the fluid. Consequently, the centrifugal compressor is less limited by boundary layer growth and separation in adverse pressure gradients than the axial compressor. The difference between the two mechanisms that increase pressure leads to noticeable differences between the two types of compressors.

As mentioned at page 294, the centrifugal compressor has several advantages over the axial compressor: (1) higher stage pressure ratio, typically between 2 and 4 compared to approximately 1.3 for the axial stage, (2) shorter length for the same overall pressure ratio, (3) wider range of stable operation between surging and choking, (4) less reduction in performance due to "dirty" blades, (5) simplicity and ruggedness of construction, (6) flow direction at discharge convenient for the installation of an inter-cooler and/or heat exchanger, (7) lighter, with smaller inertia, and (8) less expensive to manufacture.

One limitation of the centrifugal compressor is that the frontal area per mass flow rate is larger than that of an axial compressor. A larger frontal area increases the drag and weight of the engine. A second limitation of the centrifugal compressor is that is less suitable for multi-staging because the fluid discharged must be turned to the axial direction, which results

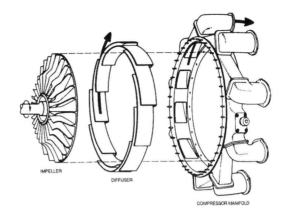

Figure 7.34 Exploded view of centrifugal compressor (reproduced with permission of Pratt & Whitney).

in additional losses that typically reduce the stage efficiency by 5 percentage points [Flack, 2005, p.374].

Centrifugal compressors are commonly used on small turboprop and turboshaft engines. More recently, centrifugal compressors have been used as the high-pressure compressors in some multiple shaft engines. This is because in high-pressure compressors, the high pressure and high density lead to small flow areas compared to those of the inlet stages. Consequently, the diameter of the centrifugal compressor is not a limiting factor in these engines.

7.3.6.1 Configuration

The centrifugal compressor consists of several major components: the air intake system, the impeller (or rotor), the diffuser, and the exit manifold, as shown in Fig. 7.34.

Impellers
The impeller is connected to the turbine through the shaft. The impeller has radially disposed blades that form convergent passages with the compressor casing. The blades can be situated on one or both sides, as shown in Fig. 7.35. The blades are either swept back or, for ease of manufacture and to reduce mechanical stresses, straight radial. Some impellers alternate blades that span the entire length of the impeller with shorter blades, called splitters (see Fig. 7.35a), which are situated at the impeller exit. The purpose of the splitter blades is to better control the flow at the exit from the impeller.

To provide a smooth transition of the axially flowing air entering the "eye" of the impeller, the blades are curved in the direction of rotation. These curved sections are often separate from the impeller, for ease of manufacture and maintenance. This component located at the inlet of the impeller is called the *inducer*.

Diffusers
The diffuser assembly consists of a number of blades located at the impeller exit. The passages between the blades are divergent to slow down the flow and increase the pressure. The size of the clearance between the impeller and the diffuser is important: if it is too large, the losses increase; if it is too small, unsteady flow and vibration could occur. Once the fluid leaves the

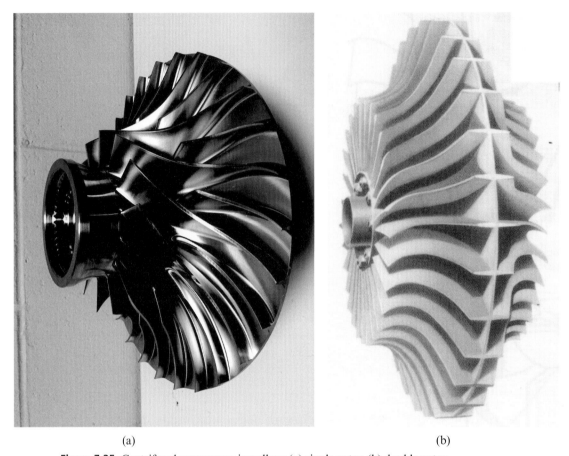

(a)   (b)

Figure 7.35 Centrifugal compressor impellers: (a) single entry, (b) double entry.

diffuser, it enters the manifold and flows in the axial direction. The centrifugal compressor of a jet engine has either a continuous annular exit or a manifold that directs the air into the burner.

7.3.6.2 Principles of Operation

The turbine rotates the impeller, which continuously induces air into the center (or "eye") of the impeller. Once it enters the impeller, the air is forced to move radially. As the radius increases, the centrifugal force increases. As a result, the pressure and velocity increase.

The air leaving the impeller enters the diffuser where it is slowed down in the divergent nozzles. The kinetic energy is reduced and the pressure and enthalpy increased. As shown later in this section, it is desirable to design the centrifugal compressor so that approximately half of the pressure rise occurs in the impeller and half in the diffuser.

Figure 7.36 Pressure and velocity variation in centrifugal compressor.

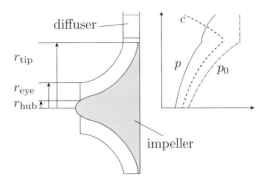

Figure 7.37 Control volume in centrifugal compressor.

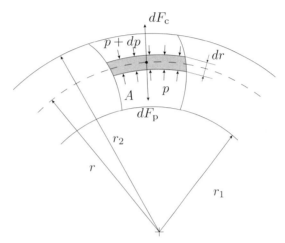

The typical variations of velocity[1] and pressure in the centrifugal compressor are shown in Fig. 7.36.

To determine the pressure variation in the centrifugal compressor, let us consider Fig. 7.37, where a small control volume is located at radius r. Radius r is bounded by r_1 and r_2, the smallest and the largest radii of the compressor, respectively. Assuming that the angular velocity is ω, the centrifugal force acting on the control volume is

$$dF_c = \omega^2 r\, dm.$$

The mass is $dm = \rho\, dA\, dr$, where ρ is the fluid density. The net pressure force acting on the control volume is

$$dF_p = (p + dp)dA - p\, dA = dp\, dA.$$

For the control volume to be in equilibrium, $dF_c = dF_p$, so that

$$\frac{dp}{\rho} = \omega^2 r\, dr. \tag{7.16}$$

---

[1] For the centrifugal compressor the absolute velocity will be denoted by c.

Integrating (7.16) between radii r_1 and r_2 yields

$$\int_1^2 \frac{dp}{\rho} = \frac{1}{2}\left(U_2^2 - U_1^2\right).$$

If the density variation is neglected, so that $\rho = \rho_{avg} = \texttt{const.}$, then

$$p_2 = p_1 + \rho_{avg}\frac{U_2^2 - U_1^2}{2}.$$

It is apparent that if ω increases, p_2 increases. Therefore, to maximize the pressure ratio and the mass flow rate, the impeller must be rotated at high speed. Modern centrifugal compressors operate at tip speeds higher than 500 m/s. To prevent losses caused by the flow leakage between the impeller and the casing, the clearance is kept as small as possible.

If instead of assuming that the density variation is small, one assumes the process in the impeller is isentropic, that is, $\rho/\rho_1 = (p/p_1)^{\frac{1}{\gamma}}$, then integrating (7.16) from the impeller inlet, 1, to the impeller exit, 2, yields

$$\left(\frac{p_2}{p_1}\right)^{\frac{\gamma-1}{\gamma}} - 1 = \frac{\gamma-1}{2}\frac{(\omega r_2)^2 - (\omega r_1)^2}{\gamma RT_1}. \tag{7.17}$$

If r_2 is much larger than r_1, the r_1^2 term can be neglected compared to r_2^2, and (7.17) reduces to [Kerrebrock, 1992, p. 267]

$$\left(\frac{p_2}{p_1}\right)^{\frac{\gamma-1}{\gamma}} - 1 = \frac{\gamma-1}{2}\mathrm{M}_t^2 \tag{7.18}$$

where

$$\mathrm{M}_t^2 = (\omega r_2)^2/(\gamma RT_1). \tag{7.19}$$

In addition, since the process is isentropic:

$$\frac{\gamma-1}{2}\mathrm{M}_t^2 = \left(\frac{p_2}{p_1}\right)^{\frac{\gamma-1}{\gamma}} - 1 = \frac{T_2}{T_1} - 1$$

so that

$$\frac{T_2}{T_1} = 1 + \frac{\gamma-1}{2}\mathrm{M}_t^2. \tag{7.20}$$

Example 7.3.5   What are the pressure ratio p_2/p_1 and the temperature ratio T_2/T_1 in an impeller if the working fluid is air and M_t has the following values: 0.6, 1, and 1.4

Solution
Substituting the M_t values in (7.18) and (7.20) yields

M_t	T_2/T_1	p_2/p_1
0.6	1.07	1.27
1	1.2	1.89
1.4	1.39	3.18

One should note that if M_t is larger than one, most likely there will be losses due to shock waves, and therefore the pressure and temperature ratios will be less than the values corresponding to isentropic flows.

The kinetic energy of the air leaving the impeller is due primarily to its tangential velocity and secondarily to its smaller radial velocity. In the diffuser, this kinetic energy is converted into thermal energy, leading to a further increase of pressure and temperature. Assuming that the process in the diffuser is isentropic yields

$$\frac{p_3}{p_2} = \left(1 + \frac{\gamma - 1}{2}M_2^2\right)^{\frac{\gamma-1}{\gamma}}, \tag{7.21}$$

where p_3 is the pressure at the exit from the diffuser and $M_2^2 = (\omega r_2)^2/(\gamma R T_2)$. Using (7.19), one obtains

$$M_2^2 = \frac{(\omega r_2)^2}{\gamma R T_1}\frac{T_1}{T_2} = M_t^2\frac{T_1}{T_2}$$

and using (7.21) yields

$$\left(\frac{p_3}{p_2}\right)^{\frac{\gamma-1}{\gamma}} - 1 = \frac{\gamma-1}{2}M_t^2\frac{T_1}{T_2} \overset{(7.20)}{=} \frac{\gamma-1}{2}M_t^2\frac{1}{1 + \frac{\gamma-1}{2}M_t^2}. \tag{7.22}$$

Using the isentropic relation between temperature and pressure ratios in (7.22) yields

$$\frac{T_3}{T_2} = \frac{1 + (\gamma-1)M_t^2}{1 + \frac{\gamma-1}{2}M_t^2},$$

and multiplying by T_2/T_1 of (7.20) gives

$$\frac{T_3}{T_1} = 1 + (\gamma-1)M_t^2 \tag{7.23}$$

which then leads to

$$\frac{p_3}{p_1} = \left[1 + (\gamma-1)M_t^2\right]^{\frac{\gamma}{\gamma-1}}. \tag{7.24}$$

Comparing (7.20) and (7.23) shows that, for high efficiency centrifugal compressors, half of the temperature rise of the stage happens in the diffuser. Therefore the pressure ratio in the impeller should be equal to the pressure ratio in the diffuser.

Example 7.3.6 The working fluid of a centrifugal compressor is air and $M_t = 0.6$.

1. Calculate the relative temperature rise in the impeller and over the stage.
2. Calculate the pressure ratio over the impeller and over the diffuser.

Solution
Using (7.20)

$$\frac{T_2 - T_1}{T_1} = \frac{\gamma - 1}{2} M_t^2 = 0.2 \times 0.6^2 = 0.072.$$

Using (7.23)

$$\frac{T_3 - T_1}{T_1} = (\gamma - 1) M_t^2 = 0.4 \times 0.6^2 = 0.144.$$

Using (7.18)

$$\frac{p_2}{p_1} = \left(1 + \frac{\gamma - 1}{2} M_t^2\right)^{\frac{\gamma}{\gamma - 1}} = (1 + 0.2 \times 0.6^2)^{3.5} = 1.27. \tag{7.25}$$

Using (7.24)

$$\frac{p_3}{p_1} = \left[1 + (\gamma - 1) M_t^2\right]^{\frac{\gamma}{\gamma - 1}} = (1 + 0.4 \times 0.6^2)^{3.5} = 1.60$$

so that

$$\frac{p_3}{p_2} = \frac{p_3}{p_1} \frac{p_1}{p_2} = 1.60/1.27 = 1.26.$$

Consequently, the pressure rise on the impeller is approximately equal to the pressure rise on the diffuser.

7.3.6.3 Energy Transfer from Impeller to Fluid

The impeller (or rotor) is the only part of the engine that transfers work to the fluid. This energy transfer is done through the blades. To determine the parameters that influence the compression, we will start by assuming first that the impeller has an infinite number of blades that have a thickness that is infinitely small, and then we will consider the case of impellers with a finite number of blades of finite thickness.

Impeller with an Infinite Number of Blades
Let us consider the following assumptions needed to determine the parameters that affect compression:

1. The fluid is a perfect gas.

2. The rotor angular velocity, ω, is constant.
3. The compression is isentropic.
4. The number of blades of the rotor is infinite, and the blades have a thickness that is infinitely small.

One could calculate the work transferred to the fluid if the pressure distribution on the blade were known. Then, the force in the tangential direction would be

$$F_u = \int_1^2 \Delta p \, dA$$

where Δp is the pressure difference between the pressure side and suction side along the blades. Then, knowing ω, one could calculate the work using the sum of forces on all the blades. Unfortunately, one does not know Δp, so one must use instead the Euler equation for turbomachinery (7.12).

Using (7.12) yields

$$\frac{P}{\dot{m}} = \frac{uF_u}{\dot{m}} = w_\infty = U_2 c_{2u\infty} - U_1 c_{1u} \tag{7.26}$$

where w_∞ is the specific work for an infinite number of blades. Using the cosine formula yields a relationship between the relative, absolute, and transport velocities shown in Fig. 7.38:

$$W^2 = U^2 + c^2 - 2Uc\cos(90 - \alpha) = U^2 + c^2 - 2Uc_u$$

and

$$2Uc_u = U^2 + c^2 - W^2 \tag{7.27}$$

Figure 7.38 Centrifugal compressor velocity triangles.

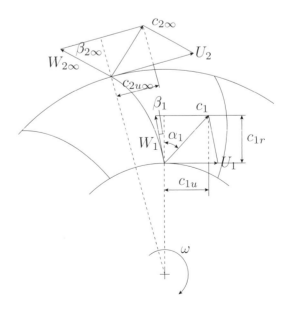

so that the specific work (7.26) can be written using (7.27) as

$$w_\infty = \frac{1}{2}\left(U_2^2 + c_{2\infty}^2 - W_{2\infty}^2 - U_1^2 - c_1^2 + W_1^2\right) =$$

$$= \frac{1}{2}(U_2^2 - U_1^2) + \frac{1}{2}(c_{2\infty}^2 - c_1^2) - \frac{1}{2}(W_{2\infty} - W_1^2). \tag{7.28}$$

Using the first law of thermodynamics (6.35) and accounting for the adiabatic process resulting from the third assumption on page 321 yields

$$w_\infty = h_{02} - h_{01} = \underbrace{h_2 - h_1}_{w_{static}} + \underbrace{\frac{1}{2}(c_{2\infty}^2 - c_1^2)}_{w_{dynamic}}. \tag{7.29}$$

This expression of the specific work allows us to identify a static and a dynamic component of the work.

Example 7.3.7 Prove that for an isentropic process

$$\int_1^2 \frac{dp}{\rho} = h_2 - h_1. \tag{7.30}$$

Solution

Using the relationship between pressure and density for an isentropic process (2.88) yields

$$\int_1^2 \frac{dp}{\rho} = \int_1^2 \frac{const \; d\rho^\gamma}{\rho} = \frac{p}{\rho^\gamma}\int_1^2 \gamma \frac{\rho^{\gamma-1} d\rho}{\rho} = \frac{p}{\rho^\gamma}\gamma \frac{\rho^{\gamma-1}}{\gamma-1}\bigg|_1^2 = \frac{\gamma}{\gamma-1}\frac{p}{\rho}\bigg|_1^2 =$$

$$= \frac{\gamma}{\gamma-1}RT\bigg|_1^2 = c_p T\big|_1^2 = h_2 - h_1.$$

Combining (7.28), (7.29), and (7.30) yields

$$h_2 - h_1 = \int_1^2 \frac{dp}{\rho} = \frac{1}{2}(U_2^2 - U_1^2) - \frac{1}{2}(W_{2\infty} - W_1^2). \tag{7.31}$$

According to (7.31), the increase in the static pressure in the impeller can be obtained by increasing ω, and implicitly U, and by slowing down the relative velocity component.

Part of $w_{dynamic}$ can be used in the diffuser to increase static pressure. This transformation, however, is done with higher losses in the diffuser than in the impeller. Therefore, for a given value of w_∞, it is better to maximize the static pressure rise in the impeller, that is, minimize $w_{dynamic}$.

By rearranging the terms in (7.31), one can identify a constant

$$\underbrace{h_1 + \frac{1}{2}(W_1^2 - U_1)}_{I_1} = \underbrace{h_2 + \frac{1}{2}(W_{2\infty}^2 - U_2)}_{I_2} = const. \tag{7.32}$$

This allows us to introduce the *rothalpy*, I

$$I := h + \frac{1}{2}(W^2 - U^2),$$
(7.33)

which is constant at any radius along the streamline, as long as there are not friction losses, that is, the process is isentropic.

When studying centrifugal compressors it is beneficial to define stagnation enthalpy based on the relative velocity,

$$h_{0W} := h + \frac{W^2}{2}$$
(7.34)

as opposed to the stagnation enthalpy based on the absolute velocity, h_0. Combining (7.33) and (7.34) yields

$$I = h_{0W} - \frac{U^2}{2}$$
(7.35)

so that h_{0W} varies with radius r.

So far, the specific work w_∞ is just a function of kinematic parameters (see (7.28)) at inlet and exit of the impeller, and not a function of geometric parameters, such as blade shape. The blade shape implicitly affects w_∞, however, by imposing modifications to the velocity triangles.

Let us consider further the influence of the kinematic parameters on the specific work transfer, w_∞. Consider a cylindrical cut at the inlet mean radius, and let us unravel this cut as shown in Fig. 7.39.

Let us consider the case when the transport velocity U_1 at inlet in the impeller is constant and the magnitude of the absolute velocity c_1 is constant. Three directions are considered for the absolute velocity: (1) axial direction, that is, $\alpha_1 = 0$; (2) positive pre-swirl, $\alpha_1 > 0$ and $c_{1u} > 0$; and (3) negative pre-swirl, $\alpha_1 < 0$ and $c_{1u} < 0$. From (7.26) one concludes that if $U_2 c_{2u\infty} = \text{const}$ then w_∞ increases if the inlet flow has a negative pre-swirl, that is, $\alpha_1 < 0$. In this case, however, the relative velocity W_1 increases and it is possible for shock waves to

Figure 7.39 Inlet impeller velocity triangles: $U_1 = \text{const}$ and $|c_1| = |c_1{}'| = |c_1{}''|$.

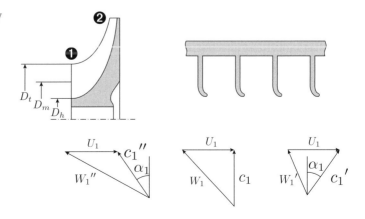

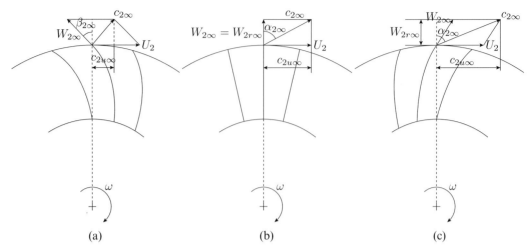

Figure 7.40 Outlet impeller velocity triangles: $U_2 = $ const and $W_{2r\infty} = $ const. (a) Backward swept, (b) radial, and (c) forward swept.

occur, which would significantly increase losses. The limit commonly used in the design is $M_{W_1} < 0.93$. This limits the magnitude of the negative pre-swirl.

Let us now consider the case when $U_1 c_{1u} = $ const, the transport velocity U_2 at exit of the impeller is constant, and the radial component of the absolute velocity $c_{2\infty}$ is constant, as shown in Fig. 7.40.

Using (7.26) one concludes that the forward-swept blades produce the largest w_∞. As shown in (7.29), the increase of w_∞ is obtained by increasing w_{dynamic}. This is not desirable because to increase static pressure one needs to slow down more the fluid in the diffuser. As a result, a larger diffuser is needed and higher losses will result. Consequently, most of the pressure increase obtained by forward-swept blades is lost due to higher losses in the diffuser. In addition, forward-swept blades are more sensitive to regime variation because of the large $c_{2\infty}$. Therefore, it is preferable to increase the specific work by increasing wheel speed instead of increasing $\alpha_{2\infty}$.

Backward-swept blades are commonly used for small centrifugal compressors. Radial flow blades have only stretching stresses and no bending stresses as occur in backward- or forward-swept blades. Radial flow blades are used for large centrifugal compressors.

It is convenient to use dimensionless parameters to describe the performance of the centrifugal compressor. The first dimensionless parameter is the *circulation coefficient*

$$\varphi_{2u\infty} := \frac{c_{2u\infty}}{U_2},\tag{7.36}$$

which is the ratio between the tangent component of the absolute velocity and the transport velocity, both at the exit of the impeller. Using the circulation coefficient, the specific work can be written as

$$w_\infty = U_2 c_{2u\infty} - U_1 c_{1u} = U_2^2 \left(\frac{c_{2u\infty}}{U_2} - \frac{U_1}{U_2} \frac{c_{1u}}{U_2} \right) = U_2^2 \left(\varphi_{2u\infty} - \frac{D_1}{D_2} \frac{c_{1u}}{U_2} \right),\tag{7.37}$$

where D_1 and D_2 are the diameters at the inlet and outlet.

The load coefficient (7.14) is

$$\overline{w}_\infty = \frac{w_\infty}{U_2^2} = \varphi_{2u\infty} - \frac{D_1}{D_2}\frac{c_{1u}}{U_2}.$$

If the flow enters the diffuser axially, that is, $c_{1u} = 0$, then

$$\overline{w}_\infty = \varphi_{2u\infty}$$

and the load coefficient is equal to the circulation coefficient.

The second dimensionless parameter specific to the centrifugal compressor is the *flow rate coefficient*

$$\varphi_{2r\infty} := \frac{c_{2r\infty}}{U_2}. \tag{7.38}$$

Using the velocity triangles of Fig. 7.40, let us find a relationship between the circulation coefficient and the flow rate coefficient. Let us start with the circulation coefficient

$$\varphi_{2u\infty} = \frac{U_2 - W_{2u\infty}}{U_2} = 1 - \frac{W_{2u\infty}}{W_{2r\infty}}\frac{W_{2r\infty}}{U_2} \overset{W_{2r\infty}=c_{2r\infty}}{=} 1 - \varphi_{2r\infty}\tan\beta_{2\infty}. \tag{7.39}$$

Typical values for the flow rate coefficient are $\varphi_{2r\infty} \in (0.08, 0.32)$.

Impeller with a Finite Number of Blades

The previous section assumed that the impeller had an infinite number of blades that have a thickness that is infinitely small. In this section we will assume that the impeller has a finite number of blades. All the other assumptions proposed on page 321 will be maintained.

Since the impeller has a finite number of blades, the fluid will no longer follow the direction imposed by the blade. As a result, the flow angle β will be different from the blade angle, β_f. The blade angle is called the fixed or metal angle, and $\beta_f = \beta_\infty$.

Let us consider the control volume shown in Fig. 7.41 and evaluate the forces acting on it. It is assumed that the trajectory of the control volume has a curvature R. In addition, the control volume is situated at radius r. The impeller is rotating with angular velocity ω.

Two centrifugal forces are acting on the control volume: one due to the rotation of the impeller

$$dF_{\text{centrif}} = \omega^2 r\,dm$$

and the other one due to the curvature $1/R$ of the control volume trajectory

$$dF_{\text{centrif}_R} = \frac{W^2}{R}dm.$$

There is also a Coriolis force acting on the control volume

$$d\vec{F}_{\text{Coriolis}} = -2(\vec{\omega} \times \vec{W})dm$$

whose magnitude is $dF_{\text{Coriolis}} = 2\omega W dm$, and the orientation is shown in Fig. 7.41. The fourth force acting on the control volume is the force due to inertia

$$dF_{\text{inertia}} = \frac{\partial W}{\partial t}dm.$$

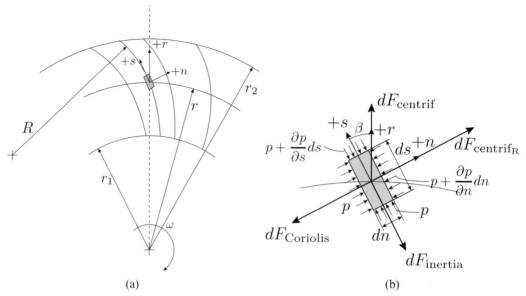

Figure 7.41 Forces acting on a control volume in an impeller with a finite number of blades. (a) Control volume in impeller and (b) detail of control volume.

The equilibrium equations along the s direction

$$\frac{\partial p}{\partial s} ds\, dn = \left(\omega^2 r \cos \beta - \frac{\partial W}{\partial t} \right) dm$$

and along the n direction

$$\frac{\partial p}{\partial n} ds\, dn = \left(\omega^2 r \sin \beta + \frac{W^2}{R} - 2\omega W \right) dm$$

can be rewritten using $dm = \rho ds\, dn$ (for a control volume of dimensions ds, dn and 1) as

$$\frac{1}{\rho}\frac{\partial p}{\partial s} = \omega^2 r \cos \beta - \frac{\partial W}{\partial t} \tag{7.40}$$

and

$$\frac{1}{\rho}\frac{\partial p}{\partial n} = \omega^2 r \sin \beta + \frac{W^2}{R} - 2\omega W. \tag{7.41}$$

The values of $\sin \beta$ and $\cos \beta$ can be written as a function of the variation of r with respect to the normal and tangential directions (see Fig. 7.42)

$$\sin \beta = \frac{\partial r}{\partial n}$$

$$\cos \beta = \frac{\partial r}{\partial s}$$

Figure 7.42 Coordinate system.

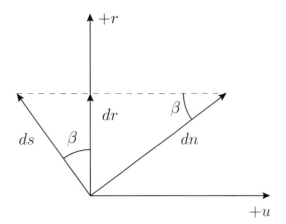

so that (7.40) and (7.41) become

$$\frac{1}{\rho}\frac{\partial p}{\partial s} = \omega^2 r\frac{\partial r}{\partial s} - \frac{\partial W}{\partial t} \tag{7.42}$$

and

$$\frac{1}{\rho}\frac{\partial p}{\partial n} = \omega^2 r\frac{\partial r}{\partial n} + \frac{W^2}{R} - 2\omega W. \tag{7.43}$$

Since it was assumed that $\omega = \texttt{const}$, the time variation of W is

$$\frac{\partial W}{\partial t} = W\frac{\partial W}{\partial s},$$

and therefore s is the only independent variable in (7.42). Consequently, (7.42) can be written as

$$\frac{1}{\rho}\frac{dp}{ds} = \omega^2 r\frac{dr}{ds} - W\frac{dW}{ds},$$

which yields

$$\frac{dp}{\rho} = \omega^2 r\,dr - W\,dW$$

and after integration gives

$$\int\frac{dp}{\rho} + \frac{W^2}{2} - \frac{U^2}{2} = \texttt{const} = I \tag{7.44}$$

where I is the rothalpy. Recall that $I = \texttt{const}$ in the entire mass if the flow is isentropic, and $I = \texttt{const}$ along the streamline if there is friction.

Let us take the partial derivative $\partial/\partial n$ of (7.44) so that

$$\frac{1}{\rho}\frac{dp}{dn} = \omega^2 r\frac{dr}{dn} - W\frac{dW}{dn} \tag{7.45}$$

and then eliminate the term $\frac{1}{\rho}\frac{dp}{dn}$ between (7.43) and (7.45)

$$\omega^2 r \frac{\partial r}{\partial n} + \frac{W^2}{R} - 2\omega W = \omega^2 r \frac{dr}{dn} - W \frac{dW}{dn}$$

to obtain

$$\frac{dW}{dn} = 2\omega - \frac{W}{R}, \tag{7.46}$$

which gives the variation of the relative velocity W along the normal direction.

For forward-swept blades, which have an opposite curvature, the sign of the centrifugal force $dF_{centrif_R}$ is changed and (7.46) becomes

$$\frac{dW}{dn} = 2\omega + \frac{W}{R}.$$

The relative velocity variation in the impeller can be determined using (7.46). In general, the radius of curvature varies along the n direction. To simplify the integration of (7.46), one will assume $R = \text{const}$. Equation (7.46) is rearranged as

$$\frac{-dW}{2\omega R - W} = \frac{-dn}{R},$$

which after integration yields

$$\ln(W - 2\omega R) + C = \frac{-n}{R}.$$

To determine the integration constant C, let us assume that $W = W_m$ at $n = 0$, which yields

$$C = -\ln(W_m - 2\omega R).$$

The relative velocity becomes

$$W = 2\omega R + (W_m - 2\omega R)e^{-n/R}.$$

Since $n/R \ll 1$, the series expansion with $x = n/R$

$$e^{-x} = 1 - x + \frac{1}{2!}x^2 - \frac{1}{3!}x^3 + \cdots$$

can be truncated to include only the first two terms, so that

$$W \approx 2\omega R + (W_m - 2\omega R)\left(1 - \frac{n}{R}\right),$$

which reduces to

$$W \approx W_m \left(1 - \frac{n}{R}\right) + 2\omega n. \tag{7.47}$$

The pressure and relative velocity variation in the channel between two adjacent blades of the impeller are shown in Fig. 7.43.

Figure 7.43 Pressure and relative velocity variation in channel between adjacent blades.

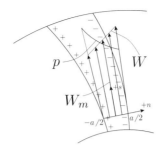

If at a given point the width of the channel between adjacent blades is a, then

$$W(n = -a/2) \approx W_m \left(1 + \frac{a}{2R}\right) - a\omega$$
$$W(n = a/2) \approx W_m \left(1 - \frac{a}{2R}\right) + a\omega.$$

For radial blades, $R \to \infty$ and the relative velocities are

$$W(n = -a/2) \approx W_m - a\omega$$
$$W(n = a/2) \approx W_m + a\omega.$$

Equation (7.47) shows that the relative velocity has two components:

$$W_I = W_m \left(1 - \frac{n}{R}\right)$$
$$W_{II} = 2\omega n.$$

W_I depends on the channel width, mass flow rate, and blade curvature. W_{II} depends on ω and represents a curl that spins in the opposite direction to the impeller.

As ω increases, the velocity W_{II} increases, the velocity $W(n = -a/2)$ decreases, and the pressure $p(n = -a/2)$ increases. The relative velocity on the pressure side of the blade, that is, at $n = -a/2$, is zero when the angular velocity is

$$\omega_{\lim} \overset{(7.47)}{=} W_m \left(\frac{1}{a} + \frac{1}{2R}\right).$$

If the angular velocity ω exceeds $\omega_{\lim}$ then backflow and separation occurs on the pressure side of the blade.

Another particular value of angular velocity is $\omega = W_m/(2R)$. In this case, the relative velocity W is constant and no longer depends on R and n:

$$W \overset{(7.47)}{=} W_m \left(1 - \frac{n}{R}\right) + 2W_m/(2R)n = W_m.$$

Figure 7.44 Vortex in relative frame of reference.

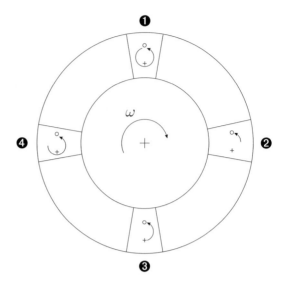

Let us consider Fig. 7.44, which shows two particles in the channel between two adjacent blades. Let us follow their position relative to the blades, assuming a frictionless flow. Because there is no friction, the fluid particles will have the same position with respect to a fixed system of reference.

Following the two particles of Fig. 7.44, they appear to rotate for an observer traveling with the impeller, while they have the same position with respect to a fixed system of reference. If the flow is viscous, the vortex will still be present, but it will be somewhat weaker.

Because of the vortex in the relative frame of reference, the angle β_2 for a finite number of blades is larger than $\beta_{2f} = \beta_{2\infty}$, as shown in Fig. 7.45. Consequently, $c_{2u} < c_{2u\infty}$, and since the specific work expressions are

$$w = U_2 c_{2u} - U_1 c_{1_u} \tag{7.13}$$

$$w_\infty = U_2 c_{2u\infty} - U_1 c_{1u} \tag{7.26}$$

then $w < w_\infty$.

The ratio between c_{2u} and $c_{2u\infty}$ is called the *slip factor*, μ

$$\mu := \frac{c_{2u}}{c_{2u\infty}} \tag{7.48}$$

where μ is always smaller or equal to 1.

The pressure ratio in the impeller can be calculated as

$$\frac{p_{02}}{p_{01}} = \left(\frac{T_{02i}}{T_{01}} \right)^{\frac{\gamma}{\gamma-1}} \overset{c_p=\text{const}}{=} \left(\frac{h_{02i}}{h_{01}} \right)^{\frac{\gamma}{\gamma-1}} = \left(\frac{h_{01} + w_i}{h_{01}} \right)^{\frac{\gamma}{\gamma-1}} \tag{7.49}$$

where the ideal work in the impeller, w_i, and the impeller efficiency, η_{12}, are related to the real work in the impeller, w

$$w_i = \eta_{12} w = \eta_{12} \left(U_2 c_{2u} - U_1 c_{1_u} \right) = \eta_{12} \left(U_2 \mu c_{2u\infty} - U_1 c_{1_u} \right).$$

Figure 7.45 Velocity triangles at impeller outlet.

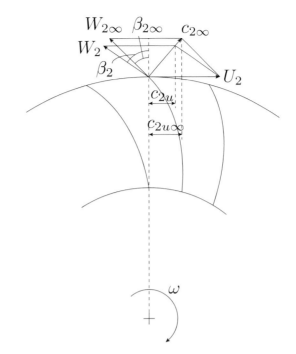

Substituting w_i in (7.49) yields the pressure ratio

$$\frac{p_{02}}{p_{01}} = \left(1 + \frac{\eta_{12}\left(U_2\mu c_{2u\infty} - U_1 c_{1u}\right)}{c_p T_{01}}\right)^{\frac{\gamma}{\gamma-1}}. \tag{7.50}$$

Equation (7.50) shows that the pressure rise in the impeller is reduced by the efficiency of the impeller and by the slip factor.

Two particular cases provide further insight in the impeller performance: (a) no preswirl, and (b) no preswirl and radial blades. If there is no preswirl, that is, $c_{1u} = 0$, then the specific work is

$$w = U_2 c_{2u} = U_2 \mu c_{2u\infty}$$

and the pressure ratio across the impeller is

$$\frac{p_{02}}{p_{01}} = \left(1 + \frac{\eta_{12} U_2 \mu c_{2u\infty}}{c_p T_{01}}\right)^{\frac{\gamma}{\gamma-1}}.$$

If the blades are radial, in addition to zero preswirl, the specific work is

$$w = U_2 c_{2u} = \mu U_2^2$$

because $c_{2u\infty} = U_2$. The pressure ratio is

$$\frac{p_{02}}{p_{01}} = \left(1 + \frac{\eta_{12}\mu U_2^2}{c_p T_{01}}\right)^{\frac{\gamma}{\gamma-1}}.$$

Figure 7.46 Impeller outlet geometry.

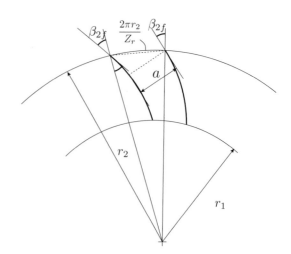

In both cases the specific work is proportional to the slip factor. Additionally, the pressure ratio is affected equally by the impeller efficiency and the slip factor.

The variation of velocity c_u due to the finite number of blades is

$$\Delta c_u = c_{2u\infty} - c_{2u} \tag{7.51}$$

so that the circulation coefficient for a finite number of blades

$$\varphi_{2u} := \frac{c_{2u}}{U_2} \stackrel{(7.36),(7.51)}{=} \varphi_{2u\infty} - \frac{\Delta c_u}{U_2}. \tag{7.52}$$

Using Stodola's assumption [Stodola, 1927], the variation of velocity c_u due to the finite number of blades is

$$\Delta c_u = \frac{1}{2} a\, \omega. \tag{7.53}$$

As shown in Fig. 7.46:

$$a = \frac{2\pi r_2}{Z_r} \cos \beta_{2f}, \tag{7.54}$$

where Z_r is the number of rotor blades. Combining (7.53) and (7.54) yields

$$\Delta c_u = \frac{\pi U_2}{Z_r} \cos \beta_{2f} \tag{7.55}$$

and the slip factor

$$\mu = \frac{c_{2u\infty} - \Delta c_u}{c_{2u\infty}} \stackrel{(7.55)}{=} 1 - \frac{\pi U_2}{Z_r c_{2u\infty}} \cos \beta_{2f}.$$

If the impeller has radial blades, then $U_2 = c_{2u\infty}$ and $\beta_{2f} = 0$, so the slip factor is

$$\mu = 1 - \frac{\pi}{Z_r}.$$

A relationship between the circulation coefficient φ_{2u} and the flow rate coefficient $\varphi_{2r\infty}$ is obtained by combining (7.39), (7.52), and (7.55):

$$\varphi_{2u} := 1 - \varphi_{2r\infty} \tan \beta_{2f} - \frac{\pi}{Z_r} \cos \beta_{2f}. \tag{7.56}$$

Note that $\beta_f \equiv \beta_\infty$. Note also that (7.56) does not show the influence of the diameter ratio D_1/D_2. This is because the influence of the D_1/D_2 ratio is rather small, so that (7.56) captures the variation of φ_{2u} quite well for average-sized centrifugal compressors. For radial blades, (7.56) simplifies to

$$\varphi_{2u} := 1 - \frac{\pi}{Z_r}.$$

Several experimental relationships were developed to improve Stodola's assumption to better match experimental results. All these relationsihps yield the slip factor as a function of β_{2f}, Z_r, and D_1/D_2:

$$\frac{1}{\mu} = 1 + \left(1.5 + 1.1\frac{\beta_{2f}}{90}\right) / [2Z_r (1 - D_1/D_2)], \qquad \beta_{2f} \text{ in degrees} \tag{7.57}$$

$$\frac{1}{\mu} = 1 + \frac{3.6}{Z_r} \frac{\cos \beta_{2f}}{1 - (D_1/D_2)^2} \tag{7.58}$$

$$\frac{1}{\mu} = 1 + \frac{1.2}{Z_r} \frac{1 + \cos \beta_{2f}}{1 - (D_1/D_2)^2}. \tag{7.59}$$

For radial blades, a fourth expression of the slip factor was proposed:

$$\frac{1}{\mu} = 1 + \frac{\frac{2}{3}\pi}{Z_r} \frac{1}{1 - \dfrac{(R_1^h)^2 + (R_1^t)^2}{2R_2^2}}, \tag{7.60}$$

where R_1^h and R_1^t are the radii of the impeller eye at hub and tip, respectively. The variation of these proposed empirical functions of the slip factor is shown in Fig. 7.47 for $Z_r = 20$ and $D_1/D_2 = 0.5$ and in Fig. 7.48 for $\beta_{2f} = 10°$ and $D_1/D_2 = 0.5$. Figure 7.47 shows that μ increases slightly when β_{2f} increases, and Fig. 7.48 shows that μ increases when the number of blades increases.

Assuming that the impeller blades have constant curvature, as shown in Fig. 7.49, the radius of this curvature can be calculated using the cosine formula applied to r_1 and r_2

$$R_0^2 := R^2 + r_1^2 - 2Rr_1 \cos(90 - \beta_{1f})$$
$$R_0^2 := R^2 + r_2^2 - 2Rr_2 \cos(90 - \beta_{2f})$$

which yields

$$R = \frac{1}{2} \frac{r_2^2 - r_1^2}{r_2 \sin \beta_{2f} - r_1 \sin \beta_{1f}}. \tag{7.61}$$

Figure 7.47 Variation of slip factor, μ with blade angle, β_{2f} for $Z_r = 20$, $D_1/D_2 = 0.5$: 1 - (7.57), 2 - (7.58), 3 - (7.59), and 4 - (7.60).

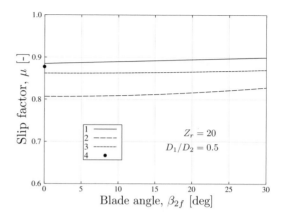

Figure 7.48 Variation of slip factor, μ with blade angle, Z_r for $\beta_{2f} = 10°$, $D_1/D_2 = 0.5$: 1 - (7.57), 2 - (7.58), 3 - (7.59), and 4 - (7.60).

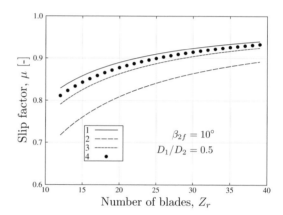

Figure 7.49 Variation of channel width.

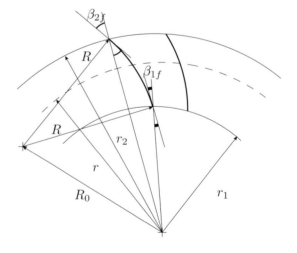

The distance between two adjacent blades at radius r is obtained similarly to that of (7.54), so that

$$a(r) = \frac{2\pi r}{Z_r} \cos \beta_f. \tag{7.62}$$

Therefore, $r \cos \beta_f$, defines how much the impeller channel diverges.

Example 7.3.8 An impeller has 20 blades, $r_1 = 10$ cm, and $r_2 = 18$ cm. The metal (or fixed) angles are $\beta_{1f} = 0$ and $\beta_{2f} = 15°$.

1. What is the variation of the distance between blades at inlet and exit?
2. Assuming that the impeller blades have constant curvature, calculate the curvature radius.

Solution

1. The gap between blades at inlet is

    $$a(r_1) = \frac{2\pi r_1}{Z_r} \cos \beta_{1f} = \frac{2\pi \, 10}{20} \cos 0 = 3.14 \text{ cm}.$$

    The gap between blades at inlet is

    $$a(r_2) = \frac{2\pi r_2}{Z_r} \cos \beta_{2f} = \frac{2\pi \, 18}{20} \cos 15 = 5.46 \text{ cm}.$$

2. The curvature radius is

    $$R \overset{(7.61)}{=} \frac{1}{2} \frac{18^2 - 10^2}{18 \sin 15 - 10 \sin 0} = 24.04 \text{ cm}.$$

Example 7.3.9 A single-stage centrifugal compressor has a mass flow rate of 35 kg/s and it rotates at 16,500 rpm. The static pressure at inlet in the impeller is 3.5 bar and the static temperature is 295 K. The inducer radius at the hub is 5 cm and at the tip is 12 cm. The impeller outlet radius is 24 cm. The blade thickness is 2 cm. The absolute inlet velocity has angle $\alpha_1 = -25°$. The impeller has backward-swept blades with an angle $\beta_{2f} = -15°$. The efficiency of the impeller is 90%. The ratio of specific heat capacities is assumed constant and equal to 1.399, and the specific heat capacity at constant pressure is constant and equal to 1005 J/(kg K).

  Calculate the Mach numbers at impeller inlet and outlet, the static and stagnation temperatures and pressures at impeller outlet, and the power needed to drive the impeller

Solution
The first step is to determine the velocity triangles at the inlet and outlet of the impeller. Let us calculate the average inlet transport velocity

$$U_1 = r_1\,\omega = \frac{1}{2}(r_1^h + r_1^t)\frac{2\pi n}{60} = \frac{1}{2}(0.05 + 0.12)\frac{2\pi\,16,500}{60} = 146.9 \text{ m/s}$$

and the average outlet transport velocity

$$U_2 = r_2\,\omega = r_2\frac{2\pi n}{60} = 0.24\frac{2\pi\,16,500}{60} = 414.7 \text{ m/s.}$$

The area at the inlet of the impeller is

$$A_1 = \pi\left[(r_1^t)^2 - (r_1^h)^2\right] = \pi\left[0.12^2 - 0.05^2\right] = 0.037385 \text{ m}^2$$

and the area at the impeller outlet is

$$A_2 = 2\pi r_2\,h = 2\pi 0.24 \times 0.02 = 0.030159 \text{ m}^2.$$

The density at the inlet of the impeller is

$$\rho_1 = p_1/(RT_1) = 350{,}000/(287.16 \times 295) = 4.132 \text{ kg/m}^3.$$

The velocity c_{1a} is then calculated from the mass flow rate

$$c_{1a} = \dot{m}/(\rho_1 A_1) = 35/(4.132 \times 0.037385) = 226.6 \text{ m/s.}$$

The speed of sound at the inlet of the impeller is

$$a_1 = \sqrt{\gamma R T_1} = \sqrt{1.399 \times 287.16 \times 295} = 344.3 \text{ m/s}$$

The absolute velocity c_1 and its tangential component c_{1u} are

$$c_1 = c_{1a}/\cos\alpha_1 = 226.6/\cos(-25) = 250.0 \text{ m/s}$$

$$c_{1u} = c_1\,\sin\alpha_1 = 250.0 \times \sin(-25) = -105.7 \text{ m/s.}$$

The angle β_1 of the relative velocity, shown in Fig. 7.50, is

Figure 7.50 Velocity triangles to scale for example 7.3.9.

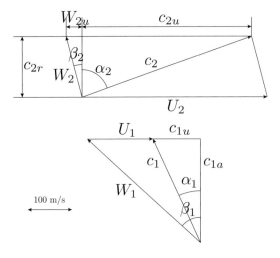

$\beta_1 = \arctan[(c_{1u} - U_1)/c_{1a}] = \arctan[((-105.7 - 146.9))/226.6] = -48.10°$.

The relative velocity at the inlet of the impeller is

$W_1 = \sqrt{c_{1a}^2 + (U_1 - c_{1u})^2} = \sqrt{226.6^2 + (146.9 - (-105.7))^2} = 339.3$ m/s.

The Mach number based on the relative velocity is

$M_{1W} = W_1/a1 = 339.3/344.3 = 0.9856$

and the Mach number based on the absolute velocity is

$M_1 = c_1/a1 = 250.0/344.3 = 0.7263$.

The stagnation temperature at the inlet is

$T_{01} = T_1 \left(1 + 0.5(\gamma - 1)M_1^2\right) = 295 \left(1 + 0.5 \times 0.399 \times 0.7263^2\right) = 326.04$ K

and the stagnation pressure at the inlet is

$p_{01} = p_1 \left(1 + 0.5(\gamma - 1)M_1^2\right)^{\frac{\gamma}{\gamma-1}} = 3.5 \left(1 + 0.5 \times 0.399 \times 0.7263^2\right)^{\frac{1.399}{0.399}} =$

$= 4.97073$ bar.

To calculate the absolute velocity in the radial direction at the impeller outlet, c_{2r}, one needs to assume a value of the density at the outlet. The computation of the velocity triangles at the impeller outlet will then allow us to calculate the outlet density. The calculated outlet density will then be used to repeat the computation of the velocity triangles. This iterative process will continue until the assumed value of the density is approximately equal to the calculated density.

Let us assume that the outlet density is $\rho_2 = 8.739$ kg/m^3, so that

$c_{2r} = \dot{m}/(\rho_2 A_2) = 35/(8.739 \times 0.030159) = 132.8$ m/s.

Assuming the split factor is 1, the relative velocity at the outlet is

$W_2 = c_{2r}/\cos\beta_{2f} = 132.8/\cos(-15) = 137.5$ m/s

and the tangent component of this velocity is

$W_{2u} = c_{2r}\tan\beta_{2f} = 132.8 \times \tan(-15) = -35.6$ m/s.

The angle of the absolute velocity at the outlet is

$\alpha_2 = \arctan((U_2 + W_{2u})/c_{2r}) = \arctan((414.7 + (-35.6))/132.8) = 70.70°$

and the absolute velocity at the outlet is

$c_2 = c_{2r}/\cos\alpha_2 = 132.8/\cos(70.70) = 401.7$ m/s.

The tangent component of the absolute velocity at the outlet is

$c_{2u} = U_2 + W_{2u} = 414.7 - 35.6 = 379.1$ m/s.

The work and the power needed to turn the impeller are

$$w = U_2 c_{2u} - U_1 c_{1u} = 414.7 \times 379.1 - 146.9 \times (-105.7) = 172,731 \text{ J/kg}$$

$$P = \dot{m}\, w = 35 \times 172,731 = 6,045,585 \text{ W} = 6.046 \text{ MW}.$$

The stagnation pressure at the outlet is

$$p_{02} = p_{01} \left[\eta_{12} w / (c_\text{p}\, T_{01}) + 1 \right]^{\frac{\gamma}{\gamma-1}} =$$
$$= 4.97073\, [0.9 \times 172,731 / (1005 \times 326.0) + 1]^{\frac{1.399}{0.399}} = 19.39376 \text{ bar}$$

so that the stagnation pressure ratio across the impeller is

$$p_{02}/p_{01} = 19.39376/4.97073 = 3.902.$$

The stagnation temperature at the impeller outlet is

$$T_{02} = T_{01} + w/c_\text{p} = 326.0 + 172,731/1005 = 497.9 \text{ K}$$

and the static temperature is

$$T_2 = T_{02} - 0.5 c_2^2 / c_\text{p} = 497.9 - 0.5 \times 401.7^2 / 1005 = 417.6 \text{ K}.$$

The speed of sound at the outlet is

$$a_2 = \sqrt{\gamma R T_2} = \sqrt{1.399 \times 287.16 \times 417.6} = 409.7 \text{ m/s}$$

and the Mach numbers are

$$M_2 = c_2 / a_2 = 401.7/409.7 = 0.9807$$

$$M_{2W} = W_2 / a_2 = 137.5/409.7 = 0.3356.$$

The static pressure at the outlet is

$$p_2 = p_{02} / \left(1 + 0.5(\gamma - 1)M_2^2\right)^{\frac{\gamma}{\gamma-1}} = 19.39376 / \left(1 + 0.5 \times 0.399 \times 0.9807^2\right)^{\frac{1.399}{0.399}}$$
$$= 10.48073 \text{ bar}.$$

so that the calculated density at the outlet is

$$\rho_2 = p_2 / (R T_2) = 10.48073 \times 10^5 / (287.16 \times 417.6) = 8.7392 \text{ kg/m}^3.$$

Since the calculated density value of 8.7392 kg/m^3 is approximately equal to the assumed density value of 8.739 kg/m^3, it is not necessary to continue the iterative process.

One should note that in this case the flow in the impeller is always subsonic. The absolute Mach number varies between 0.7263 at the inlet and 0.9807 at the outlet. In the relative frame of reference, the Mach number varies from 0.9856 at the inlet to 0.3356 at the outlet.

7.4 Combustors

The combustor is the component of the heat engine where the working fluid is heated by converting the chemical energy of the fuel into thermal energy. The combustor allows the reaction of fuel with the air at compressor outlet conditions. The fuels used in jet engines are a mixture of hydrocarbon compounds, as presented in Chapter 5. The fuel, which is liquid, is injected into the airstream, atomized, and vaporized. The fuel vapor is mixed with the air and then combustion can occur. The heat must be released so that the combustion products expand in a smooth stream of heated gas.

7.4.1 Combustor Requirements

Heat release in the combustor must satisfy several conflicting requirements: (1) high combustion efficiency to maximize heat release; (2) low pressure loss to ensure maximum overall engine performance; (3) wide range of stability to avoid flameout; (4) ability to restart the engine at high altitudes; (5) reliable ignition at low temperatures; (6) temperature profile at combustor exit that matches life requirements of the turbine; and (7) low emissions. In addition, the combustor must have (i) low weight, (ii) high durability for reliability and long life, and (iii) low cost.

The contradicting requirements make the combustor the most difficult component to design of the gas turbine engine. The intensity of the combustion, measured as the heat released per volume and time, far exceeds those of race car engines or large stationary power plant furnaces. For example, the intensity of the combustion in a jet engine combustor is of the order of 500 MW/m^3 while that of a power plant furnace is approximately 100 times smaller [Hill and Peterson, 1992, p. 242]. There are several factors that determine this difference. The most important is the pressure, which in a jet engine combustor is 20 times (or more) higher than that in a power plant furnace. Additionally, the turbulence intensity is higher and the size of the atomized fuel is smaller.

The temperature of the gases released by combustion, approximately 2400 K, exceeds the temperature at which superalloys can operate for long periods of time. Therefore, part of air coming from the compressor must be used to cool the combustor and then mix with the combustion products. Since stable combustion occurs at near stoichiometric conditions, the combustor must provide a zone where such conditions exist, in spite of the fact that the air/fuel ratio exceeds the stoichiometric value, *min L*.

The combustion efficiency, ξ_{comb}, is defined as the ratio of the enthalpy increase of the flow across the combustor and the enthalpy increase for a complete reaction of fuel and air to chemical equilibrium at the combustor exit conditions. The combustor efficiency depends on the completeness of combustion. As shown in Fig. 7.51, combustion efficiency is approximately 98% at takeoff conditions and decreases as air/fuel ratio increases.

The combustor is also required to have low emissions. This imposes stringent restrictions on the amount of carbon monoxide, CO, unburned hydrocarbons, and nitrogen oxides, that

Figure 7.51 Combustion efficiency vs. air/fuel ratio (adapted from [Anonymous, 1986]).

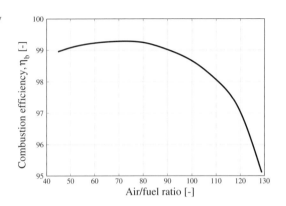

is, a mixture of NO, NO_2, and N_2O, collectively called NO_x. The largest amount of NO_x is produced at stoichiometric combustion, very close to the maximum flame temperature, which occurs at air/fuel ratios just below stoichiometric. As the temperature in the combustor increases, the NO_x increases while CO decreases. After reaching a minimum value at approximately 1850 K, the CO level starts increasing. As part of low emissions requirements, smoke must be below visible levels. Smoke consists of small carbon particles originating in the high-fuel content zones of the combustor.

7.4.2 Combustion Process

To understand how different engine parameters affect the combustion process, let us briefly review the fundamental processes that occur in the combustor. The liquid fuel is injected as an atomized spray so that it can vaporize quickly. The fuel vapor is then mixed with air and hot combustion products. Due to the high temperature of the combustion products, a high reaction rate occurs. This reaction rate can be approximated by the Arrhenius formula [Kuo, 2005, p. 122]

$$\text{Reaction rate} \propto f(T) \exp[-E_a/(\mathcal{R}T)] \tag{7.63}$$

where E_a is the activation energy, typically of the order of 250 kJ/mol. The Arrhenius formula implies that (1) two molecules must collide with some minimum energy in order to react, and (2) the number of collisions per volume and unit time in a gas at temperature T, where the energy of one molecule with respect to another one exceeds E_a, is proportional to $\exp[-E_a/(\mathcal{R}T)]$. The rate of collisions, and therefore the rate of reaction, depends on the number of molecules per unit volume, which depends on the pressure, so that

$$\text{Reaction rate} \propto p^n f(T) \exp[-E_a/(\mathcal{R}T)] \tag{7.64}$$

where $n \approx 1.9$.

  The effect of altitude on combustion can be assessed by analyzing the operation of the combustor as a component of the jet engine. Let us note that the turbine flow is choked over most of the operating range of the engine. Therefore, (3.40) shows that the mass flow

Figure 7.52 Combustion stability.

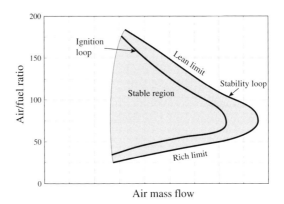

rate is proportional to $p_0/\sqrt{T_{03}}$, where p_0 is the stagnation pressure in the combustor and T_{03} is the turbine inlet temperature. Since the variation of turbine inlet temperature with altitude is small, the mass flow rate through the combustor is approximately proportional to the pressure, p_0.

For an engine operating at 15,000 meter altitude, the pressure is only 0.1189 of the sea-level pressure and therefore the mass flow rate of reactants in combustor is 11.89% of the sea-level value. The rate of reaction decreases due to the p^n term by a factor of 0.0175, that is, 1.75% of its sea-level value. As a result, the lower the pressure, the larger the reaction time and the greater the chance of flame-out. Consequently, the reaction rate can become limiting at high altitudes.

Figure 7.52 shows a typical *stability loop* obtained from experimental data. The stability loop has both a rich and a lean combustion limit. Additionally, if the mass flow rate exceeds a certain limit, flame extinction occurs. There is also an *ignition loop* that marks the region where combustion can be initiated. The ignition loop lies inside the stability loop because it is more difficult to initiate combustion under cold conditions than to maintain it.

At sea-level conditions, however, the rate of combustion is not limited by the reaction rate but by the rate at which air and fuel vapor are mixed. The fuel vapor and air mix by a combination of molecular diffusion and turbulent mixing. To further understand the combustion process, let us consider two simplified combustion cases: (1) diffusion flames and (2) combustion of premixed gases.

7.4.2.1 Diffusion Flames

In the combustion region, not all fuel is vaporized and not all the vaporized fuel is mixed with air. Therefore, some combustion takes place at the boundary between the fuel, either gas or liquid, and air. This combustion that takes place by diffusion of oxygen in one direction and fuel in the other direction is called a *diffusion flame*. The most common example of a diffusion flame is a candle flame [Kerrebrock, 1992, p. 158], where wax vaporizes from the wick and diffuses outward to meet the air that diffuses inward.

Similarly to the wax, a fuel droplet traveling in a hot gas that contains oxygen will diffuse outward to meet the inward diffusing oxygen. The droplet diameter varies in time as

$D = \sqrt{D_0^2 - kt}$, where D_0 is the initial diameter and k is of the order of 0.01 cm^2/s [Penner, 1957, p. 277]. Consequently, if the fuel droplets must be consumed in $t = 2.5 \times 10^{-3}$ s, then the initial diameter should be $D_0 = \sqrt{kt} = \sqrt{0.01 \times 2.5 \times 10^{-3}} = 0.5 \times 10^{-2}$ cm. The time t needed to consume the fuel droplet must be smaller than the residence time $t_{res} \approx A_{comb}\rho_{02}L/\dot{m}$, where A_{comb} is the flow area, ρ_{02} is the stagnation density at the compressor outlet, and L is the length of the combustor.

7.4.2.2 Combustion in Premixed Gases

Combustion in premixed gases is at the other end of the spectrum from diffusion flames. Combustion in premixed gases assumes that there is a uniform mixture of air and fuel vapor. The first question that arises is: does the mixture ignite and continue to burn? Experimental investigation in a bench-scale laboratory apparatus has shown that hydrocarbon–air mixtures ignite and burn only for a narrow range of air/fuel ratios, near the stoichiometric value [Scull, 1951]. This range decreases with decrease in pressure, as shown in Fig. 7.53. Furthermore, mixtures do not burn below a certain pressure at any air/fuel ratio.

The narrow range of excess air amenable to combustion seems problematic since all the cycle analyses done in Chapter 6 require excess air values far exceeding 1.1. The solution to this problem is to mix only a fraction of the air with the fuel and to use the rest of the air to cool the combustor.

The second question related to combustion in premixed gases is: at what speed does the flame propagate into the gas mixture? If the flow is laminar, theoretical and experimental studies have shown that for hydrocarbon–air mixtures, the flame speed is approximately 30 cm/s if the excess air, λ, is close to 1. The laminar flame speed drops rapidly for λ > 2.

If the flow is turbulent, as is the case in the combustor, the flame speed is much higher, approximately 5–8 m/s [Hill and Peterson, 1992, p. 247]. The increase in flame speed is due to the fact that turbulence increases the total area of the flame front by wrinkling it. In addition, by stretching the flame front, turbulence tends to thin the flame front, therefore increasing the gradients of temperature and concentration normal to the flame front. Since these gradients govern the thermal and species transport, the flame speed increases.

Figure 7.53 Flammability limits of gasoline/air mixture (adapted from [Scull, 1951]).

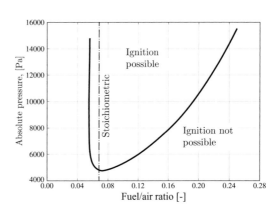

Figure 7.54 Combustion limits (adapted from [Olson et al., 1955]).

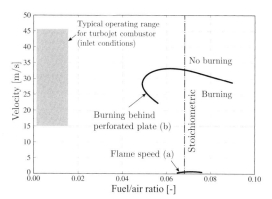

Figure 7.55 Combustor cross section (courtesy of Rolls-Royce).

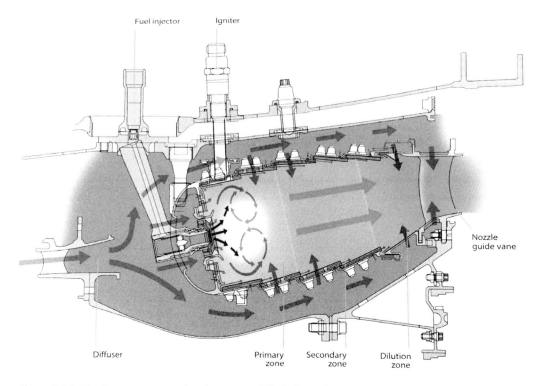

The flame speed is also dependent on the air/fuel ratio, as shown in Fig. 7.54. The curve (a) shows the flame velocities typical for hydrocarbon–air mixtures. If the flow velocity in the air/fuel mixture exceeds 60 cm/s, the flame will be extinguished [Olson et al., 1955]. Curve (b) shows the limits of combustion of vaporized hydrocarbon–air mixture burning downstream of a perforated plate positioned in a two-inch-diameter tube. Combustion is possible for air/fuel ratios and flow velocities corresponding to points located below the curve. The values of air/fuel ratio and flow velocities typical for a turbojet combustor are also shown in Fig. 7.54. Consequently, both air/fuel ratio and flow velocity must be modified in certain regions of the combustor to obtain stable combustion. The speed of the air exiting the

Figure 7.56 Individual can of a can-type combustor (courtesy of Rolls-Royce).

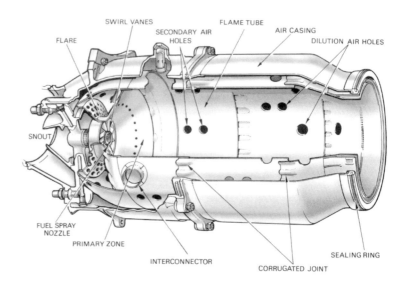

compressor and entering the combustor is of the order of 150 m/s. Since this speed is much higher than the speed needed for stable combustion, the incoming air is slowed down by a diffuser, as shown in Fig. 7.55, to approximately 100 m/s. As this air speed is still too high, a lower velocity recirculation zone is created just downstream of the fuel spray nozzle shown in Fig. 7.56. This region, called the primary zone, is where the conical fuel spray from a fuel spray nozzle intersects the recirculation vortex. The flame is initiated by an electric spark from an igniter plug. Afterward, the flame must be self-sustaining.

7.4.2.3 Combustor Cooling, Pattern Factor, and Pressure Drop

The temperature of the combustion products reaches approximately 2400 K, which is too hot for the nozzle guide vanes of the turbine. Consequently, more air is introduced in the secondary zone downstream of the primary zone. This air not only reduces the temperature of the gases but also controls emissions. The combustor should be designed so that combustion is completed in the secondary zone. The third zone of the combustor is the dilution zone. In this zone, located at the end of the combustor, more air is introduced to control the temperature profiles in both radial and circumferential directions. A typical split of the air in the combustor sets 12% through the swirl vanes, 8% through the flare, 20% for the primary zone, 40% for cooling and the secondary zone, and 20% for the dilution zone [Anonymous, 1986, p. 37].

The temperature profile in the radial direction at the exit of the combustor is sometimes, on purpose, made nonuniform, with lower temperatures at the tip and hub of the blades. Lower temperatures at the hub are desired because the blade roots have high mechanical stresses, while lower temperatures at the tip are needed because the outer wall is more difficult to cool [Schwab et al., 1983].

The temperature variation in the circumferential direction at the exit from the combustor leads to so called "hot streaks", which are not desired because they produce variations in the angle of attack in the turbine. The effect of hot streaks can be mitigated by optimizing the location of the turbine vanes with respect to the combustor hot streaks [Dorney et al., 2000], which is called airfoil clocking or indexing.

The variation of the temperature profile at the combustor outlet is characterized by a *pattern factor*, defined as the ratio between the maximum of temperature variation and the averaged temperature. A typical value for the circumferential pattern factor is 0.2.

An important parameter of the combustor performance is the stagnation pressure drop, π_{comb}. The pressure drop depends on the density change due to combustion and on the viscous losses due to the flow through the holes and slits of the burner. The stagnation pressure drop in the burner is approximated as one to two times the dynamic pressure based on the flow area of combustor, so that

$$\pi_{\text{comb}} = 1 - \epsilon \frac{\gamma}{2} M^2_{\text{comb}},$$

where $1 \leq \epsilon \leq 2$, and M_{comb} is the Mach number based on the burner flow area [Kerrebrock, 1992]. For $M_{\text{comb}} \approx 0.15$, $\epsilon \approx 2$, and $\gamma \approx 1.37$, the pressure drop is $\pi_{\text{comb}} \approx 0.97$.

7.4.3   Combustor Architecture

Three main types of combustors are used in gas turbine engines: (1) can- or tubular-type, (2) can-annular or tubo-annular type, and (3) annular type.

The *can-type combustor* consists of several individual cambers (or cans), as shown in Fig. 7.57. Each can has an inner flame tube and its individual air casing, as shown in Fig. 7.56. Can-type combustors are often used in gas turbine engines with centrifugal compressors. Ducts direct the air from the compressor diffuser into each can (or chamber). The flame tubes of the individual chambers are all interconnected so that combustion propagates around the tubes at startup. In addition, this interconnection allows all of the flame tubes to operate at the same pressure.

Figure 7.57 Can-type combustor (courtesy of Rolls-Royce).

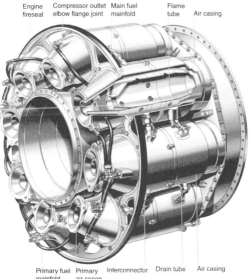

Engine fireseal   Compressor outlet elbow flange joint   Main fuel mainfold   Flame tube   Air casing

Primary fuel mainfold   Primary air scoop   Interconnector   Drain tube   Air casing

Can-type combustors have better structural strength than the other combustors. The individual cans are easy to inspect and replace, if necessary. Furthermore, they are easier to manufacture than the can-annular or the annular combustors.

The main drawback of the can-type combustor is that they use the cross-sectional area inefficiently and therefore, for the same power output, end up being longer than the other combustors. Another disadvantage of the can-type combustor is the larger temperature nonuniformity at exit, and therefore a larger pattern factor.

The *can-annular combustor* was developed from the can-type combustor by fitting the flame tubes inside a common air casing. The flow in the flame tube is similar to that of the can-type combustor except that a significant amount of air enters through the side wall as opposed to the snout (the front of the can). Consequently, the cross-sectional area is more efficiently used than in a can-type combustor. The can-annular combustor usually has a removable shroud that allows access to the cans without disassembling the engine.

The *annular combustor*, shown in Fig. 7.58, has a single flame tube contained in an inner and outer casing. As the name suggests, the flame tube has an annular form. The annular combustor is used in the majority of modern gas turbine engines.

The annular combustor has several advantages compared to the other types of combustors. The main advantage is that, for the same power output and diameter, the length of the annular combustor is approximately 75% of that of a can-annular combustor [Anonymous, 2005]. Consequently the engine is shorter and stiffer, and the weight of the engine is reduced. Another advantage of the annular combustor is that its wall area is smaller than that of a comparable can-annular combustor. Therefore the annular combustor requires approximately 15% less cooling air for the flame tube [Anonymous, 2005]. The annular combustor has also the advantage of producing an almost uniform gas mixture at inlet in the turbine. Furthermore, the annular combustor has the lowest pressure drop, π_{comb}.

Figure 7.58 Straight-through-flow annular combustor (courtesy of Rolls-Royce).

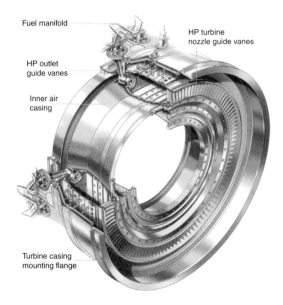

Fuel manifold

HP turbine nozzle guide vanes

HP outlet guide vanes

Inner air casing

Turbine casing mounting flange

Figure 7.59 Reverse-flow annular combustor (courtesy of Rolls-Royce).

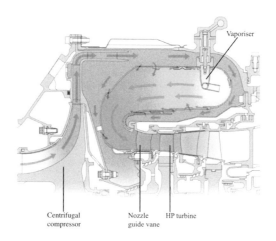

The disadvantages of the annular combustor are that is structurally weaker than the other combustors, and has a tendency to warp. The annular combustor is more complex to manufacture. In addition, to disassemble the combustor liner requires the complete removal and disassembly of the engine.

Annular combustors may have either a straight-through flow or a reverse flow. The reverse-flow combustor, shown in Fig. 7.59, is typically used in conjunction with a centrifugal compressor as it allows a shorter shaft than that required when using a straight-through flow combustor.

7.5   Axial-Flow Turbines

The turbine is the engine component that transforms a part of the kinetic energy of the hot gases released by the combustor into work. The turbine work is used to drive the compressor and other components of the engine, such as fuel and oil pumps, and electric generators. Similarly to compressors, turbines can be classified based on the flow direction as axial-flow, radial-flow, and mixed-flow turbines. Figure 7.60 shows the rotor of the axial-flow turbine of a turbojet engine. Figure 7.61 shows the radial-flow turbine of a supercharger.

Unlike radial-flow compressors, radial-flow turbines are not used for airplane propulsion although they can achieve a higher pressure ratio per stage than the axial-flow turbines. Because of the low density of the hot gases in the turbine, radial-flow turbines require a large area for the flow at small radial locations. Consequently the overall size of the engine increases, because of either the larger diameter or engine length. Radial-flow turbines are used for turbochargers, micro-gas turbines, or auxiliary power units. The rest of this section will discuss only axial-flow turbines.

As the name indicates, the flow in an axial-flow turbine is mainly in the axial direction. Axial-flow turbines of jet engines typically have between one and four stages per spool. As in the case of compressors, the stage consists of two rows of airfoils, a row of stators (also called

Figure 7.60 Axial-flow turbine (courtesy of SSPL/Getty Images).

Figure 7.61 Radial-flow turbine.

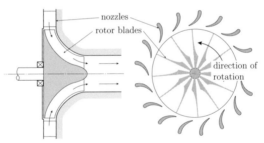

nozzles or vanes), and a row of rotors (also called blades or buckets). Unlike compressors, the turbine stage begins with a row of vanes followed by a row of blades.

To understand the differences between the turbine and compressor, one should note that the turbine flow is accelerated as the gases expand, while the compressor flow is decelerated as the air is compressed. As a result, there is a positive pressure gradient in the compressor and a negative one in the turbine. The positive pressure gradient in the compressor precludes a high flow turning on the blade, because of the risk of flow separation. Consequently, the loading of the compressor stage is low, with work per stage varying around 40 kJ/kg. The negative pressure gradient in the turbine helps delay flow separation and allows a higher flow turning, so that the work per stage varies around 250 kJ/kg. Therefore, the number of turbine stages is always smaller than the number of compressor stages. This is due to the fact that not only the stage loading but also the stage efficiency are higher in the turbine than those of the compressor.

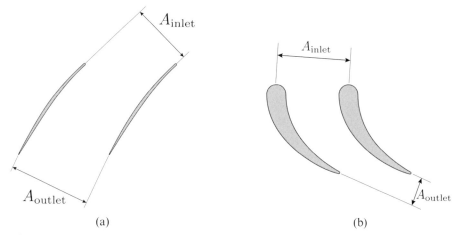

Figure 7.62 Area variation. (a) Compressor and (b) turbine.

The aerodynamic limiting factor in the compressor is separated flow, which could lead to stall and surge. The prevalent aerodynamic limiting factor in the turbine is choked flow. Having choked flow in the turbine, that is, reaching critical condition, is desirable because the mass flow rate is maximized, which leads to maximizing the turbine power. Compressor stall and surge, however, are undesirable because, as discussed in Section 7.3, they could lead to mechanical failure, not only loss of efficiency. Flow separation can also occur in turbines, however, at much larger flow turning angles.

The hot gases enter the turbine at a relatively small, subsonic Mach number and then expand in the turbine. Consequently, the passage area between the blades decreases and the turbine cascade acts like a nozzle. In contrast, in the compressor the passage areas between the blades increases and the compressor cascade acts like a diffuser.

As already mentioned, the pressure decreases as the flow expands in the turbine. Therefore, compared to the compressor, the boundary layer grows more slowly and the flow can be turned more before separation. Consequently the area variation in the turbine, $(A_{\mathrm{inlet}}/A_{\mathrm{outlet}})_{\mathrm{turbine}}$, is larger than the area variation in the compressor, $(A_{\mathrm{outlet}}/A_{\mathrm{inlet}})_{\mathrm{compressor}}$, as shown in Fig. 7.62.

While the pressure in the turbine is comparable to that in the compressor, the turbine temperature is higher than that of the compressor, which leads to a lower density in the turbine compared to that of the compressor. If the mass flow rate is written using the Mach number and the ideal gas law,

$$\dot{m} = \rho VA = \frac{p}{RT}aMA \overset{(3.20)}{=} \sqrt{\frac{\gamma}{RT}}\,pMA.$$

Taking into account that mass flow rate in the compressor and turbine are approximately equal, and the gas constant R, the ratio of specific heat capacities γ, Mach number and pressure have similar values in the turbine and the compressor yields

$$A_t = A_c \sqrt{\frac{T_t}{T_c}}$$

where A_t, T_t and A_c, T_c are the cross-sectional areas and temperatures in the turbine and the compressor, respectively. Consequently, the turbine cross-sectional area is larger than that of the compressor. As a result, turbine blades are longer than compressor blades.

Having a larger turning angle, the energy transfer in the turbine is larger than that in the compressor. For this reason the number of turbine stages in a gas turbine engine is always much smaller than the number of compressor stages.

The highest temperature in the compressor is at the exit of the compressor, and it increases with the compressor pressure ratio. Although the compressor exit temperature can exceed 900 K for compressor pressure ratios higher than 40, compressor blades are not cooled. The highest temperature in the turbine is at the inlet. This turbine inlet temperature can exceed 1800 K on modern gas turbine engines. These turbine inlet temperatures far exceed the melting temperature of the nickel-based alloys used to cast the turbine blades. For this reason high-pressure turbine blades and the nozzle guide vanes are cooled internally and externally using colder air from the compressor. The cooling air, however, can have rather high temperatures, corresponding to the temperatures at the exit from the compressor. The pressure of the cooling air from the compressor must exceed the pressure in the turbine. Therefore the cooling air from the compressor must be delivered to the turbine with minimal pressure losses. This is because the difference between the pressure at the exit of the compressor and the pressure at inlet in the turbine is rather small. Maintaining this pressure difference is critical to the cooling process.

The cooling air travels inside the blade through internal channels. Part of this air is released through holes and slots at the leading edge of the blade or nozzle guide vane. This cooling air creates a cool protective boundary layer at the airfoil surface called film cooling. Part of the cooling air is released at the trailing edge to cool the thin structure around the airfoil trailing edge. On blades, part of the cooling air is released at the blade tip to cool the top part of blade and also to reduce the flow spillage from the pressure side to the suction side, therefore improving aerodynamic efficiency. Figure 7.63 shows the cooling scheme used for blades and nozzle guide vanes.

7.5.1   Turbine Performance

The jet engine cycle analysis presented in Chapter 6 relied, among other parameters, on the stagnation pressure at outlet from the turbine, p_{04}, and turbine efficiency, η_T. Both p_{04} and η_T depend on the following parameters:

T_{03} -   inlet stagnation temperature

p_{03} -   inlet stagnation pressure

R -   gas constant

D -   characteristic dimension (usually diameter)

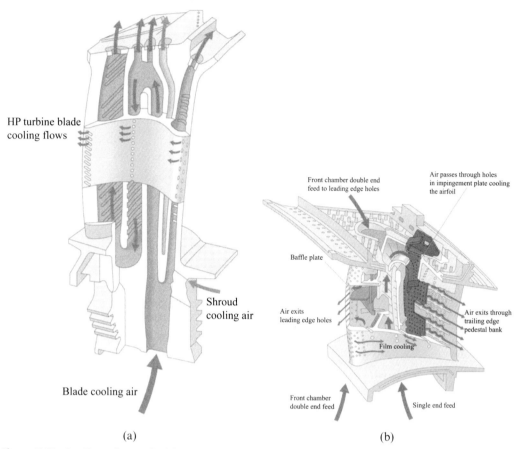

Figure 7.63 Cooling schemes for blade and nozzle guide vane (courtesy of Rolls-Royce). (a) Blade and (b) nozzle guide vane (NGV).

N - rotor rotational speed

$\dot{m}$ - mass flow rate of gas

γ - ratio of specific heat capacities

ν - kinematic gas viscosity

$\dot{m}_c$ - turbine cooling airflow

design - specification of the dimensionless geometric shape of the turbine, as a set of ratios of every important dimension to the characteristic dimension, D.

Following a dimensional analysis similar to that done for compressors, the expansion ratio p_{03}/p_{04} and the turbine adiabatic efficiency η_T can be written as

$$p_{03}/p_{04}, \eta_T = f\left(\frac{\dot{m}\sqrt{RT_{03}}}{p_{03}D^2}, \frac{ND}{\sqrt{RT_{03}}}, \frac{ND^2}{\nu}, \gamma, \text{design}\right) \tag{7.65}$$

Table 7.3 Turbine dimensionless parameters.

Dimensionless term	Significance
$\dfrac{\dot{m}\sqrt{RT_{03}}}{p_{03}D^2}$	Mass flow rate
$\dfrac{ND}{\sqrt{RT_{03}}}$	Mach number at rotor tip, $M_{\text{tip rotor}}$
$\dfrac{ND^2}{\nu}$	Reynolds number
γ	Ratio of specific heat capacities
"design"	Turbine geometric shape

where the effect of the cooling mass flow rate was neglected. The turbine dimensionless parameters are listed in Table 7.3.

The list of dimensionless parameters can be simplified further by assuming that both the variations of the Reynolds number and the ratio of specific heat capacities, γ, are small. Therefore, for a given turbine design, (7.65) simplifies to

$$p_{03}/p_{04}, \eta_{\mathrm{T}} = f\left(\frac{\dot{m}\sqrt{RT_{03}}}{p_{03}D^2}, \frac{ND}{\sqrt{RT_{03}}}\right). \qquad (7.66)$$

7.5.2 Turbine Maps

Turbine performance is illustrated by various relationships (7.66). The plots of the turbine inlet corrected mass flow rate and efficiency vs. the turbine expansion ratio and the corrected speed are called *turbine maps* or the *performance maps*. Turbine maps are typically determined experimentally or, more recently, by using numerical simulations. The turbine maps allow us to quickly determine the conditions at which the turbine is operating and to anticipate the turbine performance when the flow conditions change.

Unlike the compressor map, where typically the pressure ratio and efficiency are plotted on a single map, turbine maps are plotted on two maps, as shown in Fig. 7.64. The pressure ratio can be shown either in the abscissa or in the ordinate. The former case is illustrated in Fig. 7.64.

Figure 7.64a shows that as the expansion pressure ratio increases, the mass flow rate increases up to the point when critical conditions are reached at the throat and the flow is choked. Beyond this point, the mass flow rate remains constant even if the expansion pressure ratio increases further. As the expansion ratio changes, the work will continue to change. Choking can occur in either the stator (nozzle) or the rotor row. Figure 7.64b shows that the inlet flow parameter is only slightly dependent on rotor speed.

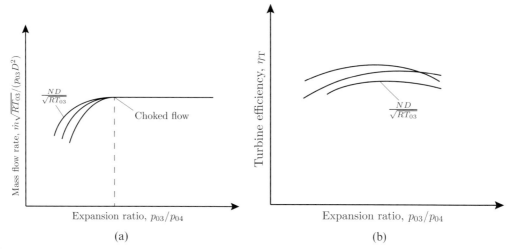

Figure 7.64 Turbine maps. (a) Mass flow rate vs. pressure ratio and (b) efficiency vs. pressure ratio.

Figure 7.65 Turbine blade notation.

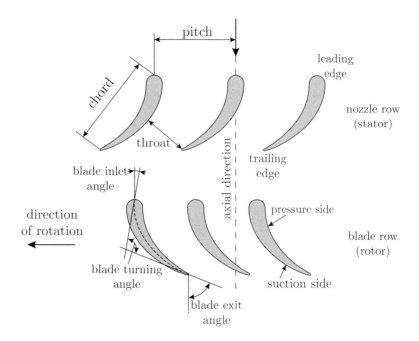

7.5.3 Turbine Stage Notation

The notation used for a turbine stage is shown in Fig. 7.65. The first row of a turbine stage is a row of nozzle guide vanes. This nozzle guide vane row is followed by a row of blades. Compared to the compressor stage, where the airfoils bring the flow closer to the axial direction, turbine airfoils push the flow away from the axial direction. In this way, while compressor airfoils reduce the flow velocity, turbine airfoils direct the flow so that velocity increases.

The throat, a notation specific to the turbine, denotes the minimum distance between two adjacent airfoils. It is at the throat of the nozzle or the blade where the flow reaches critical conditions. Solidity, the ratio between chord and pitch, has values close to 1, similarly to the values of a compressor row.

7.5.4 Turbine Velocity Diagram

The turbine stage velocity diagram is shown in Fig. 7.66. The following notation is used: state 3 at nozzle guide vane inlet, state 3.5 between the nozzle row and the blade row, and state 4 at the blade row outlet. As in the case of the compressor, the relevant velocity for stator airfoils is the absolute velocity, and the relevant velocity for the rotor airfoils is the relative velocity.

The absolute velocity V_3 at inlet in the nozzle guide vanes is turned away from the axial direction by the stator airfoils. The absolute velocity at the exit from the nozzle row is $V_{3.5}$. The relative velocity at station 3.5 is $\vec{W}_{3.5} = \vec{V}_{3.5} - \vec{U}$. This relative velocity, $W_{3.5}$, is the velocity that the blades "see" at inlet in the rotor row. The blades then turn the velocity $W_{3.5}$,

Figure 7.66 Turbine velocity diagram.

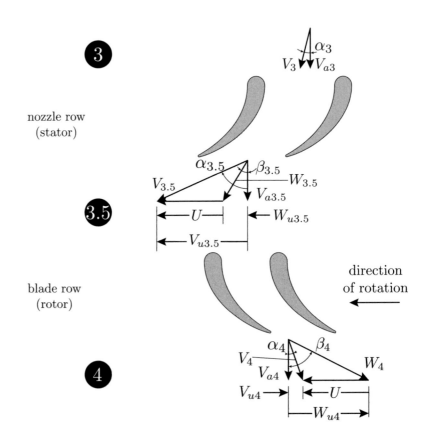

and the relative velocity at exit from the rotor row is W_4. The variation of the axial component of the velocity is typically small, so that one can approximate $V_{a3} \approx V_{a3.5} \approx V_{a4}$. In addition, since the flow is assumed axial, the transport velocity U is constant across the stage.

Example 7.5.1 The flow enters the nozzle guide vane of a turbine in the axial direction. The rotational speed is $N = 14{,}000$ rpm, the mean blade diameter is 48 cm, the velocity $V_{a3.5} = 170$ m/s, $\alpha_{3.5} = 72°$, and $\alpha_4 = 0°$. Assume that the working fluid enters and leaves the blades at the blade angle, and the axial velocity remains constant throughout the stage.

Calculate the turbine velocity diagrams at the mean blade diameter.

Solution

Let us calculate the transport velocity, U

$$U = r\omega = \frac{D}{2}\, 2\pi\, \frac{N}{60} = \frac{0.48}{2}\, 2\pi\, \frac{14000}{60} = 351.9 \text{ m/s}.$$

The tangential component of the absolute velocity at station 3.5 is

$$V_{u3.5} = V_{a3.5} \tan\alpha_{3.5} = 170 \tan 72° = 523.2 \text{ m/s}$$

so that the absolute velocity is

$$V_{3.5} = \frac{V_{a3}}{\cos\alpha_{3.5}} = \frac{170}{\cos 72°} = 550.1 \text{ m/s}.$$

The tangential component of the relative velocity at station 3.5 is

$$W_{u3.5} = V_{u3.5} - U = 523.2 - 351.9 = 171.3 \text{ m/s}$$

and the angle of the relative velocity is

$$\beta_{3.5} = \arctan\left(\frac{W_{u3.5}}{V_{a3.5}}\right) = \arctan\left(\frac{171.3}{170}\right) = 45.2°.$$

The relative velocity at station 3.5 is

$$W_{3.5} = \sqrt{W_{u3.5}^2 + V_{a3.5}^2} = \sqrt{171.3^2 + 170^2} = 241.3 \text{ m/s}.$$

Since the problem stated that the axial velocity remains constant throughout the stage,

$$V_{a4} = V_{a3} = 170 \text{ m/s},$$

and since the flow leaves the stage in the axial direction, $\alpha_4 = 0$ and therefore $V_{u4} = 0$ and $W_{u4} = U$. The relative velocity W_4 can be calculated as

$$W_4 = \sqrt{V_{a4}^2 + U^2} = \sqrt{170^2 + 351.9^2} = 390.8 \text{ m/s},$$

and finally

$$\beta_4 = \arctan\left(\frac{W_{u4}}{V_{a4}}\right) = \arctan\left(\frac{351.9}{170}\right) = 64.2°.$$

7.5.5 Energy Transfer

The specific work in the turbine is obtained using Euler's equation for turbomachinery (7.12). The expression of the turbine specific work is similar to that of the compressor specific work (7.13)

$$w_T = U_{3.5} V_{u_{3.5}} - U_4 V_{u_4}. \tag{7.67}$$

If the mean radii at locations 3.5 and 4 are equal, Eq. (7.67) becomes

$$w_T = U(V_{u_{3.5}} - V_{u_4}) \tag{7.68}$$

where $U_{3.5} = U_4 = U$. The change in the tangential component of the absolute velocity, ΔV_u, is

$$\Delta V_u := V_{u_{3.5}} - V_{u_4}.$$

Note that the specific work increases if (1) the change in the tangential component of the absolute (or relative) velocity increases, or (2) the transport velocity U increases.

Example 7.5.2 Consider the turbine velocity diagrams calculated in Example 7.5.1. Calculate the specific work and power if the turbine mass flow rate is 36 kg/s.

Solution

$$w_T = U(V_{u_{3.5}} - V_{u_4}) = 351.9 \times (523.2 - 0) = 184.1 \times 10^3 \text{ J/kg} = 184.1 \text{ kJ/kg}$$

$$\mathcal{P}_T = \dot{m} \times w_T = 36 \times 184.1 = 6620 \text{ kW}.$$

7.5.6 Turbine Parameters

The parameters introduced for the compressor in Section 7.3.5.3 – flow coefficient, work coefficient, load coefficient, hub-to-tip ratio, de Haller number, and degree of reaction – are also valid for the turbine if one substitutes the compressor stations 1, 1.5, and 2 by the turbine stations 3, 3.5, and 4. For turbines the *load factor* is defined as

$$\overline{w}_T = w_T/U^2 \tag{7.69}$$

which is half the value of the work coefficient.

In addition to these common parameters, there are specific parameters that need to be defined for the different types of turbines. This is because there are rather significant differences between the turbine of a turbojet and the turbine of a turbofan or turboprop. The goal of the turbojet turbine is to extract the minimum work from the working fluid, enough to drive the compressor, the auxiliary fuel and oil pumps, and the electric generator, and to leave the rest of energy in the exhaust gases that produce the thrust. On the other hand, the goal of the turbofan or turboprop turbine is to extract most of the energy from the working fluid, and to transfer it through the shaft to the compressor, fan, or propeller.

7.5.6.1 Turbine Efficiencies

Turbine efficiency can be expressed in different ways, depending on the type of jet engine [Dixon, 1998, p. 32]. The choice of the turbine efficiency definition is dictated by whether the exit kinetic energy is usefully utilized or wasted. For example, the exhaust kinetic energy is usefully utilized in the last turbine stage of an aircraft jet engine since it is used to generate the jet propulsive thrust. Similarly, in a multistage turbine, the exit kinetic energy of one stage is used for the downstream stage. In these cases, the ideal work of the turbine is, as shown in Fig. 7.67,

$$w_{T_i} = h_{03} - h_{04_i}$$

and the turbine efficiency is called the total-to-total efficiency:

$$\eta_{tt} := \frac{w_T}{w_{T_i}} = \frac{h_{03} - h_{04}}{h_{03} - h_{04_i}}. \tag{7.70}$$

If the useful output of the turbine is shaft power and the exhaust kinetic energy is considered a loss, the ideal work of the turbine is

$$w_{tsi} = h_{03} - h_{4_i}$$

and the relevant turbine efficiency is the total-to-static (or relative) efficiency, defined as

$$\eta_{ts} := \frac{w_T}{w_{tsi}} = \frac{h_{03} - h_{04}}{h_{03} - h_{4_i}}. \tag{7.71}$$

Figure 7.67 $h - s$ diagram of turbine expansion.

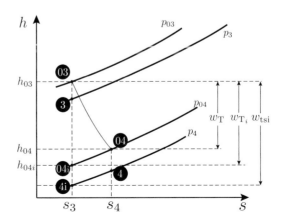

Assuming that

$$h_{04i} - h_{4i} = V_{4i}^2/2 \approx V_4^2/2 = h_{04} - h_4 \tag{7.72}$$

yields a relationship between the two efficiencies

$$\eta_{tt} = \eta_{ts} \frac{1}{1 - \frac{V_4^2}{2(h_{03} - h_{4i})}},$$

which shows that $\eta_{tt} > \eta_{ts}$. The validity of the assumption $V_{4i} \approx V_4$ will be verified in Example 7.5.3.

A third option for defining the turbine efficiency is the turbine adiabatic efficiency, $\eta_{T_{ad}}$

$$\eta_{T_{ad}} := \frac{h_{03} - h_4}{h_{03} - h_{4_i}}.$$

The pressure ratio in the turbine is reported either as the expansion ratio, $\delta_{0T} = p_{03}/p_{04}$, or its inverse, the pressure ratio, $\pi_{0T} = p_{04}/p_{03}$. The expansion ratio can be calculated using the isentropic relation

$$\delta_{0T} = \frac{p_{03}}{p_{04}} = \frac{p_{03}}{p_{04i}} = \left(\frac{T_{03}}{T_{04i}}\right)^{\frac{\gamma}{\gamma-1}}, \tag{7.73}$$

where the stagnation temperature T_{04i} is calculated from the enthalpy $h_{04i} = h_{03} - w_{T_i} = h_{03} - w_T/\eta_{tt}$.

Example 7.5.3 Air at a total temperature $T_{03} = 1400$ K and a total pressure $p_{03} = 22$ bar enters the nozzle of a turbine in the axial direction at a rate of 36 kg/s. The rotational speed is $N = 14,000$ rpm, mean blade diameter is 48 cm, velocity $V_{a3.5} = 170$ m/s, $\alpha_{3.5} = 72°$, and $\alpha_4 = 0°$. The total-to-total efficiency of the turbine is $\eta_{tt} = 90\%$.

Assume that the working fluid is air that enters and leaves the blades at the blade angle, and that the axial velocity remains constant throughout the stage.

1. Calculate the Mach number of the absolute velocity at exit from the nozzle guide vane.
2. Calculate the total-to-total expansion ratio, δ_{0T}.
3. Calculate the total-to-static efficiency, η_{ts}.
4. Calculate $\dot{m}\sqrt{T_{03}}/(p_{03}A)$.

Solution

1. The velocity triangles for this turbine were calculated in Example 7.5.1.

$$h_{03} = h_{03.5} = h_{3.5} + \frac{V_{3.5}^2}{2} \rightarrow h_{3.5} = h_{03} - \frac{V_{3.5}^2}{2}.$$

Read the enthalpy h_{03} from air tables for a known value of temperature T_{03}

$$T_{03} = 1400 \text{ K} \xrightarrow{\text{tables}} h_{03} = 1515.2 \text{ kJ/kg}$$

so that

$$h_{3.5} = 1515.2 - \frac{550.1^2}{2} 10^{-3} = 1363.9 \text{ kJ/kg} \xrightarrow{\text{tables}}$$

$$\xrightarrow{\text{tables}} \begin{cases} T_{3.5} = 1273.2 \text{ K} \\ a_{3.5} = 694.7 \text{ m/s} \end{cases}$$

$$\gamma = \frac{a_{3.5}^2}{RT_{3.5}} = \frac{694.7^2}{287.16 \times 1273.2} = 1.320$$

$$M_{3.5} = \frac{V_{3.5}}{a_{3.5}} = \frac{550.1}{694.7} = 0.792.$$

2. The expansion ratio can be calculated using

$$\frac{p_{03}}{p_{04}} = \frac{p_{03}}{p_{04i}} = \left(\frac{T_{03}}{T_{04i}} \right)^{\frac{\gamma}{\gamma-1}}. \tag{7.73}$$

Temperature T_{04i} and the ratio of specific heat capacities are not known. To calculate T_{04i} one uses the turbine total-to-total efficiency

$$\eta_{tt} = w_T / w_{T_i}$$

to find out the ideal specific work in the turbine, and then

$$h_{04i} = h_{03} - w_{T_i} = h_{03} - w_T / \eta_{tt} = 1515.2 - 184.1/0.9 = 1310.6 \text{ kJ/kg}.$$

From the tables we can then find the temperature $T_{04i} = 1228.1$ K. The next step is to find the ratio of specific heat capacities, γ, for the process between states 03 and 04i. For this we read from the tables the speeds of sound $a(T_{03}) = a(1400 \text{ K}) = 727.0$ m/s and $a(T_{04i}) = a(1228.1 \text{ K}) = 682.8$ m/s. The ratios of specific heat capacities are then calculated using $\gamma = a^2/(RT)$, which yields

$$\gamma_{03} = 727.0^2/(287.16 \times 1400) = 1.3147$$

$$\gamma_{04i} = 682.8^2/(287.16 \times 1228.1) = 1.3220.$$

The average value of γ between states 03 and 04i is $\gamma = 1.3183$, so that using (7.73) the expansion ratio is

$$\frac{p_{03}}{p_{04}} = \left(\frac{T_{03}}{T_{04i}} \right)^{\frac{\gamma}{\gamma-1}} = \left(\frac{1400}{1228.1} \right)^{\frac{1.3183}{1.3183-1}} = 1.720.$$

An alternative way of calculating the pressure at state 04i, without using (7.73), is to apply the P method. The enthalpy at state 04i was calculated above, so $h_{04i} = 1310.6$ kJ/kg.

The entropy at state 04i is equal to the enthalpy at state 03. The entropy at state 03 is obtained using the S method, knowing the pressure $p_{03} = 22$ bar and temperature $T_{03} = 1400$ K, so that $s_{03} = 7.4743$ kJ/(kg K). Using $h_{04i} = 1310.6$ kJ/kg and $s_{04i} = s_{03} = 7.4743$ kJ/(kg K) in the P method yields $p_{04i} = 12.79$ bar. Since $p_{04} = p_{04i} = 12.79$ bar, the expansion ratio in the turbine is $\delta_{0T} = p_{03}/p_{04} = 22/12.79 = 1.720$, which is identical to the value obtained using (7.73). The latter approach did not require the ratio of specific heat capacities to be calculated.

3. The total-to-static efficiency (7.71) requires finding the enthalpy h_{4i} shown in Fig. 7.67. To calculate this enthalpy we need to determine first states 04 and 4.

State 04
The pressure $p_{04} = 12.79$ bar and the enthalpy is $h_{04} = h_{03} - w_T = 1515.2 - 184.1 = 1331.1$ kJ/kg. Using the S method yields $s_{04} = 7.4908$ kJ/(kg K).

State 4
The entropy is $s_4 = s_{04} = 7.4908$ kJ/(kg K) and the enthalpy can be calculated using the velocity at the exit of the turbine, $V_4 = 170$ m/s, so that

$$h_4 = h_{04} - V_4^2/2 = 1331.1 - 170^2/2 \times 10^{-3} = 1316.6 \text{ kJ/kg.}$$

Using the P method yields $p_4 = 12.28$ bar.

State 4i
Since the pressure $p_{4i} = p_4 = 12.28$ bar and the entropy $s_{4i} = s_{03} = 7.4743$ kJ/(kg K), using the H method yields $h_4 = 1296.4$ kJ/kg. The total-to-static turbine efficiency is calculated using (7.71)

$$\eta_{ts} = \frac{h_{03} - h_{04}}{h_{03} - h_{4i}} = \frac{w_T}{h_{03} - h_{4i}} = \frac{184.1}{1515.2 - 1296.4} = 0.841.$$

Having determined h_{4i}, we can verify how approximate the assumption (7.72) was. Therefore, let us calculate V_{4i} as

$$V_{4i} = \sqrt{2(h_{04i} - h_{4i})} = \sqrt{2(1310.6 - 1296.4) \times 10^3} = 168.5 \text{ m/s}$$

and conclude that the error introduced by this approximation, $(170-168.5)/170 = 0.0088$, is acceptable.

4. To calculate the mass flow rate parameter $\dot{m}\sqrt{T_{03}}/(p_{03}A)$, one needs to find the critical area A. The critical area A is equal to $A_{3.5}$. Knowing the mass flow rate and velocity $V_{3.5}$, one needs to calculate the density $\rho_{3.5}$ in order to find $A_{3.5}$. This requires that state 3.5 be calculated. For this, one must find the entropy of state 03. For this, we apply the S method. The pressure and enthalpy at state 03 are known, so that

$$(p_{03}, h_{03}) \rightarrow s_{03}.$$

From air tables read the entropy for given h_{03}:

$s_{03p=1} = 8.3619 \text{ kJ/(kg K)}$

$s_{03} = s_{03p=1} - R \ln p_{03} = 8.3619 - 0.28716 \ln 22 = 7.4743 \text{ kJ/(kg K)}.$

Neglecting the losses in the nozzle guide vane,

$s_{3.5} = s_{03} = 7.4743 \text{ kJ/(kg K)}.$

The enthalpy $h_{3.5}$ was calculated at 1, and it is $h_{3.5} = 1363.9 \text{ kJ/kg}$. Knowing the entropy and enthalpy, one can compute the pressure:

$(s_{3.5}, h_{3.5}) \rightarrow p_{3.5}.$

For a known value of enthalpy $h_{3.5}$, read the entropy from air tables, $s_{3.5p=1} = 8.2487 \text{ kJ/(kg K)}$. Then, the pressure is

$$p_{3.5} = \exp\left(-\frac{s_{3.5} - s_{3.5p=1}}{R}\right) = \exp\left(-\frac{7.4743 - 8.2487}{0.28716}\right) = 14.83 \text{ bar}.$$

The area A is calculated as

$$A = A_{3.5} = \dot{m}/(\rho_{3.5} V_{3.5}) = \dot{m} R T_{3.5}/(p_{3.5} V_{3.5}) =$$

$$= 36 \times 287.16 \times 1273.2/(14.83 \times 10^5 \times 550.1) = 0.0161 \text{ m}^2.$$

Finally, the mass flow rate parameter is

$$\frac{\dot{m}\sqrt{T_{03}}}{p_{03} A} = \frac{36\sqrt{1400}}{22 \times 10^5 \times 0.0161} = 0.0380.$$

7.5.6.2 Turbine Degrees of Reaction

In addition to the degree of reaction defined by (7.15), which for a turbine translates into

$$R'_T := \frac{h_{3.5} - h_4}{h_3 - h_4}, \tag{7.74}$$

there is an alternative degree of reaction definition

$$\rho_T := \frac{h_{3.5} - h_4}{h_{03} - h_{04}}, \tag{7.75}$$

in which the variation of static enthalpy over the rotor is divided by the variation of stagnation enthalpy over the stage. If the velocities at inlet and exit of the stage were the same, then $R'_T = \rho_T$. Since the velocity at stage exit is larger than the velocity at inlet, then $\rho_T > R'_T$. Using the definition of turbine specific work, the degree of reaction ρ_T can be written as

$$\rho_T = \frac{\Delta h_{\text{rotor}}}{w_T},$$

where $\Delta h_{\text{rotor}} = h_{3.5} - h_4$, shown in Fig. 7.68.

Figure 7.68 $h - s$ diagram of turbine expansion.

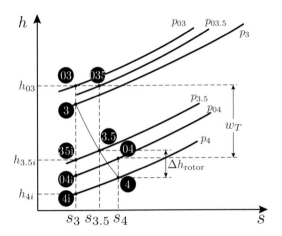

Let us derive the expression of the degree of reaction as a function of the kinematic parameters. This will allow us to connect the degree of reaction to the velocity diagram. Substituting

$$h_{3.5} = h_{03} - V_{3.5}^2/2$$

$$h_4 = h_{04} - V_4^2/2$$

$$w_T = U\Delta V_u$$

in (7.75) yields

$$\rho_T = \frac{h_{03} - h_{04} - (V_{3.5}^2 - V_4^2)/2}{h_{03} - h_{04}} = 1 - \frac{1}{2}\frac{V_{3.5}^2 - V_4^2}{U\Delta V_u}.$$

Assuming $V_{3.5a} = V_{4a}$ and using $\Delta V_u = V_{3.5u} - V_{4u}$ yields

$$\rho_T = 1 - \frac{1}{2}\frac{V_{3.5u}^2 - V_{4u}^2}{U(V_{3.5u} - V_{4u})} = 1 - \frac{V_{3.5u} + V_{4u}}{2U}. \tag{7.76}$$

Let us denote by V_m the average of the absolute velocities $V_{3.5}$ and V_4, as shown in Fig. 7.69. Similarly, W_m denotes the average of relative velocities $W_{3.5}$ and W_4. The tangent component of V_m is denoted V_{mu} and it is equal to the average of the tangent components of the absolute velocities at stations 3.5 and 4:

$$V_{mu} := \frac{V_{3.5u} + V_{4u}}{2}. \tag{7.77}$$

Let us non-dimensionalize V_{mu} by the transport velocity U

$$\overline{V}_{mu} := \frac{V_{3.5u} + V_{4u}}{2U} \tag{7.78}$$

so that we can use this dimensionless velocity in the expression of the degree of reaction (7.76)

$$\rho_T = 1 - \overline{V}_{mu}. \tag{7.79}$$

Figure 7.69 Turbine velocity diagram.

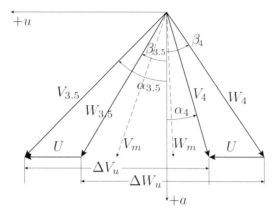

From the velocity triangles shown in Fig. 7.69 it is apparent that $U - V_{3.5u} = -W_{3.5u}$ and $U - V_{4u} = -W_{4u}$, so that using the dimensionless velocity $\overline{W}_{mu}$

$$\overline{W}_{mu} := \frac{W_{3.5u} + W_{4u}}{2U}.$$

The degree of reaction can also be written as

$$\rho_{\mathrm{T}} = 1 - \overline{V}_{mu} = -\overline{W}_{mu}. \tag{7.80}$$

Using the velocity diagram of Fig. 7.69, we can write

$$\frac{V_{3.5u}}{U} = \frac{V_a \tan \alpha_{3.5}}{V_a(\tan \alpha_{3.5} - \tan \beta_{3.5})} = \frac{\tan \alpha_{3.5}}{\tan \alpha_{3.5} - \tan \beta_{3.5}}$$

$$\frac{V_{4u}}{U} = \frac{V_a \tan \alpha_4}{V_a(\tan \beta_4 - \tan \alpha_4)} = \frac{\tan \alpha_4}{\tan \beta_4 - \tan \alpha_4}$$

so that the degree of reaction becomes

$$\rho_{\mathrm{T}} = 1 - \frac{1}{2}\left(\frac{\tan \alpha_{3.5}}{\tan \alpha_{3.5} - \tan \beta_{3.5}} + \frac{\tan \alpha_4}{\tan \beta_4 - \tan \alpha_4}\right). \tag{7.81}$$

Example 7.5.4 Calculate the degree of reaction for the turbine of Example 7.5.1.

Solution
Using (7.81) with $\alpha_{3.5} = 72°$, $\beta_{3.5} = 45.2°$, $\alpha_4 = 0$ and $\beta_4 = 64.2°$ yields

$$\rho_{\mathrm{T}} = 1 - \frac{1}{2}\left(\frac{\tan 72}{\tan 72 - \tan 45.2} + \frac{\tan 0}{\tan 64.2 - \tan 0}\right) = 0.257.$$

Turbines are often designed so that $\alpha_4 = 0$. In this case, as shown in the example above, the degree of reaction reduces to a function of $\alpha_{3.5}$ and $\beta_{3.5}$ only

$$\rho_T = 1 - \frac{1}{2} \left(\frac{\tan \alpha_{3.5}}{\tan \alpha_{3.5} - \tan \beta_{3.5}} \right). \tag{7.82}$$

Note also that if $\beta_{3.5} = 0$, then $\rho_T = 0.5$ for any $\alpha_{3.5}$. Furthermore, if $\beta_{3.5} > 0$, then $\rho_T < 0.5$.

Depending on the value of the degree of reaction, turbines are classified as *reaction* turbines, *impulse* turbines, or a combination of the two called *impulse-reaction* turbines.

Impulse Turbine

In an impulse turbine, shown in Fig. 7.70a, the pressure drop across the stage occurs only in the stator. Here the gases are expanding because of the convergent shape of the nozzle, increasing velocity while reducing pressure. The gases directed by the nozzle guide vanes impact the moving blades, which experience an impulse force. The rotor blades then only modify the flow direction without changing the speed of the gases. The impulse turbine has a zero degree of reaction

$$\rho_T = 0$$

because the enthalpies at states 3.5 and 4 are equal, $h_{3.5} = h_4$. In an ideal case, the equality of enthalpies implies the equality of pressures, $p_{3.5} = p_4$.

Since $\rho_T = 0$, the velocity $W_{mu} = 0$ and therefore the velocities $W_{3.5}$ and W_4 are symmetrical with respect to the axial direction, as shown in Fig. 7.71. For the case when V_{4u} is non-zero, the load factor (7.69) is

$$\overline{w}_T = w_T/U^2 = \Delta V_u/U = (2U + 2V_{4u})/U = 2 + 2\overline{V}_{4u}.$$

If the velocity V_{4u} is zero, the load factor is 2.

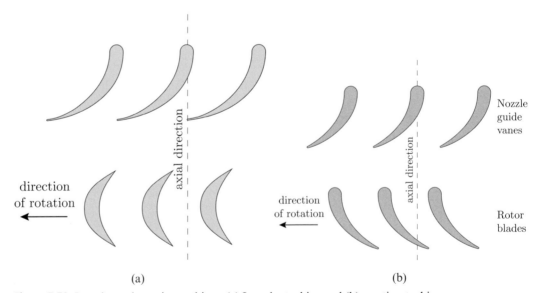

(a)   (b)

Figure 7.70 Impulse and reaction turbines. (a) Impulse turbine and (b) reaction turbine.

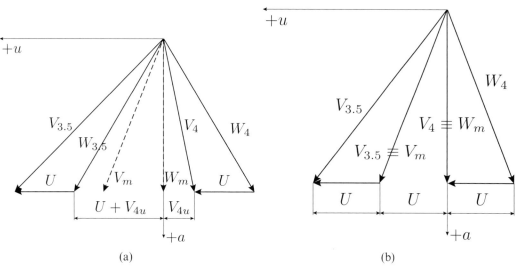

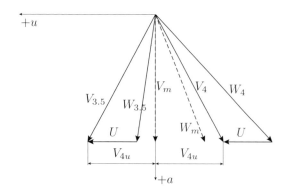

Figure 7.71 Velocity triangles for impulse turbines. (a) $V_{4u} \neq 0$ and (b) $V_{4u} = 0$.

Figure 7.72 Velocity triangles of a reaction turbine.

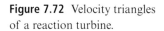

Reaction Turbine

In a reaction turbine, shown in Fig. 7.70b, the nozzle guide vanes only change the direction of the flow, without reducing the pressure. In the converging passages of the blades, the gas expands and accelerates, generating a reaction force on the blades so that

$$\rho_T = 1.$$

The velocity triangles of a reaction turbine are shown in Fig. 7.72. Since $\rho_T = 1$ the tangent component of the average absolute velocity is $\overline{V}_{mu} = 0$. Consequently, the absolute velocities $V_{3.5}$ and V_4 are symmetrical with respect to the axial direction. Therefore, the load factor is

$$\overline{w}_T = w_T/U^2 = \Delta V_u/U = 2V_{4u}/U = 2\overline{V}_{4u}.$$

Figure 7.73 Velocity triangles of an impulse-reaction turbine with $\rho_T = 0.5$.

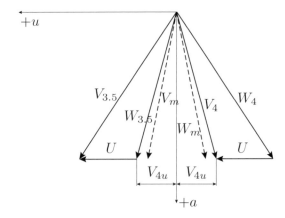

Impulse-Reaction Turbine

The impulse-reaction turbine, which is typically used on modern jet engines, is a combination between an impulse and a reaction turbine. The gases expand in both the stator and the rotor, so that

$$p_3 > p_{3.5} > p_4 \quad \text{and} \quad 0 < \rho_T < 1.$$

Figure 7.73 shows the velocity triangles for an impulse-reaction turbine with $\rho_T = 0.5$. In this case, $\overline{V}_{mu} = 0.5$ and $\overline{W}_{mu} = -0.5$, therefore the velocities V_m and W_m are symmetrical with respect to the axial direction. As a result, the velocities $V_{3.5}$ and W_4 are symmetrical with respect to the axial direction. Similarly, the velocities $W_{3.5}$ and V_4 are symmetrical with respect to the axial direction. The load factor for $\rho_T = 0.5$ is equal to $\overline{w}_T = 1 + 2V_{4u}$.

Although impulse turbines have the highest load factor, impulse-reaction turbines are typically used for jet engines because of their higher efficiency. The degree of reaction of impulse-reaction turbines used for jet engines usually varies between 0.2 and 0.5.

7.6 Exhaust Nozzles

The exhaust nozzle is the jet engine component that allows the gases discharged from the turbine to pass to the atmosphere at the desired velocity and direction. Similarly to the inlet diffuser, the complexity of the exhaust nozzle increases with maximum flight number of the vehicle. As the vehicle Mach number increases, so does the pressure ratio and the nozzle exit Mach number. For example, the nozzle pressure ratio of a jet engine can vary from between 2 and 3 at takeoff to 40 at Mach 3 [Kerrebrock, 1992, p. 139].

The exhaust nozzle of a jet engine must satisfy the following requirements: (1) match the other engine components for all engine operating conditions, (2) generate the optimum expansion ratio, (3) minimize loses at both design and off-design conditions, (4) minimize drag, (5) produce reverse thrust, if needed, and (6) reduce jet noise.

The temperature of the hot gases entering the exhaust nozzle varies between 850 and 1150 K, depending on the type of engine. For engines with afterburners, discussed in Chapter 8, the temperature can exceed 1800 K. Consequently, the nozzles are manufactured from materials that can withstand high temperatures, such as nickel alloys, titanium alloys, or ceramic composites. In addition, the construction of the nozzle should be done so that it will resist distortion and cracking. To prevent heat transfer to the aircraft structure, air is used to ventilate the nozzle. Additionally, insulating blankets are installed around the nozzle.

Several types of exhaust nozzles are used, depending on the type of jet engine. Turbofan engines have two parallel nozzles: one for the cold mass flow rate going through the fan and another one for the hot mass flow rate going through the gas generator. Turbojet, turboprop, and turboshaft engines have a single exhaust nozzle. The nozzles can be: (1) fixed-area converging nozzles, (2) fixed-area converging-diverging nozzles, (3) variable-area converging nozzles, (4) plug nozzles, and (5) two-dimensional nozzles.

Figure 7.74 shows a basic exhaust system with a fixed-area converging nozzle. The main components of this exhaust system are the exhaust cone, the jet pipe, and the convergent (or propelling) nozzle. The hot gas of the turbine enters the nozzle at high speeds, in excess of 200 m/s, producing high friction losses. The exhaust cone increases the area and decreases the flow speed by diffusion. In addition, the exhaust cone prevents the hot gas from flowing across the back face of the turbine disk. To reduce the friction losses due to the residual swirl velocity, the turbine rear struts straighten out the flow before the gases reach the jet pipe. The gases are then accelerated in the convergent or convergent-divergent nozzle.

The parameters defining the nozzle performance are the stagnation pressure and stagnation temperature at inlet, the exit static pressure, the area variation, and the efficiency. Using the $h - s$ diagram of Fig. 7.75, the nozzle efficiency is defined as

$$\eta_\mathrm{n} = \frac{h_{04} - h_5}{h_{04} - h_{5i}}. \qquad (7.83)$$

Figure 7.74 Exhaust system with fixed-area converging nozzle (courtesy of Rolls-Royce).

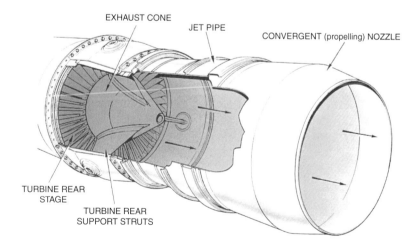

EXHAUST CONE

JET PIPE

CONVERGENT (propelling) NOZZLE

TURBINE REAR STAGE

TURBINE REAR SUPPORT STRUTS

Figure 7.75 Nozzle $h - s$ diagram.

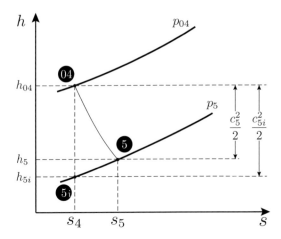

Assuming the expansion in the nozzle is adiabatic and the nozzle is fixed, so that the work is zero, yields $h_{05} = h_{04}$. Consequently, the nozzle efficiency can be written as

$$\eta_n = \frac{h_{04} - h_5}{h_{04} - h_{5i}} \overset{(3.12)}{=} \frac{c_5^2}{2} \frac{2}{c_{5i}^2} = \left(\frac{c_5}{c_{5i}}\right)^2.$$

An alternative measure of the nozzle efficiency, used in Chapter 6, was $\varphi_5 = c_5/c_{5i}$. The two parameters, η_n and φ_5, are linked by $\eta_n = \varphi_5^2$.

The ratio between the stagnation pressure at inlet and the static pressure at exit is

$$\frac{p_{04}}{p_5} = \left(\frac{T_{04}}{T_{5i}}\right)^{\frac{\gamma}{\gamma-1}}. \tag{7.84}$$

The temperature ratio is

$$\frac{T_{04}}{T_5} = \frac{T_{05}}{T_5} = 1 + \frac{\gamma - 1}{2} M_5^2. \tag{7.85}$$

Assuming that the heat capacity at constant pressure is constant, and combining (7.83) and (7.85) yields

$$T_{5i} = T_{04} - \frac{T_{04} - T_5}{\eta_n}. \tag{7.86}$$

Substituting (7.86) and (7.85) in (7.84) yields

$$\frac{p_{04}}{p_5} = \left(\frac{\eta_n}{\dfrac{1}{1 + \frac{\gamma-1}{2}M_5^2} - 1 + \eta_n}\right)^{\frac{\gamma}{\gamma-1}} \tag{7.87}$$

which at sonic condition becomes

$$\frac{p_{04}}{p_{cr}} = \left(\frac{\eta_n}{\frac{2}{1+\gamma} - 1 + \eta_n} \right)^{\frac{\gamma}{\gamma-1}} = \left(\frac{\eta_n}{\frac{1-\gamma}{1+\gamma} + \eta_n} \right)^{\frac{\gamma}{\gamma-1}}. \tag{7.88}$$

To calculate the ratio between the nozzle exit area and the critical area, let us use mass conservation

$$\dot{m}_{cr} = \dot{m}_5$$

so that

$$\rho_{cr} a_{cr} A_{cr} = \rho_5 c_5 A_5$$

which yields

$$\frac{A_5}{A_{cr}} \stackrel{(2.66)}{=} \frac{\rho_{cr} a_{cr}}{\rho_5 c_5} = \frac{p_{cr}}{T_{cr}} \frac{T_5}{p_5} \frac{a_{cr}}{c_5} \stackrel{(3.20)}{=} \frac{p_{cr}}{p_5} \frac{T_5}{T_{cr}} \frac{\sqrt{T_{cr}}}{\sqrt{T_5}} \frac{1}{M_5} = \frac{1}{M_5} \frac{p_{cr}}{p_5} \sqrt{\frac{T_5}{T_{cr}}} = \frac{1}{M_5} \frac{p_{cr}}{p_{04}} \frac{p_{04}}{p_5} \sqrt{\frac{T_5}{T_{04}}} \sqrt{\frac{T_{04}}{T_{cr}}} =$$

$$\stackrel{(7.85),(7.87),(7.88)}{=} \frac{1}{M_5} \left(\frac{\frac{1-\gamma}{1+\gamma} + \eta_n}{\frac{1}{1+\frac{\gamma-1}{2}M_5^2} - 1 + \eta_n} \right)^{\frac{\gamma}{\gamma-1}} \sqrt{\frac{1}{1+\frac{\gamma-1}{2}M_5^2} \frac{\gamma+1}{2}}. \tag{7.89}$$

7.6.1   Converging Nozzle

The converging nozzle is the simplest and most common type of exhaust nozzle. This nozzle is used for both the hot turbine discharged gases and for the cool by-pass airflow of the fan. Depending on the nozzle exit pressure relative to the atmospheric pressure, the nozzle can operate under two conditions.

At low thrust regimes, the flow is subsonic. During most operating conditions, however, the flow reaches the speed of sound and the convergent nozzle is choked. In this case, the pressure at exit of the nozzle, p_e, is equal to the atmospheric pressure, *i.e.*, $p_e = p_a$, and the critical pressure is less than or equal to the atmospheric pressure, $p_a \geq p_{cr}$. This is the first operating condition.

According to (3.40), if the flow is choked and the stagnation pressure is constant, the mass flow rate, and therefore the velocity, cannot be increased unless the stagnation temperature is decreased. If the stagnation temperature is constant but the upstream stagnation pressure is increased above the pressure that caused sonic flow in the convergent nozzle, the static pressure at exit exceeds the atmospheric pressure. As a result, the pressure thrust term $A_e(p_e - p_a)$ of thrust equation (6.7) becomes nonzero, increasing the jet engine thrust due to the gas stream momentum change.

In this case, the atmospheric pressure is smaller than the critical pressure, $p_a < p_{cr}$. Since the exit pressure is higher than the atmospheric pressure, the gases continue to expand and

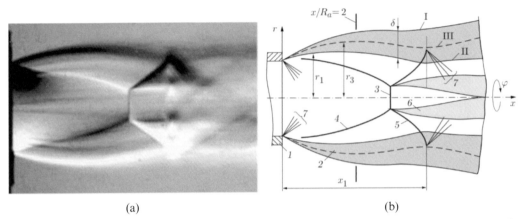

(a)   (b)

Figure 7.76 Underexpanded jet downstream of converging nozzle with exit Mach number, $M_e = 1$ and $p_e/p_a = 2.64$: (a) Schlieren picture of initial jet segment, (b) sketch of flow features: (1) converging nozzle exit, (2) jet mixing layer, (3) Mach disk, (4,5) hanging an reflected shock waves, (6) shear layer behind the point of interaction of shock waves 3-5, (7) expansion fan [Zapryagaev et al., 2004]. Reprinted by permission from Springer Nature Journal of Applied Mechanics and Technical Physics, Effect of Streamline Curvature on Intensity of Streamwise Vortices in the Mixing Layer of Supersonic Jets, V. I. Zapryagaev, N. P. Kiselev, and A. A. Pavlov, ©2004.

accelerate for a short distance beyond $M = 1$. This small region of supersonic flow ends with a normal shock and/or several oblique shocks that decelerate the flow to subsonic. This operating condition is called underexpanded or supercritical and is illustrated in Fig. 7.76.

7.6.2 Converging-Diverging Nozzle

There are losses in the converging nozzle operating at supercritical condition because the gases do not expand fast enough to immediately reach atmospheric pressure. High-pressure ratio engines can recover some of these losses by replacing the converging nozzle by a converging-diverging nozzle. The converging-diverging nozzle, also called the de Laval nozzle, has a cross-sectional area that decreases and then increases, as shown in Section 3.4.3.2.

Depending on the ratio between the exit static pressure and the inlet stagnation pressure, the converging-diverging nozzle has seven operating conditions, analyzed in Section 3.4.3.2. The first two operating conditions correspond to an exit pressure equal or larger than the pressure p_2 of Fig. 3.14. In these cases the flow is subsonic in the nozzle, reaching at most $M = 1$ at the throat if the exit pressure is p_2. The nozzle is not typically operated in these regimes, except during engine startup.

The nozzle is operating at the design condition if the nozzle exit pressure is equal to the atmospheric pressure and the flow transitions smoothly from subsonic in the convergent to supersonic in the divergent. No shocks are present in the nozzle or at the exit of the nozzle. The design operating condition, denoted as case 6 in Fig. 3.14, yields the maximum thrust.

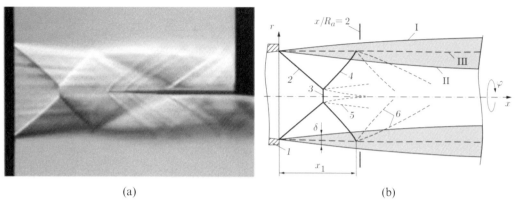

(a)   (b)

Figure 7.77 Overexpanded jet downstream of converging nozzle with exit Mach number, $M_e = 2$ and $p_e/p_a = 0.643$: (a) Schlieren picture of initial jet segment, (b) sketch of flow features: (1) converging-diverging nozzle exit, (2) compression shock, (3) Mach disk, (4) reflected shock, (5) shear layer behind the point of interaction of shock waves 2-4, (6) expansion fan [Zapryagaev et al., 2004]. Reprinted by permission from Springer Nature Journal of Applied Mechanics and Technical Physics, Effect of Streamline Curvature on Intensity of Streamwise Vortices in the Mixing Layer of Supersonic Jets, V. I. Zapryagaev, N. P. Kiselev, and A. A. Pavlov, ©2004.

Figure 7.78 Shock diamonds in a Schlieren picture of underexpanded jet, $M = 1.15$, $p_e/p_a = 0.83$ [André et al., 2014].

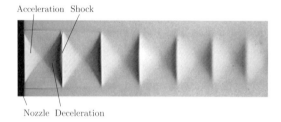

In the third operating condition, the exit pressure is less than p_2 but higher than p_6. In this case the flow is subsonic in the converging part, reaches sonic condition at the throat, becomes supersonic in the divergent part, and then is slowed down by a normal shock or a series of oblique shocks. The flow leaves the nozzle at subsonic speeds, which is not advantageous for thrust generation. The flow downstream of the nozzle consists of series of compression waves and expansion waves. The result of these waves is a series of "shock diamonds" (or Mach diamonds), as shown in Figs. 7.77–7.79.

The fourth operating condition corresponds to exit pressure p_4 in Fig. 3.14. In this case the normal shock is located at the nozzle exit.

In the fifth operating condition of Fig. 3.14, the exit pressure p_5 is just above the design exit pressure. Similar to operating conditions 3 and 4, the flow downstream of the nozzle consists of series of compression waves and expansion waves, as shown in Fig. 7.77. The operating cases 3, 4, and 5 are all called overexpanded cases. None of these overexpanded conditions are desirable for thrust generation.

Figure 7.79 Shock diamonds in
the exhaust of a Pratt & Whitney
J58 turbojet, the engine of
Lockheed SR-71 aircraft (courtesy
of NASA).

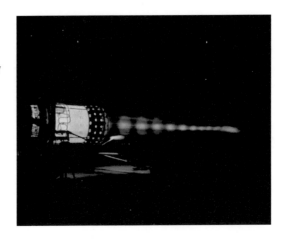

In the seventh operating condition, the pressure at exit of the nozzle, p_7 in Fig. 3.14, is slightly lower than the design exit pressure. The flow continues to expand outside of the nozzle. A series of expansion and compression waves are generated downstream of the nozzle. This results in a series of shock diamonds opposite in order to those of the operating cases 3, 4, and 5. This operating case 7 is an underexpanded or supercritical condition.

Example 7.6.1 The inlet stagnation pressure and stagnation temperature of a convergent-divergent nozzle are 25 bar and 600 K. The nozzle exit area is 0.4 m^2 and the minimum area is 0.25 m^2. The nozzle efficiency is $\eta_n = 0.97$. The excess air is $\lambda = 3$. Determine the operating condition of the nozzle and the mass flow rate if the atmospheric pressure is 0.7 bar. One can assume that the ratio of specific heat capacities, γ, is constant in the nozzle.

Solution
Let us first calculate γ. For $T_{04} = 600$ K, read from the air tables B the speed of sound $a_a = 486.8$ m/s and from the stoichiometric combustion products tables C $a_{\lambda=1} = 477.6$ m/s. The ratio of specific heat capacities for air, γ_a, is

$$\gamma_a = \frac{a_a^2}{RT_{04}} = \frac{486.8^2}{287.16 \times 600} = 1.3754$$

and for stoichiometric combustion products, $\gamma_{\lambda=1}$, is

$$\gamma_{\lambda=1} = \frac{a_{\lambda=1}^2}{RT_{04}} = \frac{477.6^2}{287.16 \times 600} = 1.3239$$

so that the ratio of specific heat capacities of the mixture is

$$\gamma_\lambda = \frac{1 + min\,L}{1 + \lambda min\,L}\gamma_{\lambda=1} + \frac{(\lambda - 1)min\,L}{1 + \lambda min\,L}\gamma_a = 0.3481 \times 1.3239 + 0.6518 \times 1.3754 = 1.3575.$$

To determine the operating condition of the nozzle, let us calculate the exit pressures that correspond to conditions 2, 4, and 6.

Using (7.89), where $A_5/A_{cr} = 4/2.4 = 1.6667$, $\gamma = \gamma_\lambda = 1.3575$ and $\eta_n = 0.97$ yields

$$1.6667 = \frac{1}{M_5}\left(\frac{\frac{-0.3575}{2.3575} + 0.97}{\frac{1}{1 + \frac{0.3575}{2}M_5^2} - 1 + 0.97}\right)^{\frac{1.3575}{0.3575}}\sqrt{\frac{2.3575}{2}\frac{1}{1 + \frac{0.3575}{2}M_5^2}}.$$

Using an iterative process, two solutions are obtained for M_5: 0.3712 and 1.8975. The subsonic solution $M_5 = 0.3712$ corresponds to condition 2 and the supersonic solution $M_5 = 1.8975$ corresponds to condition 6.

The pressure p_5 at the exit of the nozzle is calculated using (7.87). For the subsonic case, the pressure is

$$p_5 = p_{04}\left(\frac{\frac{1}{1 + \frac{0.3575}{2}0.3712^2} - 1 + 0.97}{0.97}\right)^{\frac{1.3575}{0.3575}} = 25 \times 0.9091 = 2.2728 \text{ bar.}$$

For the supersonic case corresponding to condition 6, the nozzle pressure exit is

$$p_5 = p_{04}\left(\frac{\frac{1}{1 + \frac{0.3575}{2}1.8975^2} - 1 + 0.97}{0.97}\right)^{\frac{1.3575}{0.3575}} = 25 \times 0.1404 = 0.35105 \text{ bar.}$$

Next, let us calculate the pressure for condition 4, when a normal shock is located at the nozzle exit. Using (3.77), the pressure downstream of the normal shock is

$$p_{5d} = p_5\left(\frac{2\gamma}{\gamma + 1}M_1^2 - \frac{\gamma - 1}{\gamma + 1}\right) = 0.35105\left(\frac{2 \times 1.3575}{2.3575}1.8975^2 - \frac{0.3575}{2.3575}\right) =$$
$$= 0.35105 \times 3.995 = 1.4023 \text{ bar.}$$

The atmospheric pressure of 0.7 bar is between conditions 4 and 6. Consequently, the imposed condition by the atmospheric pressure corresponds to condition 5, where the nonisentropic shocks and expansions occur outside of the nozzle. For this condition the flow is choked at the throat. Therefore, the mass flow rate can be calculated using condition 6, so that

$$\dot{m} = \rho_5 A_5 c_5.$$

The temperature T_5 at the exit of the nozzle is needed to calculate the density. The temperature is obtained using (7.85)

$$T_5 = T_{04}\frac{1}{1 + \frac{\gamma - 1}{2}M_5^2} = 600\frac{1}{1 + \frac{0.3575}{2}1.8975^2} = 365.05 \text{ K}$$

so that the density is

$$\rho_5 = \frac{p_5}{RT_5} = \frac{0.35105}{287.16 \times 365.05} = 0.3349 \text{ kg/m}^3.$$

The velocity c_5 is calculated using the Mach number

$$c_5 = \mathrm{M}_5 a_5 = M_5\sqrt{\gamma_\lambda RT_5} = 1.8975\sqrt{1.3575 \times 287.16 \times 365.05} = 715.8 \text{ m/s}$$

so that the mass flow rate is

$$\dot{m} = \rho_5 A_5 c_5 = 0.3349 \times 0.4 \times 715.8 = 95.88 \text{ kg/s}.$$

7.6.3   Variable Nozzle

Ideally the exhaust nozzle should always operate at design conditions, irrespective of the engine operating point and atmospheric pressure. One way of achieving this is by using a nozzle with variable area. The variable nozzle can either have a constant fixed exit area with a variable throat area or a constant throat area with a variable exit area. Two solutions are used for variable nozzles: a plug nozzle, shown in Fig. 7.80, and an iris nozzle, shown in Fig. 7.81.

Figure 7.80 Plug nozzle on Jumo 004.

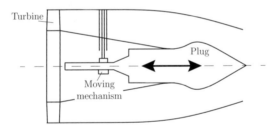

Figure 7.81 Iris variable nozzle on a General Electric F110 engine of an F-16 aircraft (courtesy of Emmanuel Dunand/AFP/Getty Images).

In the former case, the nozzle area changes by translating the plug along the axis. The iris nozzle can change either the throat (minimum) area or the exit area by adjusting its iris-like petal design.

Variable nozzles are not typically used in civil aviation because in this case the airplane is designed to operate at one altitude and flight condition. The only exception was the Concorde airplane whose Rolls-Royce/Snecma Olympus 593 engines had variable nozzles. Many military jet engines, including all with afterburning (see Section 8.2), use a variable nozzle because they operate under diverse conditions. Example 8.2.1 illustrates how to determine the nozzle area change as the engine switches from a dry run (operating without afterburning) to a wet run (operating with afterburning).

Variable nozzles are also used for thrust vectoring and as thrust reversers. Thrust vectoring, or gas-dynamic steering, refers to the ability to control the direction of the thrust of the engine. Thrust vectoring was initially used for providing upward vertical thrust for vertical and short takeoff and landing. Currently thrust vectoring is also used in combat situations for performing various maneuvers that are not possible on an airplane powered by conventional engines. Thrust vectoring is shown in Fig. 7.82 on the Pratt & Whitney F119 engine and in Fig. 7.83 on the General Electric F404 engine installed in an F-18 airplane.

Figure 7.82 Thrust vectoring on Pratt & Whitney F119 engine (reproduced with permission from Pratt & Whitney).

Figure 7.83 Thrust vectoring on General Electric F404 turbofan (courtesy of NASA).

Figure 7.84 Thrust reverse systems (courtesy of Rolls-Royce).

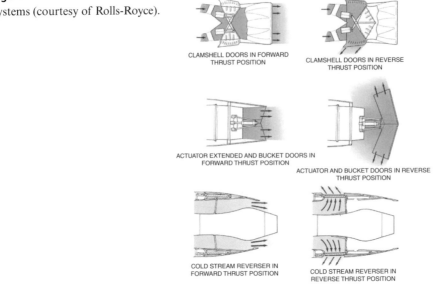

CLAMSHELL DOORS IN FORWARD
THRUST POSITION

CLAMSHELL DOORS IN REVERSE
THRUST POSITION

ACTUATOR EXTENDED AND BUCKET DOORS IN
FORWARD THRUST POSITION

ACTUATOR AND BUCKET DOORS IN REVERSE
THRUST POSITION

COLD STREAM REVERSER IN
FORWARD THRUST POSITION

COLD STREAM REVERSER IN
REVERSE THRUST POSITION

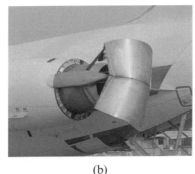

(a)   (b)   (c)

Figure 7.85 Thrust reversers: (a) Clamshell door (Crown copyright 2020), (b) bucket deployed (courtesy of Stellan Hilmerby), and (c) cold stream (courtesy of Eric Piermont/AFP/Getty Images).

Thrust reversers are used to change the direction of the exhaust flow while the airplane is on the ground, to assist with airplane deceleration. This is needed either after the airplane touches down or in the case of a rejected takeoff. There are several types of thrust reversers, as shown in Fig. 7.84: (1) clamshell door system, (2) bucket target system, and (3) cold stream reverser system. The first two thrust reversers are used on small to medium thrust engines, while the last one is used on large turbofan engines. Figure 7.85a shows a clamshell door system. Figure 7.85b shows the bucket target system being deployed on the Pratt & Whitney JT8D engine of an MD-80 airplane. Figure 7.85c shows the cold stream reverser system being deployed on an Airbus 320 airplane.

Problems

1. The flow coefficient for a compressor stage is $\phi = 0.5$. The flow enters and leaves the stage in the axial direction. The turning angle of the stator is 20 deg.
   1. What is the turning angle of the rotor?
   2. Draw the velocity triangles.
2. The flow enters the fan of a turbofan engine axially with a velocity of 180 m/s. What should the value of the flow coefficient be such that the flow is subsonic on the fan blade? (Hint: assume the speed of sound is 340 m/s.)
3. The flow coefficient of the stage shown in Fig. 7.86 is $\phi = 1$. Assume axial inlet velocity. Note that the exit angle of the rotor is 30 deg and the turning angle of the stator is 12 deg.

Figure 7.86 Compressor stage.

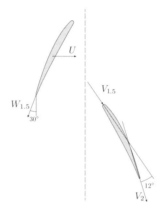

   1. Draw the stage velocity diagram.
   2. Is this a compressor or a turbine stage? Justify.
   3. What should the turning angle of the stator be such that the flow leaves the stage axially? Is this value realistic? Justify.
4. The flow coefficient for a compressor stage is $\phi = 0.4$. The flow enters and leaves the stage in the axial direction. The turning angle of the stator is 25 deg.
   1. What is the turning angle of rotor?
   2. Draw the velocity triangles.
5. Consider the compressor velocity triangles shown in Fig. 7.87, where the incoming velocity V_1 is in the axial direction, the angle $\beta_1 = 45°$ and the velocity $W_{U1.5} = 0.5\,U$. What is the specific work of the compressor if the transport velocity $U = 160$ m/s?
6. The incoming velocity V_1 in an axial compressor is 160 m/s, the turning angle on the rotor is $\theta_r = 22°$, the shaft rotates at $n = 10,000$ rpm, and the radius is 0.2 m. What should the turning on the stator be such that the velocity leaves the stage in the axial direction?
7. Air at 101.3 kPa and 288 K enters an axial-flow compressor with a velocity of 160 m/s. There are no inlet guide vanes. The stage has a tip diameter of 0.68 m and a hub diameter

Figure 7.87 Compressor velocity triangle.

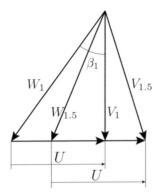

of 0.46 m. The angular velocity of the rotor is 8500 rpm. The air enters the rotor axially and leaves the stator as close to the axial direction as possible. The turning angle is 15° in the rotor and up to 12° in the stator.

Assume:

1. constant specific heat and $\gamma = 1.4$;
2. air enters and leaves the blades at the blade metal angles;
3. the axial components of the velocities are constant in the stage.

Divide the blade into three equal span-wise sections.

1. Construct the velocity diagrams at the middle of each section, for both the rotor and stator.
2. Plot the radial variation of the camber line for the rotor and stator.
3. Calculate the power required by the compressor.
4. Calculate and plot the radial variation of the total enthalpy at the exit of the stage.
5. Calculate and plot the radial variation of the total-to-total pressure ratio for the stage.
6. Calculate and plot the radial variation of the degree of reaction, flow coefficient, work coefficient, Mach number, and de Haller number.

8. Derive a formula for the compressor ideal work of a jet engine flying at velocity V and temperature T_H, as a function of the following parameters:

1. compressor pressure ratio, π_{oc};
2. ambient temperature, T_H;
3. flight velocity, V;
4. ratio of specific heats, γ;
5. specific heat a constant pressure, c_p (assume c_p constant).

9. The mass flow rate of gases in a one-stage turbine is 32 kg/s, the turbine inlet temperature is $T_{03} = 1350$ K, and the pressure is $p_{03} = 20$ bar. The flow enters and exits the turbine axially. The rotational speed is 12,000 rpm and the mid-span radius is 25 cm. The axial velocity at the exit from the stator is $V_{a3.5} = 160$ m/s and the angle of the absolute velocity is $\alpha_{3.5} = 70°$. Assume that the flow leaves the airfoil blades at the metal blade angle and the axial velocity remains constant throughout the stage.

1. Draw the velocity diagrams at the mean blade diameter.
2. Calculate the Mach number of the absolute velocity at the exit from the nozzle.

3. Calculate the specific work and power of the turbine.
4. Calculate the total-to-total expansion rate.
5. Calculate the inlet flow parameter $\frac{\dot{m}\sqrt{T_{03}}}{p_{03}A}$.

Assume the excess air is $\lambda = 1$.

10. A one-stage axial free turbine has a flow coefficient $\phi = 0.5$. The gases enter the turbine axially. The turning angle on the vane is $68°$.
    1. What should the turning angle on the blade be such that the flow leaves the turbine axially?
    2. What should the wheel speed U be such that the free turbine generates 2000 hp using a mass flow rate of 10 kg/s?

Bibliography

B. André, T. Castelain, and C. Bailly. Experimental exploration of underexpanded supersonic jets. *Shock Waves*, 24:21–32, 2014. 371

Anonymous, editor. *The Jet Engine*. Rolls-Royce plc, 1986. 340, 344

Anonymous, editor. *The Jet Engine*. Rolls-Royce plc, 2005. 346

W. W. Bathie. *Fundamentals of Gas Turbines*. John Wiley & Sons, Inc., Second edition, 1996. 286, 292

C. J. Carter, S. A. Guillot, W. F. Ng, and W. W. Copenhaver. Aerodynamic performace of a high-turning compressor stator with flow control. In *37th AIAA/ASME/SAE/ASEE Joint Propulsion Conference and Exhibit*, AIAA-2001-3973, pages 1–6. American Institute of Aeronautics and Astronautics, July 2001. 302

L. J. Cheshire. The design and development of centrifugal compressors for aircraft gas turbines. *Proc. I. Mech. E.*, 153, 1945. 300

D. E. Culley, M. M. Bright, P. S. Prahst, and A. J. Strazisar. Active flow separation control of a stator vane using embedded injection in a multistage compressor experiment. *Transactions of the ASME - Journal of Turbomachinery*, 126(1):24–34, January 2004. 302

N. A. Cumpsty. *Compressor Aerodynamics*. Longman, 1989. 300

I. J. Day. Review of stall, surge and active control in axial compressors. In *Eleventh International Symposium on Air Breathing Engines*, pages 97–105, Tokyo, Japan, September 1993a. AIAA. 301, 302

I. J. Day. Active suppression of rotating stall and surge in axial compressors. *ASME Journal of Turbomachinery*, 115(1):40–47, January 1993b. 302

S. L. Dixon. *Fluid Mechanics and Thermodynamics of Turbomachinery*. Butterworth-Heinemann, forth edition, 1998. 357

D. J. Dorney, D. L. Sondak, and P. G. A. Cizmas. Effects of hot streak/airfoil ratios in a high-subsonic single-stage turbine. *International Journal of Turbo & Jet-Engines*, 17(2):119–132, 2000. doi: 10.1515/TJJ.2000.17.2.119. 345

H. W. Emmons, C. E. Pearson, and H. P. Grant. Compressor surge and stall propagation. *Transactions of the ASME*, 77(4):455–469, May 1955. 300, 301

A. H. Epstein, J. E. Ffowcs-Williams, and E. M. Greitzer. Active suppression of compressor instabilities. *Journal of Propulsion and Power*, 5(2):204–211, March-April 1989. 301

R. D. Flack. *Fundamentals of Jet Propulsion with Applications*. Cambridge University Press, 2005. 315

R. Florea, K. C. Hall, and P. G. A. Cizmas. Eigenmode analysis of unsteady viscous flows in turboma-chinery cascades. In Torsten H. Fransson, editor, *Proceedings of the 8th International Symposium on*

Unsteady Aerodynamics and Aeroelasticity of Turbomachines, pages 767–782, Stockholm, Sweden, 1997. 301

R. Florea, K. C. Hall, and P. G. A. Cizmas. Reduced-order modeling of unsteady viscous flow in a compressor cascade. *AIAA Journal*, 36(6):1039–1048, 1998. 301

E. M. Greitzer. Surge and rotating stall in axial flow compressors. *ASME Journal of Engineering for Power*, 98:190–217, April 1978. 301

P. Hill and C. Peterson. *Mechanics and Thermodynamics of Propulsion*. Addison Wesley, second edition, 1992. 290, 339, 342

T. Iura and W. D. Rannie. Experimental investigations of propagating stall in axial-flow compressors. *Transactions of the ASME*, 76(3):463–471, April 1954. 300, 301

S. Johnson, F. L. Carpenter, O. K. Rediniotis, and P. G. A. Cizmas. Experimental validation of a pulse modulated flow control actuator for turbomachinery. *Journal of Propulsion and Power*, 33:1358–1368, 2017. 303

J. L. Kerrebrock. *Aircraft Engines and Gas Turbines*. The MIT Press, second edition, 1992. 283, 286, 291, 318, 341, 345, 366

K. K. Kuo. *Principles of Combustion*. John Wiley & Sons, second edition, 2005. 340

N. M. McDougall. Stall inception in axial flow compressors. PhD thesis, University of Cambridge, 1988. 300

A. B. McKenzie. *Axial Flow Fans and Compressors, Aerodynamic Design and Performance*. Cranfield Series on Turbomachinery Technology. Ashgate, Aldershot, 1997. 300

F. K. Moore and E. M. Greitzer. A theory of post-stall transients in axial compression systems: Part i - development of equations. *Transactions of the ASME - Journal of Engineering for Gas Turbines and Power*, 108(1):68–76, January 1986. 301

W. K. Olson, J. H. Childs, and E. J. Jonash. The combustion problem of the turbojet at high altitude. *Transactions of the American Society of Mechanical Engineers*, 1955. 343

J. D. Paduano, A. H. Epstein, L. Valavani, J. P. Longley, E. M. Greitzer, and G. R. Guenette. Active control of rotating stall in a low-speed axial compressor. *Transactions of the ASME - Journal of Turbomachinery*, 115:48–56, 1993. 301, 302

S. S. Penner. *Chemistry Problems in Jet Propulsion*. Pergamon Press, 1957. 342

J. R. Schwab, R. G. Stabe, and W. J. Whitney. Analytical and experimental study of flow through and axial turbine stage with a nonuniform inlet radial temperature profile. In *19th AIAA/SAE/ASME Joint Propulsion Conference*, AIAA Paper 83-1175, Seattle, WA, June 1983. 344

W. E. Scull. Relation between inflammables and ignition sources in aircraft environments. Technical Report Report 1019, National Advisory Committee for Aeronautics, 1951. 342

J. Seddon and E. L. Goldsmith. *Intake Aerodynamics*. AIAA educational series, second edition, 1999. 291

P. D. Silkowski. Aerodynamic design of moveable inlet guide vanes for active control of rotating stall. Master's thesis, Massachussetts Institute of Technology, 1990. 302

A. Stodola. *Steam and Gas Turbines*, volume 1 and 2. McGraw Hill, 1927. 332

K. L. Suder, M. D. Hathaway, S. A. Thorp, A. J. Strazisar, and M. B. Bright. Compressor stability enhancement using discrete tip injection. *Journal of Turbomachinery. Transactions of the ASME*, 123(1):14–23, January 2001. 302

V. I. Zapryagaev, N. P. Kiselev, and A. A. Pavlov. Effect of streamline curvature on intensity of streamwise vortices in the mixing layer of supersonic jets. *Prikladnaya Mekhanika i Tekhnicheskaya Fizika*, 45(3):32–43, May-June 2004. 370, 371

8   Thrust Augmentation

Thrust augmentation is usually needed for a short time period at (1) takeoff, (2) climb, (3) combat, and (4) high speed performance. Thrust augmentation allows us to avoid using a bigger (and heavier) engine that would penalize the performance of the aircraft when the additional thrust is not necessary. In other words, instead of utilizing a heavier and more powerful engine whose maximum power is only needed for a short period of time, it is often better to use a smaller engine that produces the required short-duration thrust by power augmentation. This section presents three methods of thrust augmentation: (1) water injection, (2) afterburning, and (3) inter-turbine combustion.

8.1   Water Injection

Water injection is a technology that was developed in the late 1940s and early 1950s. Water injection increased engine thrust and allowed heavily-loaded aircraft to take off from short runways. Figure 8.1 shows the Boeing B-52 bomber powered by eight Pratt & Whitney J57-P-1W turbojet engines climbing using water injection. Water injection increased the thrust of the J57-P-1W by 10%. However, the amount of smoke produced by engines using water injection is significant. Currently, B-52 bombers are equipped with a new generation of more powerful Pratt & Whitney TF33-P-3 turbofan engines than do not use water injection. The Boeing KC-135 aerial refueling aircraft powered by Pratt & Whitney J57-P-59W turbojet engines also used water injection.

    Water injection was also used on commercial transport aircraft. Boeing 707-120 aircraft with Pratt & Whitney JT3C-6 turbojet engines, Boeing 747-100 and 200 aircraft with Pratt & Whitney JT9D-3AW and -7AW turbofan engines, and the BAC One-Eleven with Rolls-Royce Spey turbofan engines used water injection. Water injection is currently used to augment the thrust for supersonic flight. Figure 8.2 shows the water injection device used by the Japan Aerospace Exploration Agency on a Teledyne YJ-69-T-406 engine, which generated a 8% thrust increase for a 3.5% water/air ratio and a 17.1% fuel increase. Water injection is currently reassessed in the aviation sector because of its potential as a disruptive technology to reduce airplane emissions and maintenance costs.

Figure 8.1 B-52 bombers climbing using water injection for thrust augmentation (courtesy of USAF/Handout/Getty Images).

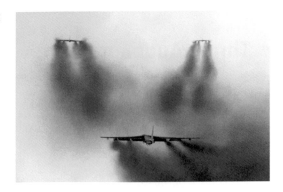

Figure 8.2 Water injection on a Teledyne YJ-69-T-406 engine (courtesy of Japan Aerospace Exploration Agency (JAXA), ©JAXA).

Water can be injected either at the inlet of the compressor or upstream of the combustor. As shown in the following, water injection at the inlet of the compressor produces approximately twice the amount of thrust increase compared to the case when water is injected upstream of the combustor. In both cases, thrust augmentation increases when the airplane speed increases.

8.1.1 Compressor Inlet Water Injection

One way of achieving thrust augmentation is by injecting water at the inlet of the compressor. Since the air temperature in the compressor increases rapidly as the air advances through the compressor stages, the injected water quickly evaporates. As a result of the evaporation

Figure 8.3 Variation of the position of the operation point on the compressor map when water is injected at the inlet of the compressor.

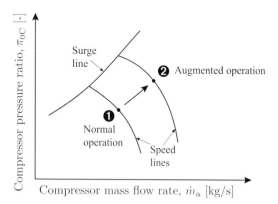

process, the air temperature decreases. The temperature decrease is proportional to the latent heat of water and the ratio between the mass flow rate of water and the mass flow rate of air. Since the inertia of the spool is large, the speed of the spool and implicitly the compressor speed, N, do not change significantly immediately after water is injected. Consequently, the corrected compressor speed, $N/\sqrt{\theta_0}$ increases since θ_0 decreases. Recall that $\theta_0 = T_{01}/288.15$, where the inlet compressor stagnation temperature, T_{01}, is specified in Kelvin. As shown in Fig. 8.3, the operating point moves from the normal operation point, 1, to the augmented operation point, 2.

As a result, the mass flow rate of air through the compressor increases. Consequently, the total mass flow rate through the compressor, $\dot{m}_{cps}$, increases because of (1) the increase of the air mass flow rate, $\dot{m}_{air\ cooled}$, due to the increase of air density and (2) the addition of the mass flow rate of water injected, $\dot{m}_{water}$. Consequently, $\dot{m}_{cps} = \dot{m}_{air\ cooled} + \dot{m}_{water}$. The pressure ratio also increases when the operation point moves to 2. Assuming the pressure drop in the combustor is not affected by water injection, the inlet pressure turbine increases. As a result, the speed of the gases leaving the exit nozzle, c_5, increases. Consequently, the engine thrust

$$\mathcal{T} = c_5 \cdot \dot{m}_g - u \cdot \dot{m}_a + A_e(p_5 - p_H) \tag{6.7}$$

increases since both $\dot{m}_g$ and c_5 increased. Note that the mass flow rate of gases is equal to the total mass flow rate through the compressor, $\dot{m}_{cps}$, plus the mass flow rate of fuel, $\dot{m}_f$; that is, $\dot{m}_g = \dot{m}_{cps} + \dot{m}_f$.

The influence of the ratio between the mass flow of water injected and the mass flow of air on (1) compressor pressure ratio, (2) ratio of total mass flow and air mass flow before injection, and (3) augmented thrust ratio is shown in Fig. 8.4 [Lundin, 1949]. Two values of the compressor inlet temperature are considered: (1) 59° F and (2) 99° F. The figure also shows the values of the injected water/air ratio that produce saturation at compressor inlet and outlet. The compressor pressure ratio increases as the mass flow rate of water is increased. The pressure increment is approximately the same for either 59° F or 99° F inlet temperature. The relative increase is, however, larger for 99° F inlet temperature than for 59° F inlet temperature.

Figure 8.4 Variation of compressor pressure ratio, mass flow rate to normal mass flow rate ratio, and augmented thrust ratio as a function of the ratio between the injected mass flow rate of water and the mass flow rate of air [Lundin, 1949].

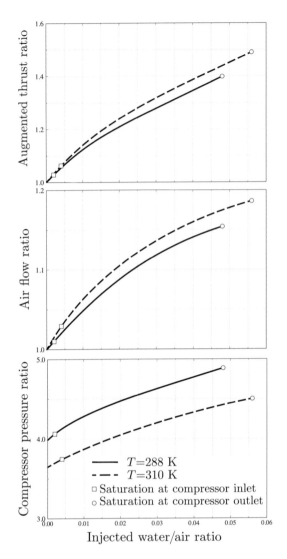

Consequently, water injection at the inlet in the compressor is more efficient when the ambient temperature is high.

8.1.2   Water Injection Upstream of the Combustor

Another location for water injection is upstream of the combustor. In this case the phenomena that lead to thrust augmentation are different from the case when water is injected at the compressor inlet. The effectiveness of water injection upstream of the combustor is lower than injection at the compressor inlet.

Let us assume that the flow through the turbine is choked. That is the case for most of the operating points of the jet engine. Since the inlet temperature in the engine, T_{01},

Figure 8.5 Variation of the position of the operation point on the compressor map when water is injected upstream of the combustor [Lundin, 1949].

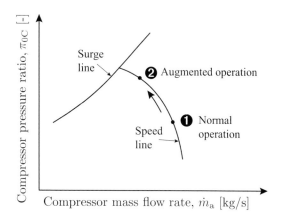

and the area at the turbine inlet are constant, the mass flow rate through the turbine depends on the turbine inlet pressure (and to a lesser degree on the properties of the fluid) since $\dot{m}_g \sqrt{T_{03}}/(p_{03}A) = \mathtt{const}$, as shown in Fig. 7.64. When water is injected upstream of the combustor, the mass flow rate in the compressor decreases since the pressure at the turbine inlet does not change. Due to the inertia of the spool, the wheel speed does not change and consequently the operating point moves toward smaller compressor mass flow rates, as shown in Fig. 8.5. The compressor pressure ratio at the new operating point increases, and as a result, turbine pressure increases and therefore the mass flow rate in the turbine increases. This results in a new equilibrium point of the engine, having a lower compressor air flow, a higher compressor pressure ratio, and a higher total mass flow rate.

The thrust augmentation is the result of both the increased total mass flow and the higher jet velocity provided by the increased pressure ratio. The water injected in the combustor increases the mass flow rate through the turbine and lowers the combustor inlet temperature. Consequently, more fuel can be burned without exceeding the maximum turbine inlet temperature. The amount of water injected is, however, limited by the compressor surge. A comparison between the water injection in the compressor and in the combustor, shown in Fig. 8.6, illustrates the fact that injection in the compressor leads to higher augmentation than injection in the combustor. Water injection in the combustor provides a smaller thrust increase than water injection in the compressor. In addition, for the same thrust augmentation, the amount of water injected in the combustor must be approximately twice the amount of water injected in the compressor.

8.1.3 Water Injection for Reducing Emissions and Maintenance Costs

As more powerful engines have been designed and built, water injection has been almost abandoned for thrust augmentation in the aviation sector, the exception being the Teledyne turbojet engine used for supersonic flight mentioned on page 381. Aeroderivative industrial engines, however, have continued to use and improve upon water injection. The use of water

Figure 8.6 Comparison between water injection in compressor and combustor (specific liquid consumption in lb/(hr lb thrust)) as a function of flight Mach number [Lundin, 1949].

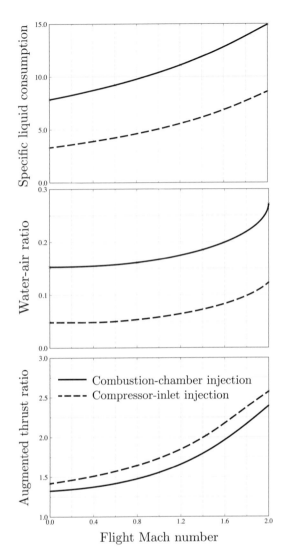

injection in the aviation sector is currently being reassessed because of the potential benefit of reducing emissions and maintenance costs [Daggett et al., 2004].

As opposed to old style water injection schemes designed before the 1970s, *water misting* has the potential benefits of greatly reduced turbine temperature and improved TSFC, in addition to reduced emissions [Daggett et al., 2010]. As presented in Section 8.1.1, when water is atomized and sprayed in the compressor inlet, the temperature at the exit of the compressor decreases. Consequently, for a given burner temperature rise, the lower burner inlet temperature yields a lower mean reaction temperature that reduces NO_x formation [Kerrebrock, 1975]. A 2.2% water to core air flow ratio during takeoff and climbout reduces NO_x by 50% while using approximately 1130 kg of water for a large-sized airplane [Daggett et al., 2004]. If water is added inside the burner during the takeoff and climbout at a water-to-fuel ratio

Table 8.1 Variation of TSFC and turbine inlet temperature (TIT) corresponding to approximately 50% NOx reduction at standard day conditions (temperature 288 K, pressure 101,325 Pa) (adapted from [Daggett, 2004]).

	NO_X	TSFC	TIT [K]
LPC	-47%	-4%	-242
Burner	-50%	+2%	-45

Figure 8.7 Turbine inlet temperature variation during water misting (adapted from [Daggett et al., 2010]).

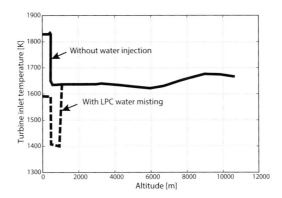

of 0.5:1, a 50% NO_X reduction is obtained while using approximately 510 kg of water. Table 8.1 compares the results of water injection in the low pressure compressor (LPC) and in the burner for a similar reduction of NO_X formation.

Injecting water in the compressor reduces TSFC, while injecting water in the burner slightly increases TSFC. Furthermore, the reduction of turbine inlet temperature is significantly larger when misting water in the compressor. An additional benefit of water injection in the compressor is the reduction of engine noise by approximately 0.6 dBA due to the reduction of the velocity at the exit nozzle.

Let us now consider using water injection from takeoff to top of climb (TOC) at 10,500 m. The water flow rate will be adjusted to reduce NO_X by 50–65% during takeoff and climb to a 900 m altitude. Between 900 m and TOC, the water flow will be adjusted to improve engine life by reducing turbine inlet temperature by 28 K [Daggett et al., 2004]. Figure 8.7 shows the variation of turbine inlet temperature for a modern 385,000 N (85,000 lbf) turbofan engine while the airplane climbs to 10,500 m.

Figure 8.7 shows that the LPC water misted engine has a peak turbine inlet temperature 242 K lower than the baseline engine. The water injected in the burner reduces the peak turbine inlet temperature by 64 K. The reduction of turbine inlet temperature by LPC misting, however, uses more water than when injecting in the burner, as shown in Table 8.2. Once the aircraft is operating in freezing conditions or reaches 1500 m altitude, the misting is switched from the LPC to the high-pressure compressor to avoid water droplet freezing.

Table 8.2 Water consumption and fuel savings/increase as function of flight schedule (adapted from [Daggett, 2004]).

	Water [kg]	Fuel [kg]
Takeoff to 3000 m		
Compressor (50% NO_x reduction)	1089	−32
Burner (65% NO_x reduction)	671	27
Takeoff to TOC		
11 K TIT reduction		
Compressor	1615	−39
Burner	2028	79
28 K TIT reduction		
Compressor	2404	−49
Burner	4069	159

Table 8.3 High-pressure turbine service life [Halila et al., 1982].

Condition	% Life used	Time [hours]
Takeoff	36	300
Max. Climb	49	3300
Max. Cruise	15	7200
Balance	<0.2	7200
Total	100	18 000

Water injection in the burner is more efficient at reducing flame temperature than compressor misting. Therefore, the amount of water needed to reduce NO_x by 65% from takeoff to 900 m is 418 kg less. However, the flame temperature quenching by injecting in the burner resulted in a 27 kg increase in fuel use due to losses in thermal efficiency. The compressor efficiency improvement obtained by misting resulted in 32 kg of fuel savings as opposed to a 27 kg increase when injecting in the burner. Table 8.2 shows that while the weight penalty of carrying water for takeoff is acceptable, the amount of water needed for injection from takeoff to TOC is a prohibitive payload penalty for the aircraft.

Reducing flame temperature yields a reduction of NO_x. At the same time, however, there is a tendency to increase hydrocarbon and CO emissions. This tendency can be managed by increasing the compressor overall pressure ratio [Bahr, 1973].

Reducing the temperature in the burner and turbine is also beneficial because it increases the service life of the hot section of the engine. Typically the service life of the hot section of the engine, that is, the burner and the high-pressure turbine, is much lower than that of the rest of the engine. For example, on a General Electric CF6 turbofan engine, the service life of the low and high-pressure compressors is 36,000 hours, while the service life of the burner and the high-pressure turbine is 18,000 hours [Davis and Stearns, 1985].

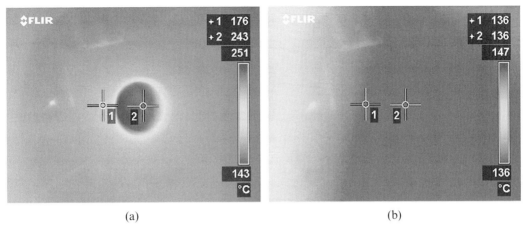

(a)   (b)

Figure 8.8 Aft view thermal image of turbojet before water injection and seconds after injection begins [Guarnieri and Cizmas, 2008]. (a) Before water injection and (b) after water injection.

Table 8.3 shows the design practices and service life experience of the CF6 turbofan engine [Halila et al., 1982]. As the maximum turbine inlet temperature occurs during takeoff, this operating regime reduces the turbine life at the fastest pace. At standard day conditions, 36% of the 18,000-hour design life is consumed during takeoff in only 300 hours. For hot day takeoff conditions, the entire turbine blade's life is consumed in only 250 hours of continuous operation. As the turbine life increases logarithmically with the decrease of turbine inlet temperature, reducing the hot section by water injection lowers engine maintenance costs.

Another benefit of water injection is the reduction of the jet engine thermal signature [Guarnieri and Cizmas, 2008]. Water injection is a simple method for shielding the infrared signature of a jet engine from heat seeking missiles. Figure 8.8 shows the thermal image of a small (1 kN thrust) turbojet before and after water injection. The temperatures reported by the FLIR camera were lower than the temperatures measured by the thermocouples because the thermal imaging measurements were taken indirectly via an infrared mirror. For example, thermocouples in the turbine measured turbine outlet temperature at 825 K (552°C) compared to 516 K (243°C) reported by the FLIR camera [Guarnieri and Cizmas, 2008].

8.2   Afterburning (Reheat)

Afterburning (or reheat) is another method for increasing the thrust of turbojet and turbofan engines. Afterburning can be used to improve airplane takeoff, climb, and combat performance. A turbojet engine equipped with an afterburner is called an *afterburning turbojet*, while a turbofan that has an afterburner is called an *augmented turbofan*. Figure 8.9 shows drawings of turbojet and turbofan engines with afterburning. Figure 6.29 shows cutaway drawings of the General Electric F404 and Pratt & Whitney F135 augmented turbofans.

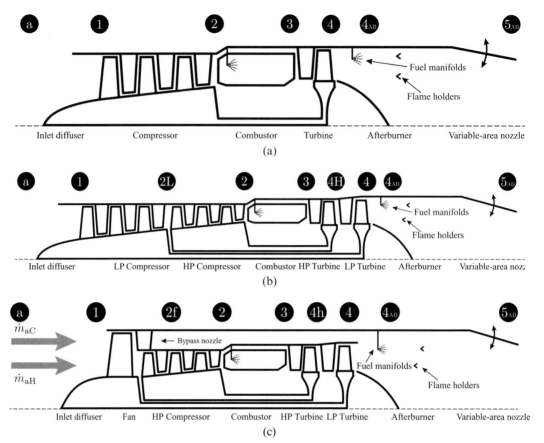

Figure 8.9 Types of engines with afterburning. (a) Single-spool with afterburning, (b) two-spool turbojet with afterburning, and (c) two-spool augmented turbofan.

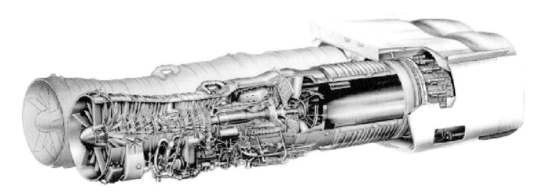

Figure 8.10 Two-spool turbojet with afterburning: Rolls-Royce/SNECMA Olympus 593 (courtesy of Rolls-Royce).

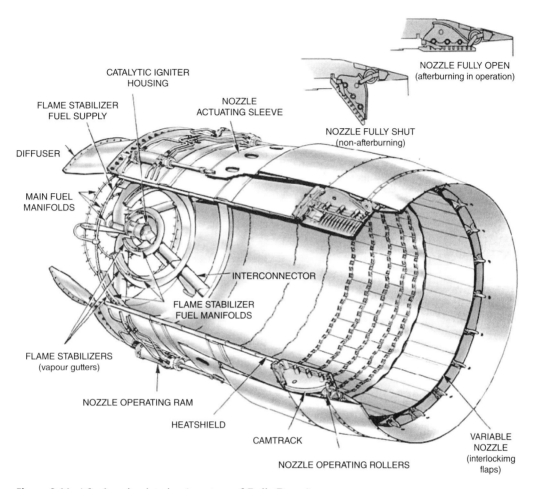

NOZZLE FULLY OPEN
(afterburning in operation)

NOZZLE FULLY SHUT
(non-afterburning)

CATALYTIC IGNITER
HOUSING

NOZZLE
ACTUATING SLEEVE

FLAME STABILIZER
FUEL SUPPLY

DIFFUSER

MAIN FUEL
MANIFOLDS

INTERCONNECTOR

FLAME STABILIZER
FUEL MANIFOLDS

FLAME STABILIZERS
(vapour gutters)

NOZZLE OPERATING RAM

HEATSHIELD

CAMTRACK

NOZZLE OPERATING ROLLERS

VARIABLE
NOZZLE
(interlockimg
flaps)

Figure 8.11 Afterburning jet pipe (courtesy of Rolls-Royce).

Figure 8.10 shows a cutaway drawing of the Rolls-Royce/Snecma Olympus 593 two-spool turbojet with afterburning, the engine that powered the Concorde airplane.

Fuel is injected downstream of the turbine, in the nozzle jet pipe, as shown in Fig. 8.11. The fuel is delivered by a spray bar. To maintain the combustion, flame holders are required downstream of the spray bar. The role of the flame holders is to create regions of reduced velocity. Ideally, the flow velocity behind the flame holders should be equal to the flame speed, that is, on the order of few m/s. The difference between the velocity of gases exiting the turbine, typically 300 m/s or more, and the typical flame speed is large. Consequently, the aerodynamic losses generated by the flame holders are important. Therefore, a compromise is necessary between aerodynamic efficiency and stable combustion.

Because the temperature of the flame in the afterburner can exceed 2000 K, the flame holders are positioned so that the flame is concentrated along the axis of the jet pipe. This allows a fraction of the turbine discharge gases and, in the case of augmented turbofans, the bypass flow to circulate along the jet pipe walls and maintain a safe wall temperature.

To reduce the velocity of the gases in the afterburner, the area of the afterburning jet pipe is larger than the jet pipe of the same engine without afterburning. For this reason, the afterburning jet pipe must have a propelling nozzle with variable area. The nozzle is closed during engine operation without afterburning, also called *dry operation*. The nozzle is open during engine operation with afterburning, called *wet operation*. By opening the nozzle during the afterburning operation, one avoids the pressure increase in the jet pipe, which otherwise would hamper the functioning of the engine. The timing of the nozzle opening and closing is critical. If the fuel is injected in the jet pipe and the nozzle opens late, the turbine temperature could exceed the normal limits. If the nozzle opens too early, flameout could happen.

Example 8.2.1 The stagnation pressure and stagnation temperature at exit from a turbojet with afterburning are $p_{04} = 3.5$ bar and $T_{04} = 1000$ K. The stagnation temperature in the afterburning jet pipe during a wet run is 1800 K. The ambient pressure is 1.013 bar. The pressure losses in the afterburning jet pipe can be neglected. Calculate the ratio between the nozzle throat area during wet and dry operation.

Solution
Let us check whether the flow is choked in the nozzle by calculating

$$\frac{p_{04}}{p_a} = \frac{3.5}{1.013} = 3.455.$$

Since this value far exceeds $[(\gamma + 1)/2]^{\gamma/(\gamma-1)}$, the flow is choked and

$$\frac{\dot{m}\sqrt{T_{05}}}{p_{05}A_5} = \text{const.}$$

Assuming an adiabatic process during dry operation, and since no work is being done in the nozzle, energy conservation yields $T_{05\text{dry}} = T_{04}$. Because the mass flow rate and the stagnation pressure are constant during wet and dry operation, then

$$\frac{A_{5\text{wet}}}{A_{5\text{dry}}} = \sqrt{\frac{T_{05\text{wet}}}{T_{05\text{dry}}}} = \sqrt{\frac{1800}{1000}} = 1.34.$$

Consequently, the nozzle area during wet operation is 34% larger than the nozzle area during dry operation.

The afterburner increases the weight of the engine. In addition, the combustion efficiency in the afterburner, ξ_{AB}, is smaller than the combustion efficiency in the combustor, ξ_{C}. The thrust increase obtained by using the afterburner is, however, significant, as it can reach

Figure 8.12 Reheat cycle.

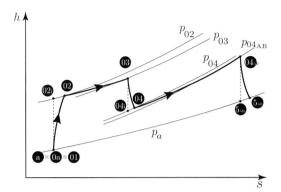

70% [Anonymous, 1986, p.176]. The fuel consumption increases significantly, so that the afterburner must only be used for a short duration to avoid running out of fuel.

The reheat that takes place in the afterburner between states 04 and 04_{AB} increases the enthalpy of the gases, as shown in Fig. 8.12. As a result, the temperature of the gases in the afterburner can be several hundred degrees Kelvin higher than that in the combustor.

Example 8.2.2 The excess air in the combustor is $\lambda_C = 3$. At the exit from the turbine, the temperature is $T_{04} = 1000$ K and the pressure is $p_{04} = 2.8$ bar. The mass flow rate of gases at the turbine exit is 40 kg/s. The enthalpy increase in the afterburning process is 600 kJ/kg. The pressure loss in the afterburning process is 4%. The combustion efficiency in the afterburner is $\xi_{AB} = 80\%$. Standard fuel is used in the engine so that $LHV = 43,500$ kJ/kg and $minL = 14.66$.

1. What is the mass flow rate of fuel injected in the combustor?
2. What is the mass flow rate of air available at the inlet in the afterburner?
3. What is the mass flow rate of fuel injected in the afterburner?
4. What is the temperature at the end of afterburning?
5. What is the entropy variation in the afterburning process? Draw the Mollier diagram.

Solution

1. The mass flow rate of air is obtained from

$$\dot{m}_g = \dot{m}_a(1 + f) = \dot{m}_a \left(1 + \frac{1}{\lambda_C minL}\right)$$

so that

$$\dot{m}_a = \dot{m}_g / \left(1 + \frac{1}{\lambda_C minL}\right) = 40 / \left(1 + \frac{1}{3 \times 14.66}\right) = 39.11 \text{ kg/s.}$$

2. The mass flow rate of fuel used in the combustor is

$$\dot{m}_{f_C} = \dot{m}_g - \dot{m}_a = 40 - 39.11 = 0.89 \text{ kg/s}.$$

The mass flow rate of air needed for stoichiometric combustion in the combustor is

$$\dot{m}_{a_{stoic}} = \dot{m}_{f_C} \times minL = 0.89 \times 14.66 = 13.04 \text{ kg/s}.$$

The mass flow rate of air available for combustion in the afterburner is

$$\dot{m}_{a_{avail}} = \dot{m}_a - \dot{m}_{a_{stoic}} = 39.11 - 13.04 = 26.07 \text{ kg/s}.$$

3. Neglecting the enthalpy of the fuel compared to LHV, the energy balance in the afterburner is

$$\dot{m}_g h_{04} + \dot{m}_{f_{AB}} \xi_{AB} LHV = (\dot{m}_g + \dot{m}_{f_{AB}}) h_{04_{AB}}. \tag{8.1}$$

The enthalpy at the exit from the turbine, h_{04}, is calculated using the stagnation temperature, T_{04}, and the excess air, λ_C. Using the thermodynamic properties of air tables B and the thermodynamic properties of stoichiometric combustion products tables C yields $h_{04} = 1071.87 \text{ kJ/kg}$.

The maximum enthalpy in the afterburner is

$$h_{04_{AB}} = h_{04} + \Delta h = 1071.87 + 600 = 1671.87 \text{ kJ/kg}.$$

The mass flow rate of fuel injected in the afterburner is obtained using (8.1), which yields

$$\dot{m}_{f_{AB}} = \dot{m}_g \frac{h_{04_{AB}} - h_{04}}{\xi_{AB} LHV - h_{04_{AB}}} = 40 \frac{1671.87 - 1000}{0.8 \times 43,500 - 1671.87} = 0.81 \text{ kg/s}.$$

Combustion of the fuel injected in the afterburner occurs with an excess air, λ_{AB}

$$\lambda_{AB} = \frac{\dot{m}_{a_{avail}}}{\dot{m}_{f_{AB}} minL} = \frac{26.07}{0.81 \times 14.66} = 2.19.$$

The overall excess air in the afterburner, considering the fuel injected in the combustor and in the afterburner, is

$$\lambda_{ovrl} = \frac{\dot{m}_a}{(\dot{m}_{f_C} + \dot{m}_{f_{AB}}) minL} = \frac{39.11}{(0.89 + 0.81) \times 14.66} = 1.57.$$

4. The maximum temperature of the gases in the afterburner is calculated using stagnation enthalpy, $h_{04_{AB}}$, and the overall excess air in the afterburner, λ_{ovrl}, which yields $T_{04_{AB}} = 1456.7 \text{ K}$.

5. The entropy at the exit of the turbine is determined by the excess air, $\lambda_C = 3$, the stagnation temperature $T_{04} = 1000 \text{ K}$, and the pressure $p_{04} = 2.8 \text{ bar}$. Using the air and stoichiometric combustion products thermodynamic tables, one obtains $s_{04} = 7.7806 \text{ kJ/(kg K)}$.

State 04_{AB} is determined by the pressure $p_{04_{AB}} = 0.96 \times p_{04} = 2.688$ bar, the enthalpy $h_{04_{AB}} = 1671.87$ kJ/kg, and the overall excess air, $\lambda_{ovrl} = 1.57$. Using the air and stoichiometric combustion products thermodynamic tables, one obtains $s_{04_{AB}} = 8.3620$ kJ/(kg K). Therefore, the entropy variation in the afterburner is $s_{04_{AB}} - s_{04} = 8.3620 - 7.7806 = 0.5814$ kJ/(kg K).

8.3   Inter-turbine Combustion

The vast majority of gas turbine engines operate following the Brayton cycle, which is not capable of maximum efficiency because of the irreversibility that is introduced by the temperature varying process during compression and expansion. All thermodynamic cycles that are capable of theoretical maximum efficiency, such as Carnot, Stirling, and Ericsson cycles, use constant temperature compression and expansion. A comparison of the Brayton and Carnot cycles shown in Fig. 6.3, and the Ericsson and Isothermal expansion, isentropic compression cycles, shown in Fig. 8.13, should also include the variation of efficiency and net work per unit flow vs. the pressure ratio. Assuming a ratio of specific heat capacities $\gamma = 1.4$, and temperatures $T_1 = 288$ K and $T_3 = 1152$ K, Fig. 8.14 shows a comparison of the efficiency and net work for these cycles. Note that the net work per unit flow is defined as w_{net}/RT_1, where w_{net} is the net work and R is the gas constant.

Figure 8.13 (a) Ericsson and (b) Isothermal expansion, isentropic compression cycles. Circled symbols **s, p, T** denote isentropic, isobaric, and isothermal processes, respectively. Symbol ⇗ denotes heat addition or rejection.

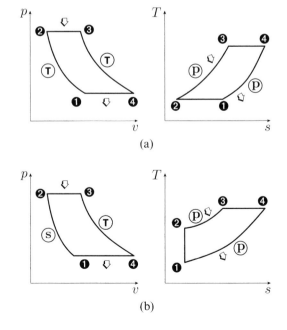

(a)

(b)

It is apparent that the thermal efficiency of the Brayton cycle

$$\eta_B = 1 - 1/(p_2/p_1)^{(\gamma-1)/\gamma} = 1 - T_1/T_2$$

is smaller than the efficiency of the Ericsson cycle, $\eta_E = 1 - T_1/T_3$, because $T_2 < T_3$. In the limit, when the efficiency of the Brayton cycle reaches that of the Ericsson (and Carnot) cycle, the net work of the Brayton cycle is zero, making it impractical. Similarly, the regenerative Brayton cycle reaches the thermal efficiency of the Ericsson cycle as the pressure ratio tends to zero and the net work tends to zero, making it impractical at this limit. The Carnot cycle must have a pressure ratio larger than $(T_3/T_1)^{\gamma/(\gamma-1)}$ (which for the given temperatures and ratio of specific heats is 128) so that the net work is positive. Such large pressure ratios in the compressor vastly exceed current and future compressor pressure ratios of gas turbine engines. Figure 8.14 shows that the Ericsson cycle has the highest thermal efficiency and highest dimensionless net work; therefore it should be the prime candidate for the gas turbine engine cycle. Achieving the isothermal compression is challenging, however. The second-best option is the isothermal expansion, isentropic compression cycle whose net work and thermal efficiency are better than those of the Brayton and regenerative Brayton cycles. Both the Ericsson and the isothermal expansion, isentropic compression cycles use an isothermal expansion. The present challenge is achieving an as-close-as-possible isothermal expansion by using *in situ* reheat in the turbine (aka intra-turbine combustion).

For a set power (or net work) of the gas turbine engine, the use of a turbine combustor allows a decrease in the thermal load in the main combustor, therefore reducing the maximum temperature of the cycle and the temperature variation throughout the combustion process. This is shown in Fig. 8.15, where the two gray areas are equal, so that the work generated by the Brayton cycle is equal to that of the isentropic compression, isothermal expansion cycle. Decreasing the combustor temperature also diminishes the need for costly combustor materials and thermal barrier coatings on the combustor liner, enabling more resource-efficient manufacturing. If the temperature in the combustor is kept unchanged, the work increases by using inter-turbine combustion, as shown in Fig. 8.16. In this case, the

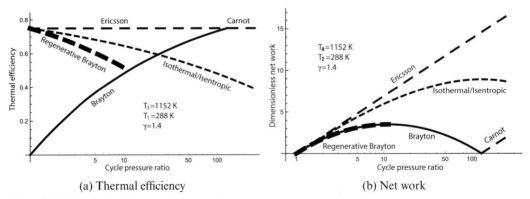

(a) Thermal efficiency (b) Net work

Figure 8.14 Thermal efficiency and net work vs. pressure ratio.

Figure 8.15 Brayton and isentropic compression, isothermal expansion cycles.

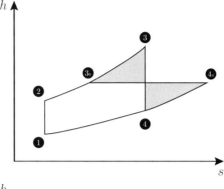

Figure 8.16 Real cycle approximating isothermal expansion with inter-turbine combustion.

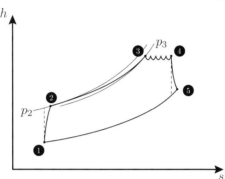

inter-turbine combustion is approximated as a series of small expansions followed by heat additions occurring in each stage of the turbine.

Due to the distributed fuel injection and combustion, both in the main combustor and in the turbine combustor, the amount of fuel to be burned at each location is smaller, thereby allowing a more complete and efficient combustion, decreasing the amount of unburned hydrocarbons and also the emission of solid particles (e.g. soot), creating the premises for a greener powerplant. This concept can be applied not only to stationary powerplants but also to aviation gas turbine engines, therefore promising to be a major contributor to the target of 90% NO_x reduction by 2050 per passenger and per kilometer relative to the year 2000 [Advisory Council for Aviation Research and Innovation in Europe].

The turbine-combustor concept is not only beneficial for land-based gas turbine engines used for power generation but also for aviation jet engines. Using cycle analysis, it was shown [Sirignano and Liu, 1999] that a 20% increase in gas turbine engine thrust can be achieved by a 50% increase in the TSFC by using an afterburner or by a 10% increase in the TSFC by using turbine combustion. Therefore, *in situ* reheat can become an option for increasing jet engine thrust. Recall that the classical approaches currently used to increase engine thrust are: (1) water injection either at inlet the compressor or, less efficiently, at inlet in the combustor (Section 8.1), and (2) combusting fuel downstream of the turbine, in an afterburner, where heat is added into the gas turbine engine system by a secondary combustion downstream of the turbine (Section 8.2). Thrust increase by water injection is significantly smaller than that obtained in an afterburner. The afterburner, however, increases

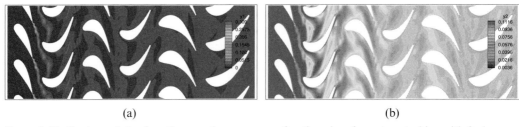

(a)   (b)

Figure 8.17 Contour plots of methane and oxygen mass fractions in a four-stage turbine with fuel injected in the first nozzle guide vane [Chambers et al., 2006]. (a) Methane mass fraction and (b) oxygen mass fraction.

the length and the weight of the engine and yields a high specific fuel consumption, due to the low efficiency of the low-pressure secondary combustion. Cycle studies [Witt, 1967; Liu and Sirignano, 2001] compared the performance of a jet engine using traditional compression together with isothermal expansion, that is, an isothermal expansion, isentropic compression cycle – a hybrid between the Ericsson and Brayton cycles shown in Fig. 8.16. It was observed that the turbine combustion benefits increase as the flight Mach number increases. At high Mach numbers (above 2.2), the compressor inlet temperature increases, and so does the compressor outlet temperature, increasing the compressor consumed work, while the amount of fuel injected in the engine is limited for the classical gas turbine engine by the material temperature constraints. This leads to a decrease in the available engine energy and, consequently, a decrease in engine thrust, especially since friction losses are higher at high Mach numbers.

An inter-turbine combustion concept was developed by Westinghouse Electric Corporation. The concept proposes fuel injection through the airfoils in order to reheat the expansion-cooled gas to achieve a quasi-isothermal expansion. In this manner, the power output increases, the NO_x emissions decrease, and the cycle thermal efficiency increases towards the ideal limit of the isothermal expansion, isentropic compression cycle. However, the ability to achieve stable and complete combustion in the turbine flow channels between the blades and vanes has not yet been demonstrated. A computational fluid dynamics investigation of *in situ* reheat in a four-stage turbine-combustor, shown in Fig. 8.17, identifies the high-strain regions where quenching could occur. In addition, the very short residence time is often lower than the auto-ignition delay [Isvoranu and Cizmas, 2003], and stabilization requires additional means, such as a cavity feeding the combustor with burned gases.

Problems

1. The gas at the exit from the turbine of a turbojet engine has a temperature of $T_{04} = 900$ K and a pressure of $p_{04} = 1.5$ bar. The excess air in the turbine is $\lambda = 2.7$. The engine has an afterburner that raises the temperature to $T_{04AB} = 1900$ K. The pressure drop in the afterburner is 7% and the excess air in the afterburner is $\lambda_{AB} = 1.3$. The engine is at takeoff conditions and the atmospheric pressure is 1 bar. To simplify computation, assume that the expansion in the exit nozzle is isentropic, whether the afterburner is on or off. The entropy correction $\Delta s'$ can be neglected. Standard fuel is used in the engine so that $LHV = 43,500$ kJ/kg and $minL = 14.66$.

1. What is the velocity c_5 of the gases at the exit from the engine nozzle when the afterburner is off?
2. What is the specific heat added in the nozzle when the afterburner is on?
3. What is the velocity c_{5AB} of the gases at the exit from the engine nozzle when the afterburner is on?
4. By what fraction does the thrust increases when the afterburner is on? (one can neglect in a first approximation the thrust increase due to the variation of mass flow rate of gases).

2. The excess air in the combustor is $\lambda_C = 3$. At the exit from the turbine the temperature is $T_{04} = 1000$ K and the pressure is $p_{04} = 2.8$ bar. The mass flow rate of gases at the turbine exit is 40 kg/s. The enthalpy increase in the afterburning process is 600 kJ/kg. The pressure loss in the afterburning process is 4%. The combustion efficiency in the afterburner is $\xi_{AB} = 80\%$. Standard fuel is used in the engine so that $LHV = 43,500$ kJ/kg and $minL = 14.66$.

1. What is the mass flow rate of fuel injected in the combustor?
2. What is the mass flow rate of air available at the inlet in the afterburner?
3. What is the mass flow rate of fuel injected in the afterburner?
4. What is the temperature at the end of afterburning?
5. What is the entropy variation in the afterburning process? Draw the Mollier diagram.

3. The air mass flow rate of a jet engine is 12 kg/s. The stagnation enthalpy at the exit of the turbine is 918 kJ/kg and the excess air is 4.5. Standard fuel is used in the engine so that $LHV = 43,500$ kJ/kg and $minL = 14.66$.

1. What should the mass flow rate of fuel injected in the afterburner be such that the stagnation enthalpy at the end of the afterburner is 1600 kJ/kg?
2. What is the excess air in the afterburner?
3. What is the maximum temperature in the afterburner?

4. The air mass flow rate of a jet engine is 9 kg/s. At the exit of the turbine the stagnation enthalpy is 600 kJ/kg and the excess air is $\lambda_C = 4.5$. Standard fuel is used in the engine so that $LHV = 43,500$ kJ/kg and $minL = 14.66$.

1. What is the mass flow rate of fuel injected in the combustor?
2. What is the mass flow rate of air available at the inlet of the afterburner?
3. What should the mass flow rate of fuel injected in the afterburner be such that the stagnation enthalpy at the end of the afterburner is 1600 kJ/kg if the combustion efficiency in the afterburner is $\xi_{AB} = 0.8$?
4. What is the excess air λ_{AB} in the afterburner?
5. What is the maximum temperature in the afterburner?

5. The air mass flow rate at the inlet of a jet engine is 10 kg/s. At the exit of the turbine the stagnation enthalpy is 600 kJ/kg. The mass flow rate of fuel in the jet engine combustor is $\dot{m}_{fC} = 0.04$ kg/s and in the afterburner is $\dot{m}_{f_{AB}}$ is 0.06 kg/s. The combustion efficiency

in the afterburner is $\xi_{AB} = 0.5$. The mass flow rate of air that is left at the exit from the afterburner is 6 kg/s. The maximum temperature in the afterburner is 1650 K.

1. What is the excess air λ_{AB} in the afterburner?
2. What is the excess air λ_C in the jet engine combustor?

Note: for standard fuel, $LHV = 43,500$ kJ/kg and $minL = 14.66$.

Bibliography

Advisory Council for Aviation Research and Innovation in Europe. Strategic Research & Innovation Agenda, 2017 Update, Volume 1. http://www.acare4europe.org,2017. 397

Anonymous, editor. *The Jet Engine*. Rolls-Royce plc, 1986. 393

D. W. Bahr. Technology for the reduction of aircraft turbine engine exhaust emissions. In *Atmospheric Pollution by Aircraft Engines*, volume Paper 29. Advisory Group for Aerospace Research & Development, 1973. 388

S. B. Chambers, H. C. Flitan, P. G. Cizmas, D. Bachovchin, T. Lippert, and D. Little. The influence of in situ reheat on turbine-combustor performance. *Journal of Engineering for Gas Turbines and Power*, 128(3):560–572, July 2006. 398

D. L. Daggett. Water misting and injection of commercial aircraft engines to reduce airport NO_x. Technical Report NASA CR–2004-212957, NASA, Glenn Research Center, 2004. 387, 388

D. L. Daggett, S. Ortanderl, D. Eames, C. Snyder, and J. Berton. Water injection: Disruptive technology to reduce airplane emissions and maintenance costs. In *2004 World Aviation Congress*, Reno, NV, November 2004. SAE International. doi: 10.4271/2004-01-3108. 386, 387

D. L. Daggett, L. Fucke, R. C. Hendricks, and D. J. H. Eames. Water injection on commercial aircraft to reduce airport nitrogen oxides. Technical Report NASA TM–2010-213179, NASA, Glenn Research Center, 2010. 386, 387

D. Y. Davis and E. M. Stearns. Energy efficient engine flight propulsion system final design and analysis. Technical Report NASA CR-168219, NASA, NASA Lewis Research Center, August 1985. 388

Jason A. Guarnieri and Paul G. A. Cizmas. A method for reducing jet engine thermal signature. *International Journal of Turbo & Jet-Engines*, 25:1–11, 2008. doi: 10.1515/TJJ.2008.25.1.1. 389

E. E. Halila, D. T. Lenahan, and T. T. Thomas. High pressure turbine test hardware detailed design report. Technical Report NASA CR-167955, NASA, NASA Lewis Research Center, June 1982. 388, 389

D. D. Isvoranu and P. G. A. Cizmas. Numerical simulation of combustion and rotor-stator interaction in a turbine combustor. *International Journal of Rotating Machinery*, 9(5):363–374, 2003. also published at ISROMAC-9. 398

J. L. Kerrebrock. Effect of compression ratio on NO_x production by gas turbines. *Journal of Aircraft*, 12(9):752–753, September 1975. doi: 10.2514/3.44489. 386

F. Liu and W. A. Sirignano. Turbojet and turbofan engine performance increases through turbine burners. *Journal of Propulsion and Power*, 17(3):695–705, 2001. 398

B. T. Lundin. Theoretical analysis of various thrust-augmentation cycles for turbojet engines. Technical Report 981, National Advisory Committee for Aeronautics, September 1949. 383, 384, 385, 386

W. A. Sirignano and F. Liu. Performance increases for gas-turbine engines through combustion inside the turbine. *Journal of Propulsion and Power*, 15(1):111–118, 1999. 397

Stewart H. De Witt. Reheat gas turbine power plant with air admission to the primary combustion zone of the reheat combustion chamber structure. Unites States Patent Office: 3,315,467, Westinghouse Electric Corporation, April 1967. 398

Part III

Rocket Engines

9 Classification of Rocket Propulsion Systems

9.1 Introduction

Rocket propulsion is a form of jet propulsion where mass (or matter) is accelerated from storage to high exit velocities. Rockets differ from typical air-breathing jet propulsion in that the rocket vehicle itself supplies <u>all</u> the propellant for the rocket motor. The exception to this is the mixed-mode (or multi-mode) engine that will be discussed later in this chapter.

The energy source in a rocket engine can be chemical, nuclear, or solar. The momentum can be imparted to the mass that leaves the rocket engine by pressure, electromagnetic forces, or electrostatic forces. Consequently, there is a multitude of possible rocket engines. The goal of this chapter is to present the essential components of rocket engines, the criteria for their classification, and to briefly present the main types of rocket engines.

9.1.1 Essential Components of Rocket Engines

This section presents the main components of a rocket engine, followed by the definition of total and specific impulse. These introductory concepts are needed to understand the classification of rocket propulsion systems.

The main components of a rocket engine vary depending on the type of engine. Figure 9.1 shows the simplified schematic diagrams of three rocket engines. Two of those are chemical rockets: a liquid propellant rocket engine, Fig. 9.1a, and a solid propellant rocket motor, Fig. 9.1b. Both chemical rockets have a combustion chamber and a nozzle. The liquid propulsion rocket engine has a propellant supply system that includes the propellant tanks. The solid rocket motor has also the solid propellant inside the combustion chamber.

In a chemical rocket engine, the propellant supply system provides the fuel and oxidizer to the combustion chamber. The parameter associated with the propellant supply system is the mass flow rate, $\dot{m}$. In the combustion chamber, energy is added to the working

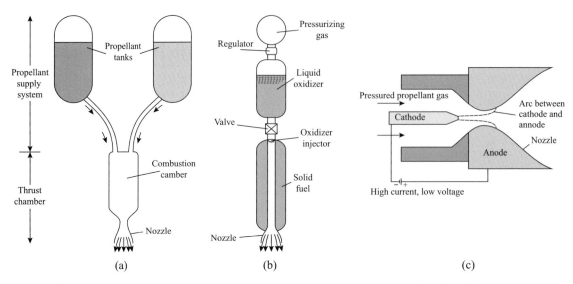

Figure 9.1 Main components of a rocket engine. (a) Liquid propellant engine, (b) solid propellant motor, and (c) electrothermal propulsion thruster.

fluid. Stagnation enthalpy, h_{0c}, and stagnation pressure, p_{0c}, define the state in the combustion chamber. The third major component of the chemical rocket engine is the nozzle. The nozzle converts the thermal energy of the working fluid into kinetic energy and thrust. The efficiency of the nozzle is a crucial parameter that affects the rocket engine performance. The main parameters that define the nozzle are the area ratio of the nozzle exit area A_e to the throat area A_t, $\epsilon := A_e / A_t$, and the Mach number M_e or the velocity u_e at the exit of the nozzle.

The electric propulsion system shown in Fig. 9.1c is an electrothermal propulsion thruster.[1] This consists of a propellant delivery system, a nozzle that is the anode, and a cathode. The anode and cathode create an arc that ionizes a small column of gas and heat the remainder of the propellant.

The specific details of each component depends on the type of rocket engine. These details will be presented in Sections 9.3 – 9.7.

9.1.2   Total and Specific Impulse

Before presenting the classification of rocket engines, let us define the total and specific impulse. The latter will often be mentioned when comparing different propulsion systems.

The *total impulse* I_t is defined as the thrust force $\mathcal{T}$ integrated over the burning time t

$$I_t = \int_0^t \mathcal{T} dt. \tag{9.1}$$

---

[1] The word "thruster" is commonly used for a small rocket engine

Table 9.1 Range of specific impulse.

Name	Type	Specific impulse
Saturn V, first stage	chemical	263 s (2.58 km/s)
Saturn V, second and third stages	chemical	421 s (4.13 km/s)
N1-L3,[2] first stage	chemical	330 s (3.24 km/s)
N1-L3, second stage	chemical	346 s (3.39 km/s)
N1-L3, third stage	chemical	353 s (3.46 km/s)
Timberwind 250	nuclear	780 s (7.65 km/s)
Xenon Hall thruster	electrical	1800 s (17.66 km/s)

If the thrust force is constant, then the total impulse becomes

$$I_t = \mathcal{T} t. \tag{9.2}$$

The *specific impulse* I_{sp} is defined as the total impulse per unit mass or weight of fuel. The former definition yields

$$I_{sp} = \frac{I_t}{\int_0^t \dot{m}_f \, dt} \tag{9.3}$$

with the units m/s, while the latter definition yields

$$I_{sp} = \frac{I_t}{g \int_0^t \dot{m}_f \, dt} \tag{9.4}$$

where g is the gravitational acceleration, $g = 9.8$ m/s^2. For the definition (9.4), the units of the specific impulse are seconds. The specific impulse gives the variation of impulse obtained when using a given amount of fuel. The higher the specific impulse, the more efficient the rocket engine is. Table 9.1 shows the range of specific impulse for several rocket engines. Note that the specific and total impulse are parameters used for all types of rocket engines.

9.2   Criteria for Classification of Rocket Engines

Several criteria can be used to classify rocket propulsion systems, such as the power source, the number and type of propellants, the engine mode, the basic function, and the type of vehicle.

9.2.1 Power Source

Rocket propulsion systems are classified based on the power source as (1) chemical, (2) nuclear, or (3) electric rocket propulsion.

---

[2] Soviet competitor to Saturn V

Figure 9.2 Liquid propellant rocket Saturn V (courtesy of AFP/Getty Images).

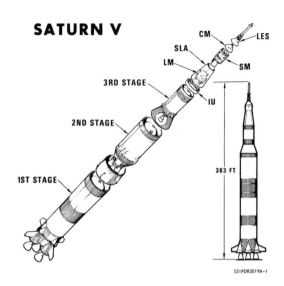

Chemical Rocket Propulsion

Chemical rocket propulsion systems are used as either the boosters or the main engines of space vehicles. As the name suggests, chemical rocket propulsion utilizes the energy generated by high-pressure combustion reactions. As a result of the chemical reaction between a fuel and an oxidizing substance, the gaseous products of this reaction reach high temperatures that vary between 2800 and 4400 K. The gases leaving the combustion chamber are then expanded in the exit nozzle where they reach velocities varying between 1.5 to 4.3 km/s. Figures 9.2 and 9.3 show two examples of chemical rocket propulsion: Saturn V, a liquid propellant rocket, and the Space Shuttle, which has two large solid rocket booster motors and three main engines, which use liquid propellants.

Nuclear Rocket Propulsion

Nuclear rocket propulsion can be divided into two categories: *nuclear thermal* and *nuclear electric*. In nuclear thermal propulsion, the working fluid, commonly hydrogen, is heated using the energy resulting from a nuclear reaction. This energy can be produced using either a fission reactor or a fusion reactor. Consequently, the power source is separate from the propellant. This is different from chemical rockets, where the energy is generated as a reaction that occurs within the propellants. The nuclear rocket does not require an oxidizer. Typically the propellant expands through the nozzle at approximately twice the speed of a chemical rocket. The specific thrust of a nuclear thermal engine can be as high as 1000 s.

Nuclear electric propulsion uses the energy generated by a nuclear reactor to produce electricity. This electricity is utilized to produce thrust as presented in the following paragraph.

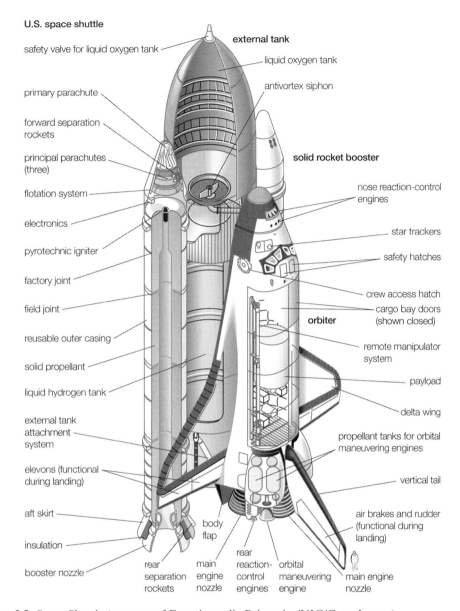

Figure 9.3 Space Shuttle (courtesy of Encyclopaedia Britannica/UIG/Getty Images).

Electric Rocket Propulsion

Electric propulsion uses energy from a nuclear reaction, batteries, or solar radiation receivers to yield or augment the exhaust velocity of the propulsion device. Depending on how the energy is used, electric propulsion is divided into *electrothermal*, *electrostatic*, and *electromagnetic*. The advantage of electric propulsion is that the exhaust velocity can be much higher than in chemical propulsion; therefore the mass flow rate of propellant can be much smaller for the same thrust value. The increased exhaust velocity is due to the fact that, in principle, any amount of energy can be added to the propellant. Consequently, the specific impulse does not have a hard upper limit as long as an energy source is available.

In spite of the high specific impulse, electric rockets are very limited in thrust. They use large amounts of electric power of the order of kilowatts and megawatts, and produce thrust values of 1 N or less. As a result, electric propulsion is used for satellite maneuvering and potentially for interplanetary travel.

9.2.2   Number of Propellants

Based on the number of propellants, rocket engines are classified as (1) monopropellant, (2) bipropellant, and (3) tripropellant. A more in-depth presentation of these propellants is given in Section 9.4.

Monopropellant

There are two types of monopropellant rocket engines: those that depend on a chemical reaction of the propellant, and those that do not. Hydrazine (N_2H_4) is the propellant commonly used for the former monopropellant engines, while nitrogen is often used for the latter.

Bipropellant

The bipropellant rocket uses two propellants: the fuel and the oxidizer. Commonly used propellants are liquid oxygen with liquid kerosene or liquid oxygen with liquid hydrogen. The oxidizer and the fuel are stored separately and mixed in the combustion chamber.

Tripropellant

The tripropellant rocket uses three propellants: one oxidizer and two fuels. The tripropellant rocket can either burn all propellants simultaneously or use an oxidizer with two fuels that are burned in sequence during the flight. A rocket engine that uses liquid hydrogen fuel delivers the highest specific impulse among the possible rocket fuels.

9.2.3   Type of Propellant

Based on the type of propellant, there are several types of rocket engines: solid, liquid, hybrid, and gaseous. This section will briefly present each of them.

Solid Propellant

The solid propellant provides the thrust for solid rocket motors.[3] The solid propellant typically accounts for approximately 90% of the total mass of the solid rocket motor. Solid propellant contains all the elements needed for its combustion. The solid propellants

---

[3] Rocket propulsion systems that do not have turbopumps are commonly called motors instead of engines.

are classified according to the distribution of fuel and oxidant as *homogeneous* and *composite* propellants. The former has the fuel and oxidant contained within the same molecule while the latter is a heterogeneous mixture of oxidizing crystals in an organic plastic-like fuel binder.

Liquid Propellant

Liquid propellants are the working fluids of liquid rocket engines. There are several types of liquid propellants:

1. Fuel; *e.g.*, kerosene, liquid hydrogen, alcohol.
2. Oxidizer; *e.g.*, liquid oxygen, nitric acid (HNO_3), nitrogen tetroxide (or amyl, N_2O_4).
3. Mixture of fuel and oxidizer capable of self-decomposition; *e.g.*, hydrazine (N_2H_4).

The liquid propellant engine allows for repetitive firing and throttling operations. The liquid propellant engine, however, is a complex system that includes storage tanks, piping, turbopumps, and control valves. Additionally, some liquid propellants (liquid oxygen, liquid hydrogen) can only be stored in liquid form by cryogenic means. This results in additional mass disadvantages for liquid propellant engines.

Hybrid Propellant

Hybrid propellant is a mix of liquid and solid propellants. There are multiple ways to combine the liquid and solid propellants: classical, inverse, and mixed hybrid. In the *classical* configuration, the oxidizer is liquid and the fuel is solid. The *inverse* or *reverse* configuration has the fuel as a liquid and the oxidizer as a solid. The *mixed hybrid* configuration has a solid fuel with a small amount of solid oxidizer combined with a liquid oxidizer that is injected in the combustion chamber.

Gaseous Propellant

Gaseous propellants, such as air, nitrogen, or helium, are stored at high temperature and are mainly used for low thrust maneuvers and attitude control. For certain applications, the gaseous propellant is heated electrically or by combustion to improve performance.

9.2.4 Engine Mode

Rocket propulsion systems can be classified as single-mode or mixed-mode engines, depending on whether or not they have all the oxidizer on board.

Single-Mode Engine

Single-mode engines are the vast majority of rocket propulsion systems. They carry on board all the oxidizer needed for propulsion.

Mixed-Mode Engine

Mixed-mode-mode engines (or multi-mode engines) operate as air-breathers to save takeoff weight and then switch to on-board oxidizers when the air density decreases at high altitude.

Figure 9.4 Solid rocket motor [Evans, 1988].

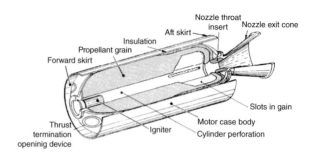

9.3   Solid Rocket Motor

The solid rocket motor is a simple, relatively inexpensive, and reliable rocket propulsion system. Solid rocket motors provide rapid response and are storable for typically 5 to 20 years, depending on the type of propellant. Solid rocket motors do not require a propellant delivery system, turbopumps, or a metering system. Solid rocket motors are used in many applications, such as: (1) strap-on boosters or stages for space launch vehicles, (2) missile propulsion systems, (3) upper-stage propulsion systems for orbital transfer vehicles, and (4) gas generators for starting liquid engines [Heister et al., 2019, p. 229].

The solid propellant of a solid rocket motor contains all the elements needed for its combustion. The solid propellant, called *grain*, is contained in the combustion chamber, called the *case*, as shown in Fig. 9.4. The grain is bonded to the case. An insulation layer protects the case and the exhaust nozzle.

The solid propellants are classified according to the distribution of fuel and oxidant as *homogeneous* or *composite* propellants. The homogeneous propellant has the fuel and oxidant contained within the same molecule. Homogeneous propellants contain enough chemically bonded oxygen to sustain the reaction of combustible elements in an intramolecular decomposition reaction. The most common homogeneous propellants are combinations of nitroglycerin and nitrocellulose, called *double-base* propellants. The double-base propellants also include a small fraction of additives whose purpose and weight fraction are shown in Table 9.2 for the JP-N propellant.

Diethyl pthalate, a nonexplosive plasticizerm,[4] supplements the nitroglycerin, an explosive plasticizer. The nonexplosive plasticizer is added for safety reasons. Potassium sulfate, a flash suppressor, promotes burning at low temperatures. Ethyl centralite prevents the autocatalytic decomposition of the two major constituents, in order to avoid propellant self-ignition. The carbon black prevents the transmission of radiant energy to an almost transparent propellant. This radiant energy could otherwise cause internal ignition around impurities and voids. The wax is added for propellants formed by extrusion.

Composite propellants are heterogeneous mixtures of oxidizing crystals in an organic plasticlike fuel binder [Hill and Peterson, 1992, p. 590]. The fuel binders are made of synthetic rubbers, such as hydroxyl terminated polybutadiene. The fuel binder contains the oxidizer

---

[4] A plasticizer is a substance added to the propellant to increase its plasticity.

Table 9.2 Typical double-base propellant (JP-N) [Geckler and Klager, 1961, p. 19-4].

Material	Weight %	Purpose
Nitrocellulose (13.25%N)	51.40	Polymer
Nitroglycerin	42.93	Explosive plasticizer
Diethyl pthalate	3.20	Nonexplosive plasticizer
Ethyl centralite	1.00	Stabilizer
Potassium sulfate	1.20	Flash suppressor
Carbon black (added)	0.20	Opacifying agent
Candelilla wax (added)	0.07	Die lubricant

which consists of ground crystals of nitronium, ammonium, or potassium perchlorate or ammonium or potassium nitrate.

Once ignited, the propellant burns smoothly, commonly at a rate of approximately 1 cm/s, although the rate for a specific propellant can vary between 0.1 and 5 cm/s. The thrust and performance of the solid rocket motor depends on the shape of the propellant. Depending on the geometry of the grain, the thrust can increase, be constant, or decrease in time [Shafer, 1959, p. 16-22].

9.4 Liquid Propellant Rocket Engine

The liquid propellant rocket engine is one of the most common rocket propulsion systems. The liquid propellant rocket engine has a specific thrust higher than that of a solid rocket motor. In addition, the liquid propellant rocket engine can vary the thrust over time, that is, it has throttling capabilities. Some liquid propellant rocket engines allow for multiple starts during a mission.

Liquid propellant rocket engines can be divided into two categories: boosters and auxiliary propulsion. Boost propulsion is used for imparting a significant velocity increase to a payload. Auxiliary propulsion is used for trajectory adjustments and attitude control.

Boost propulsion is used for booster stages and upper stages of launch vehicles. It can also be used for large missiles. The thrust typically varies between 4.5 kN and 8 MN, while the chamber pressure varies between 24 and 210 bar (2.4–21 MPa).

Auxiliary propulsion is used for satellites, spacecraft, and space rendezvous. The thrust level is low, less than 4.5 kN, while the chamber pressure varies between 1.4 and 21 bar (0.14–2.1 MPa). An auxiliary propulsion unit can be fired several thousand times during a single mission.

As presented in Section 9.2.3, liquid propellant rocket engines can use monopropellants, bipropellants, or tripropellants. Some liquid propellants are *cryogenic propellants* that consist of liquefied gas at low temperature. Two commonly used cryogenic propellants are liquid oxygen (LOX) and liquid hydrogen (LH2). Liquid oxygen has a freezing point of 54 K

Table 9.3 Liquid propellant characteristics.

| Propellant | Application | Specific Impulse[1] | | Bulk Density [kg/m^3] | Optimum Mixture Ratio | Combustion Temperature [K] |
		Sea Level $\epsilon = 8$ [s]	In Space $\epsilon = 25$ [s]			
Liquid O_2 and kerosene	Booster and sustainer	261	324	1009	2.25	3477
Fluorine (F_2) and ammonia (NH_3)	Booster and sustainer	301	368	1137	2.77	4255
F_2-H_2	Sustainer	364	447	304	4.0	2866
F_2 and hydrazine (N_2H_4)	Booster and sustainer	303	372	1281	1.75	4811
Liquid O_2 and H_2	Sustainer	357	441	256	3.5	2755
Nitric acid (HNO_3) and UDMH	Terminal propulsion	246	304	1217	2.4	3089
Monopropellant 90% hydrogen peroxide (H_2O_2)	Trajectory correction	137	167	1394	-	1014

and a boiling point of 90 K. Liquid hydrogen has a melting point of 14 K and a boiling point of 20 K.

Some liquid propellants, such as nitric acid and hydrogen peroxide (H_2O_2), do not need special insulation or other provisions for storage in the vehicle. For this reason they are called *storable propellants*. Storable propellants are suited for vehicles that use them after being carried on board for a long period of time, such as for landing.

Table 9.3 shows the characteristics of several common liquid propellants. Some of the propellants are high-energy propellants that generate higher specific impulse and therefore allow for a reduction in size or weight for the same thrust. However, the active nature of these chemicals, such as fluorine (F_2), hydrazine, and liquid hydrogen, generates challenges in engine design, testing, and operation.

Monopropellants

As mentioned in Section 9.2.2, there are two types of monopropellant rocket engines: those that depend on a chemical reaction of the propellant, and those that do not. A monopropellant that depends on a chemical reaction does not need any other ingredients to release its thermochemical energy. This monopropellant is a fluid that can undergo an exothermic reaction that yields gaseous products with enough heat generation to sustain the chemical reaction. The monopropellant is either a single substance, *e.g.*, hydrazine (N_2H_4), hydrogen peroxide (H_2O_2), or propyl nitrate ($C_3H_7NO_3$), or a mixture of compounds, *e.g.*, hydroxylammonium nitrate (NH_3OHNO_3), also known as AF-M315E [Spores et al., 2013].

   Monopropellants are relatively simple to use compared to other liquid propellants. They impose minimal constraints on the injection location in the reaction chamber and have relatively low reaction temperatures, *e.g.*, under 1500 K for hydrazine. The specific impulse of engines using monopropellants, *e.g.*, 220 s for hydrazine, is smaller than that of engines using bipropellants or solid propellants, as shown in Table 9.1. Therefore, monopropellants are used primarily for attitude control thrusters and gas generators.

   Hydrazine is the most used monopropellant in space applications. At room temperature hydrazine requires a solid catalyst, Shell 405, composed of 30% by weight iridium deposited on aluminum oxide Al_2O_3 [Heister et al., 2019, p. 359]. Above 450 K several materials decompose hydrazine, including iron, nickel, and cobalt. The catalyst decomposes the hydrazine into ammonia (NH_3), nitrogen gas, and hydrogen gas

$$3N_2H_4 \rightarrow 4NH_3 + N_2 \qquad\qquad -112 \text{ kJ/mol } N_2H_4$$
$$2NH_3 \rightarrow N_2 + 3H_2 \qquad\qquad +46 \text{ kJ/mol } NH_3$$

   The first reaction is exothermic, generating 112 kJ/mol of hydrazine, while the second reaction is endothermic, requiring 46 kJ/mol of ammonia. The endothermic reaction reduces the temperature of the overall reaction, and therefore has a negative impact on the performance. On the other hand, the endothermic reaction reduces the molecular mass of products, which has a positive impact on performance.

   Multiplying the endothermic reaction by $2x$ and adding it to the exothermic reaction according to Hess' law, the overall reaction can be written as

$$3N_2H_4 \rightarrow 4(1-x)NH_3 + (1+2x)N_2 + 6xH_2.$$

   The reaction parameter $0 \leq x \leq 1$ is a measure of the amount of ammonia dissociation, and is dependent on the catalyst, the operating pressure, and the hydrazine residence time in the catalyst bed.

   The specific impulse and the characteristic velocity c^* (see Section 10.4) are affected by x. The specific impulse varies between a demonstrated value of 220 s and a maximum theoretical value of 260 s with no ammonia decomposition [Schmidt, 1984]. The maximum c^* value is approximately 1350 m/s when 30% of ammonia decomposed.

   The decomposition reaction is highly exothermic, and the mixture of nitrogen, hydrogen, and ammonia products has a temperature of approximately 1300 K. Hydrazine was first used during World War II by the Germans to power the Messerschmitt Me 163 Komet rocket-powered aircraft.

   Another monopropellant is hydrogen peroxide, which was also first used during World War II by the Germans in the V-2 rocket. In this case, the hydrogen peroxide, using sodium permanganate as a catalyst, powered the turbines needed to drive the alcohol and liquid oxygen pumps.

   A monopropellant that does not depend on a chemical reaction is nitrogen. Nitrogen gas released from storage under pressure is used for attitude and speed control on monopropellant thrusters. The thrust values achieved in this case are small.

Bipropellants

Most liquid propellant rocket engines use bipropellants, that is, two propellants: the fuel and the oxidizer. The difference between a fuel and an oxidizer consists in the way the electrons are transferred: fuels donate electrons while oxidizers accept electrons.

The bipropellant system must use separate tanks and pumping systems to deliver the propellants into the combustion chamber where they are mixed and combusted. Some propellant combinations need ignition devices, such as spark plugs or chemical pyrotechnic igniters. Other propellant combinations ignite spontaneously when they mix. These propellant combinations are called *hypergolic*. Common hypergolic fuels, hydrazine (N_2H_4), monomethylhydrazine, and unsymmetrical dimethylhydrazine (UDMH), and oxidizer, nitrogen tetroxide, are all liquid at ordinary temperatures and pressures. The hypergolic bipropellants can be fired any number of times by simply opening and closing the propellant valves. Therefore, hypergolic bipropellants are well suited for spacecraft maneuvering. However, hypergolic bipropellants present a serious risk of explosion from accidental mixing of the propellant components.

A *nonhypergolic* bipropellant combination, such as, liquid hydrogen/liquid oxygen or liquid methane/liquid oxygen, needs an energy source to be ignited, *e.g.*, an electric or heat discharge.

Commonly used bipropellants can be either storable or cryogenic. Storable propellants are liquids over a wide range of temperatures and pressures, typical for ambient conditions. For military applications, the temperature range extends between 223 K and 342 K. The properties of some storable (noncryogenic) and cryogenic liquid propellants are given in Tables 9.4 and 9.5, respectively.

Table 9.4 Properties of storable (noncryogenic) liquid propellants.

Propellant	Molecular Mass	Density [kg/m³]	Freeze pt. [K]	Boil pt. [K]	Vapor press. [atm abs.]	Usage
RP-1[1]	165	810	229–219	446–537	0.02 @ 344 K	Fuel, Coolant
JP-4	128	750-830	213	406–517	0.5 @ 344 K	Fuel, Coolant
UDMH[2]	60	790	296	337	1.2 @ 344 K	Fuel, Coolant
MMH[3]	16	880	291	359	0.6 @ 344 K	Fuel, Coolant
Ethyl alcohol	46	830	156	352	0.8 @ 344 K	Fuel, Oxidant Coolant
Hydrazine	32	1000	275	386	0.2 @ 344 K	Fuel, Oxidant Coolant
Hydrogen peroxide 95%	33	1400	268	419	0.003 @ 298 K	Monopropellant, Oxidant, Coolant
IRFNA[4]	56	1600	287	339	1.2 @ 344 K	Oxidant, Coolant
WFNA[5]	60	1420–1500	231	359	6.7 @ 344 K	Oxidant, Coolant

[1] rocket propellant-1 or refined petroleum-1;   [2] unsymmetrical dimethyl hydrazine;
[3] monomethyl hydrazine.   [4] inhibited red fuming nitric acid;   [5] white fuming nitric acid.

Table 9.5 Properties of cryogenic liquid propellants.

Propellant	Molecular Mass	Density[1] [kg/m^3]	Freeze point [K]	Boil point [K]	Critical press. [atm abs.]	Critical temp. [K]	Usage
Hydrogen (LH2)	2.016	71	14	20	13	33	Fuel, Coolant
Ammonia	17.03	683	196	240	111	406	Fuel, Coolant
Oxygen (LOX)	32.00	1142	54	90	50	155	Oxidizer

[1] at boiling point

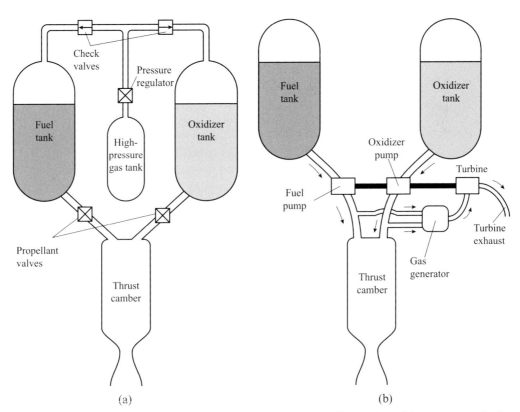

Figure 9.5 Simplified diagrams of bipropellant rockets. (a) Bipropellant rocket with gas pressure feed system and (b) bipropellant rocket with turbopump feed system.

The liquid propellants are supplied under pressure from tanks to the combustion chamber. A simplified diagram of a bipropellant rocket engine is shown in Fig. 9.5a. This rocket engine uses a gas pressure feed system to deliver the fuel and oxidizer to the *thrust chamber*, which consists of the combustion chamber, injectors, and nozzle.

Figure 9.5b shows an alternative configuration of a bipropellant rocket that uses a turbopump feed system with a separate gas generator. The turbine of the turbopump captures the energy of the gas generator flow and powers both the fuel pump and the oxidizer pump.

Two examples of liquid bipropellant rocket engines are the V-2 rocket engine and the Space Shuttle main engine. The fuel of the V-2 rocket engine was alcohol (75% ethanol and 25% water) and the oxidizer was liquid oxygen. The engine used 58 kg/s of alcohol and 68 kg/s of liquid oxygen to produce a thrust of 249 kN. The 589 hp turbopump that delivered the propellants was powered by the steam generated by hydrogen peroxide that used sodium permanganate as a catalyst.

Figure 9.6 shows the RS-25 engine, the Space Shuttle main engine built by Rocketdyne. The fuel was liquid hydrogen and the oxidizer was liquid oxygen. The vacuum specific impulse was 452.3 s and the sea level specific impulse was 366 s. The vacuum thrust was 2279 kN and the sea level thrust was 1,860 kN. The nozzle ratio was $\epsilon = 69.1$ and the thrust to weight ratio was 73.1. The pressure in the combustion chamber was 206.4 bar (20.64 MPa).

Tripropellants

As mentioned in Section 9.2.2, some liquid propulsion rocket engines use three propellants: one oxidizer and two fuels. The tripropellant rocket can either burn all propellants simultaneously or use an oxidizer with two fuels that are burned in sequence during the flight.

The *simultaneous tripropellant* rocket uses high energy density metals, such as lithium or beryllium, with a bipropellant system. The fuel and oxidizer of the bipropellant system provide the activation energy needed for initiating a more energetic reaction between the metal and the oxidizer.

Although theoretical modeling predicts better performance compared to bipropellant rockets, simultaneous tripropellant rocket operation is challenged by several factors: (1) solid metal injection into the thrust chamber, (2) metal combustion initiation and sustainment, and (3) mass, momentum, and heat transport in a multiphase system. A successful test of a simultaneous tripropellant rocket by Rocketdyne used liquid lithium, gaseous hydrogen, and liquid fluorine [Clark, 1972]. This rocket engine produced a specific impulse of 542 s.

The *sequential-burn tripropellant* rocket changes the fuel that is combusted during flight. Consequently, a dense fuel like kerosene can be used early in flight, followed by a lighter fuel like liquid hydrogen. The sequential burn of different fuels provides some of the benefits of a multi-stage rocket.

The rocket engine that uses liquid hydrogen fuel delivers the highest specific impulse among the possible rocket fuels. The drawback of using hydrogen is that, due to its low density, the storage tanks are large. As a result, the structures needed for storing hydrogen have a significant weight that offsets to some degree the light weight of the fuel. The kerosene engine has a smaller specific impulse than that of hydrogen engine. However, the higher density of kerosene results in smaller storage tanks, which reduces the atmosphere drag of the vehicle. In addition, a rocket engine using kerosene provides the high thrust that is needed at takeoff. Therefore, there is an optimal altitude range for transitioning from mainly burning kerosene to hydrogen.

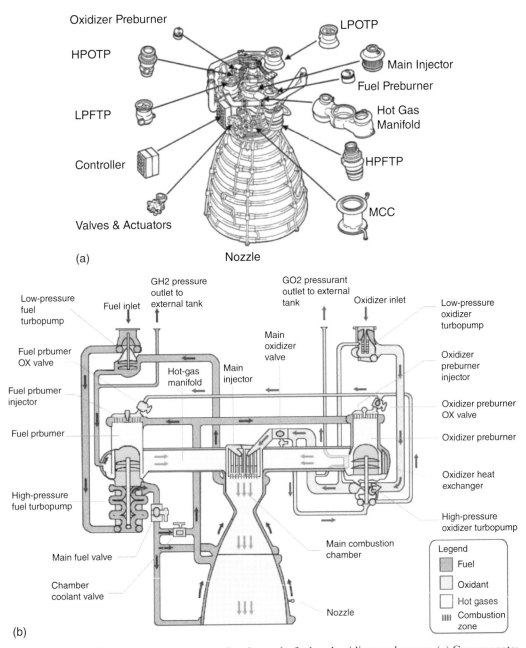

Figure 9.6 Space Shuttle main engine powerhead: nozzle, fuel and oxidizer preburners. (a) Components: LPOTP, HPOTP (low- and high-pressure oxidizer turbopump), LPFTP, HPFTP (low- and high-pressure fuel turbopump), MCC (main combustion chamber) and (b) diagram.

At liftoff, the sequential-burn tripropellant rocket typically burns both fuels, kerosene and hydrogen. This is called *mode 1* of operation. The fraction of hydrogen fuel is gradually augmented as the altitude increases, eventually burning only hydrogen once the kerosene has be consumed. Past the altitude where the kerosene was used up, the sequential-burn tripropellant rocket becomes a hydrogen/oxygen bipropellant rocket. This is called *mode 2* of operation.

Energomash, a Russian rocket engine manufacturer, built a sequential-burn tripropellant rocket, RD-701, used liquid oxygen, kerosene, and liquid hydrogen. The RD-701 has a thrust in vacuum of 4 MN in mode 1 and 1.6 MN in mode 2. The sea level thrust is 3.2 MN. The nozzle diameter is 2.4 m and the dry mass of the engine is 1923 kg. The specific impulse in vacuum is 415 s in mode 1 and 460 s in mode 2. The specific impulse at sea level is 330 s.

9.5   Hybrid Propellant Rocket Engine

The hybrid propellant rocket engine combines solid and liquid propellants. As presented in Section 9.2.3, three propellant configurations are used: classical, inverse, and mixed hybrid. The classical configuration is a liquid propulsion system with a solid fuel or a solid propulsion system with a liquid oxidizer. Figure 9.7 shows a hybrid propellant rocket engine with a classical configuration.

The oxidizer is first introduced in a vaporization or precombustion chamber before it reaches the solid fuel. Combustion occurs near the numerous fuel ports of the fuel grain. The solid fuel absorbs the heat and vaporizes or pyrolyzes, generating gaseous fuel. An aftermixing or afterburning chamber, downstream of the grain, facilitates the complete combustion of the fuel and oxidizer before reaching the nozzle.

The oxidizers used for hybrid propellant can be either storable, *e.g.*, chlorine trifluoride (ClF_3), hydrogen peroxide (H_2O_2), oxygen-difluoride (OF_2), or cryogenic, *e.g.*, liquid oxygen.

Several options are available for the fuel of the hybrid propellant, such as: hydroxyl-terminated polybutadiene (HTPB), carboxyl-terminated polybutadiene (CTPB), polybutadiene acrylonitrile (PBAN), hydroxyl-terminated polybutadiene and polycyclopentadiene (PCPD), lithium hydride/polybutadiene, and metallized solid fuels. The metallized solid fuels, lithium, beryllium, boron, and aluminum are expected to increase fuel performance, although they have some limitations. Boron is difficult to ignite, beryllium is highly toxic, and lithium has a low heat of reaction.

Common fuel-oxidizer combinations used in hybrid propellants include: hydroxyl-terminated polybutadiene-liquid oxygen, hydroxyl-terminated polybutadiene-hydrogen peroxide, hydroxyl-terminated polybutadiene-chlorine trifluoride, and lithium hydride/polybutadiene-oxygen-difluoride. Figure 9.8 shows the theoretical vacuum specific impulse of hydroxyl-terminated polybutadiene reacting with several oxidizers vs. the mixture ratio. The specific impulse is calculated at 68.9 bar (1000 psi) and a nozzle expansion ratio $\epsilon = 10$.

The hybrid propellant rocket engine has a specific impulse higher than that of the solid rocket motor but smaller than that of the liquid bipropellant rocket engine. The hybrid

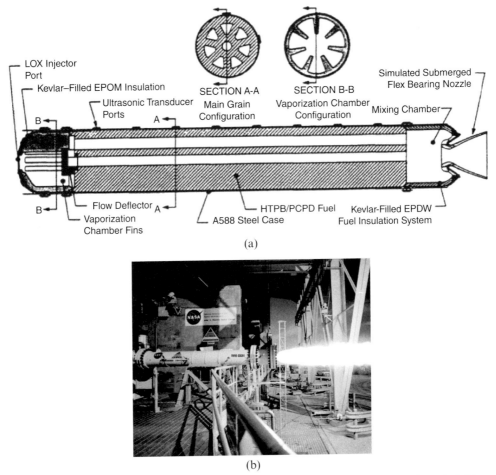

(a)

(b)

Figure 9.7 AMROC 250k hybrid propellant rocket engine, a 250,000 pound vacuum thrust hybrid booster that used liquid oxygen and hydroxyl-terminated polybutiene. (a) Layout [Story et al., 2003] and (b) Testing (courtesy of NASA).

propellant rocket engine, however, has a density specific impulse, ρI_{sp}, higher than that of the liquid bipropellant rocket engine.

The oxidizer of the hybrid propellant rocket engine is loaded on the rocket at the launching pad. Therefore, the hybrid propellant rocket engine is safe from explosion and detonation during fabrication and storage, especially when the solid fuel is hydroxyl-terminated polybutadiene (HTPB), which is nothing other than the rubber used for tires.

Other advantages of the hybrid propellant rocket engine include its relative simplicity, especially when turbopumps are not used to deliver the oxidizer, and the possibility of stopping and restarting the engine. In addition, cracks, voids, and debonds in the fuel grain are not as problematic as for a solid rocket motor, because the combustion in the hybrid propellant rocket engine is limited by the oxidizer [Heister et al., 2019, p. 440].

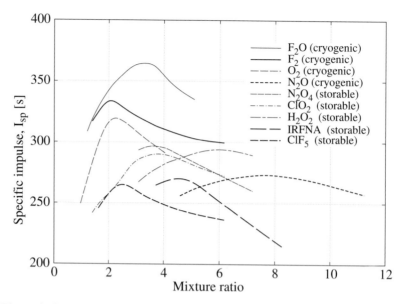

Figure 9.8 Theoretical vacuum specific impulse of hydroxyl-terminated polybutadiene-chlorine with different oxidizers vs. mixture ratio at 68.9 bar (1000 psi) and a nozzle expansion ratio $\epsilon = 10$.

Some of the disadvantages of the hybrid propellant rocket engine include: (1) a rather complicated grain geometry needed to mitigate the low fuel regression rates, (2) the specific impulse my vary during steady-state operation due to variations in the mixture ratio, and (3) a tendency to large-amplitude, low-frequency pressure fluctuations, called chugging.

9.6   Nuclear Thermal Rocket Engine

The energy needed for a nuclear thermal rocket engine can be produced using a fission reactor, or, in principle, a fusion reactor. This power source is separate from the propellant, which is commonly hydrogen. No oxidizer is needed in a nuclear thermal rocket engine.

The *nuclear fission reactor rocket* generates the heat by the fission of uranium. This heat is then transferred to the working fluid. The nuclear fission reactor rocket is typically a high thrust rocket whose thrust exceeds 50 kN.

Nuclear fission reactor rockets were designed, built, and tested in the United States and the Soviet Union starting in the 1950s. Although their thrust to weight ratio is currently approximately 30, as opposed to 70 or more for chemical rocket engines, the nuclear fission reactor rockets can have a specific impulse as high as 1000 s, which is double that of the chemical rocket engines. The Timberwind 250 nuclear fission reactor rocket[5] specifications

---

[5] Project Timberwind was part of the Strategic Defense Initiative, during late 1980s and early 1990s.

list the vacuum specific impulse at 1000 s, the sea level thrust at 780 s, the vacuum thrust at 2.4 MN, and the sea level thrust at 1.9 MN for a nozzle of 8.7 m diameter and an engine mass of 8300 kg.

Several concepts of *nuclear fusion reactor rocket* have been studied. However, none of them are currently feasible or practical [Sutton and Biblarz, 2010, p.11].

9.7 Electric Rocket Engine

Electric propulsion uses energy from any source except that stored in the chemical bonds of the propellants. This energy is used to provide and augment the exhaust velocity of the propellant. The electric rocket engine can use different energy sources: nuclear reaction, batteries, or solar radiation. A significant advantage of the electric rocket engine is that, in principle, any amount of energy can be added to the propellant. As a result, the specific impulse of electric rocket engines can be as high as 5000 s (see Table 9.6), that is, an order of magnitude higher than that of chemical propulsion rockets. Consequently, for a given thrust value, the mass flow rate of propellant of an electric rocket engine is much smaller than that of a liquid propellant rocket engine.

Solar energy is an option for the energy source of the electric rocket engine. The limitation is the low energy density of solar energy. The alternative is to use nuclear energy, which has a much higher energy density. While nuclear energy could be used directly as the heat source for the propellant, the heat transferred to the propellant would have to pass through walls. Therefore, the maximum temperature of the propellant is limited by the maximum allowable temperature of the walls [Lanin, 2013]. In spite of this heat transfer limitation, the specific impulse of an electric propulsion rocket using pure hydrogen as propellant is higher than that of a chemical rocket using oxygen and hydrogen as propellants. Consequently, adding energy electrically to the propellant in space appears to be the most promising approach. As of 2019, more than 500 spacecraft operating throughout the solar system used electric propulsion for primary propulsion, station keeping, or orbit raising [Lev et al., 2019]. It is expected that by 2020, half of the new satellites will carry full electric propulsion [Forrester, 2016]. Electric propulsion, however, is not appropriate for launching a vehicle from Earth's surface because the thrust generated, typically less than 1 N, is too small.

Table 9.6 Range of specific impulse for electric propulsion.

Electric Propulsion Type	Specific Impulse [s]
Electrothermal (arc heating)	280–1200
Electrostatic (ion propulsion)	1200–5000
Electromagnetic	
Hall effect	1000–1700
Pulsed plasma	700–2500

Figure 9.9 Electrostatic propulsion thruster.

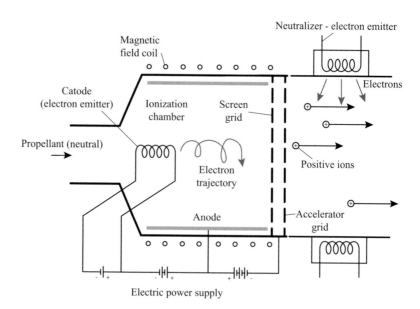

A typical electric propulsion thruster has the following subsystems [Sutton and Biblarz, 2010, p. 622]: (i) an energy source (solar, batteries, or nuclear) with its supplemental devices: heat conductors, pumps, radiators, panels, concentrators, and controls; (ii) conversion devices that transform the energy into electricity at the desired voltage, frequency, and current needed for the electric propulsion system; (iii) a propellant system for storing, metering, and delivering the propellant; and (iv) a thruster that converts the electric energy into kinetic energy of the propellant.

Electric propulsion can be divided into three basic types:

1. Electrothermal – systems where electrical energy is used to add heat to the propellant, which expands thermodynamically so that the gas is accelerated through a nozzle, like in a chemical rocket. The propellant is heated electrically by heat resistors or electric arcjets, as shown in Fig. 9.1c. The exhaust velocity ranges between 1 and 5 km/s. Hydrogen, ammonium, nitrogen, or hydrazine are used as propellants.

2. Electrostatic or ion propulsion – systems where electrical energy is used to accelerate the propellant particles. In the electrostatic engine, the working fluid is commonly xenon. The working fluid is ionized by stripping off electrons. Then the electrically charged ions are accelerated by the electrostatic fields to 2–60 km/s. To avoid buildup of charge on the engine, the ions are electrically neutralized as they leave the engine by combining them with electrons, as shown in Fig. 9.9.

3. Electromagnetic or magnetoplasma – systems where electrical energy is used to accelerate the propellant by the interaction of electric and magnetic fields within a plasma. A plasma is an energized hot gas containing ions, electrons, and neutral particles. The propellant

exhaust velocity ranges between 1 and 50 km/s. While there are several types and geometries of magnetoplasma engines, the Hall-effect thruster developed and patented by NASA has a good flight record in the Soviet Union (now Russia). The electromagnetic engine, like the electrostatic engine, works only in a vacuum.

Bibliography

J. Clark. *Ignition! An Informal History of Liquid Rocket Propellants*. Rutgers University Press, 1972. 416

P. R. Evans. Composite motor case design. In *Design Methods in Solid Rocket Motors*, number 150 in AGARD Lecture Series, chapter 4A. North Atlantic Treaty Organization, revised edition, 1988. 410

C. Forrester. *Beyond Frontiers*. Broadgate Publications, 2016. 421

R. D. Geckler and K. Klager. Solid-propellant rocket engines. In H. H. Koelle, editor, *Handbook of Astronautical Engineering*. McGraw-Hill, 1961. 411

S. D. Heister, W. E. Anderson, T. L. Pourpoint, and R. J. Cassady. *Rocket Propulsion*. Cambridge University Press, 2019. 410, 413, 419

P. Hill and C. Peterson. *Mechanics and Thermodynamics of Propulsion*. Addison Wesley, second edition, 1992. 410

A. Lanin. *Nuclear Rocket Engine Reactor*, volume 170. Springer Series in Materials Science, 2013. 421

D. Lev, R. M. Myers, K. M. Lemmer, J. Kolbeck, H. Koizumi, and K. Polzin. The technological and commercial expansion of electric propulsion. *Acta Astronautica*, 159:213–227, 2019. 421

E. W. Schmidt. *Hydrazine and Its Derivatives: Preparation, Properties, Applications*. J. Wiley, 1984. 413

J. I. Shafer. Solid rocket propulsion. In H. S. Seifert, editor, *Space Technology*. John Wiley and Sons, 1959. 411

R. A. Spores, R. Masse, S. Kimbrel, and C. McLean. GPIM AF-M315E propulsion system. In *49th AIAA/ASME/SAE/ASEE Joint Propulsion Conference & Exhibit*, AIAA 2013–3849. San Jose, CA, July 2013. 412

G. Story, T. Zoladz, J. Arves, D. Kearney, T. Abel, and O. Park. Hybrid propulsion demonstration program 250k hybrid motor. In *39th AIAA/ASME/SAE/ASEE Joint Propulsion Conference and Exhibit*, number AIAA 2003-5198, Huntsville, AL, July 2003. 419

G. P. Sutton and O. Biblarz. *Rocket Propulsion Elements*. Wiley, eighth edition, 2010. 421, 422

10   Chemical Rocket Performance

In Chapter 9, the total and specific impulse were introduced since they were used in comparing different types of rocket engines. It was illustrated there that the specific impulse is an important indicator of efficiency and overall system performance. This chapter presents the general equations and parameters that measure chemical rocket performance as a function of the propellant and chamber characteristics, nozzle design, and operating altitude.

10.1   Thrust Equation

The thrust equation (6.7) of air-breathing propulsion systems will be used to obtain the thrust equation for a chemical rocket engine by setting the incoming mass flow rate to zero, so that

$$T = \dot{m}\, u_e + (p_e - p_a)A_e, \tag{10.1}$$

where $\dot{m}$ is the mass flow rate of propellant leaving the engine at velocity u_e, p_e is the nozzle exit pressure, A_e is the nozzle exit area, and p_a is the atmospheric pressure.

10.2   Rocket Equation

Several simplifying assumptions are needed to obtain a simple analytical expression for u_e:

1. The working fluid is a homogeneous, perfect gas of constant composition.
2. There is no heat transfer across the rocket walls, that is, the flow is adiabatic.
3. The viscous effects are neglected.
4. There are no shock waves in the flow.
5. The flow is steady, isentropic, and one dimensional.

Figure 10.1 Thrust chamber.

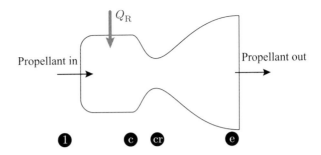

6. Expansion through the nozzle is fast enough so that the chemical equilibrium is unchanged, that is, one assumes frozen equilibrium.

The *frozen equilibrium* assumption is valid as long as the residence time is smaller than the time needed for recombination reactions to occur. For example, if the average gas velocity is 2000 m/s, the residence time per unit length is 500 μs/m. If the nozzle is large, the frozen equilibrium is not valid. In this case, a new time-independent equilibrium can be reached at all points in the nozzle. This condition is called *shifting equilibrium* or *equilibrium flow* [Archer and Saarlas, 1996, p. 435]. The recombination process may release extra energy, which will increase the rocket performance. Therefore, using the frozen equilibrium assumption will yield a conservative result, the engine performing somewhat better than predicted.

As shown in Fig. 10.1, the propellant enters the thrust chamber at state "1", and the heat per unit mass Q_R is added at constant pressure so that the propeller reaches state "c" at inlet in the de Laval nozzle. The propellant enters the de Laval nozzle and reaches critical condition at the throat. Then it expands in the divergent part of the nozzle, reaching state "e" at the nozzle exit.

Since the flow is assumed adiabatic and there are no moving/deforming parts, the energy conservation equation (3.10) yields

$$h_{0c} = h_e + \frac{u_e^2}{2} \tag{10.2}$$

where h_{0c} denotes the stagnation enthalpy in the chamber. Equation (10.2) yields the nozzle exit velocity:

$$u_e^2 = 2c_p(T_{0c} - T_e) = 2\frac{\gamma}{\gamma - 1}RT_{0c}\left(1 - \frac{T_e}{T_{0c}}\right). \tag{10.3}$$

Assuming isentropic flow in the nozzle, the exit velocity u_e can be written as a function of the chamber stagnation temperature T_{0c}, pressure ratio p_e/p_{0c}, and the molecular mass $\mathcal{M}$:

$$u_e = \sqrt{\frac{2\gamma}{\gamma - 1}\frac{\mathcal{R}}{\mathcal{M}}T_{0c}\left[1 - \left(\frac{p_e}{p_{0c}}\right)^{\frac{\gamma-1}{\gamma}}\right]} \tag{10.4}$$

Figure 10.2 Expansion efficiency, E, vs. temperature ratio, T_{0c}/T_e.

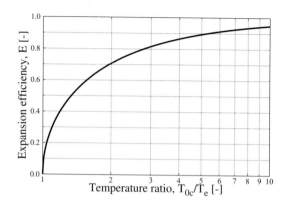

where $\mathcal{R}$ is the universal gas constant, $\mathcal{R} = 8.314$ J/(mol K) = 8314 J/(kmol K). Equation (10.4) is called the *rocket equation*.

For a given temperature T_{0c}, the exit velocity is maximum if the exit pressure is zero, that is, the rocket operates in vacuum, so that

$$u_{max} := \sqrt{\frac{2\gamma}{\gamma - 1} \frac{\mathcal{R}}{\mathcal{M}} T_{0c}} = \sqrt{\frac{2\gamma}{\gamma - 1} R T_{0c}}. \tag{10.5}$$

10.2.1 Expansion Correction (Expansion Efficiency), E

Using (10.5), the exit velocity u_e can be written as

$$u_e = u_{max} \cdot E \tag{10.6}$$

where E is called the *expansion correction* or the *expansion efficiency* and is equal to

$$E := \sqrt{1 - \left(\frac{p_e}{p_{0c}}\right)^{\frac{\gamma-1}{\gamma}}} = \sqrt{1 - \frac{T_e}{T_{0c}}}. \tag{10.7}$$

The expansion efficiency is always less than 1, except for when $p_e = 0$, which yields $E = 1$. Figure 10.2 shows the variation of the expansion efficiency, E, as a function of temperature ratio, T_e/T_{0c}. The variation of E can also be plotted as a function of pressure ratio p_e/p_{0c}, but in this case γ must also be taken into account because $p_e/p_{0c} = (T_e/T_{0c})^{\gamma/(\gamma-1)}$.

While the maximum velocity in the nozzle, u_{max}, is a function of stagnation temperature in the combustion chamber and the molecular mass of the propellant, the expansion efficiency is a function of the expansion ratio p_e/p_{0c}.

10.3 Design Thrust, $\mathcal{T}_D$

Design thrust is defined as the thrust for the case when nozzle exit pressure is equal to the atmospheric pressure, $p_e = p_a$. In this case (10.1) reduces to

$$\mathcal{T}_D = \dot{m} u_e. \tag{10.8}$$

Example 10.3.1 Calculate the design thrust T_D and the nozzle exit velocity u_e for a rocket engine with the following parameters:

$\gamma = 1.2$ $p_{0c} = 60$ bar $\mathcal{M} = 15$ kg/kmol   design altitude $(p_e = p_a)$
$p_e = 0.4$ bar $T_{0c} = 3900$ K $\dot{m} = 200$ kg/s

Solution

$$u_{max} = \sqrt{\frac{2\gamma}{\gamma - 1} \cdot \frac{\mathcal{R}}{\mathcal{M}} \cdot T_{0c}} = \sqrt{\frac{2 \times 1.2}{1.2 - 1} \cdot \frac{8314}{15} \cdot 3900} = 5093 \text{ m/s}$$

$$E = \sqrt{1 - \left(\frac{p_e}{p_{0c}}\right)^{\frac{\gamma-1}{\gamma}}} = \sqrt{1 - \left(\frac{0.4}{60}\right)^{\frac{0.2}{1.2}}} = 0.752$$

$$u_e = u_{max}E = 5093 \times 0.752 = 3832 \text{ m/s}$$

$$T_D = \dot{m}\,u_e = 200 \times 3832 = 766 \times 10^3 \text{ N}.$$

The design thrust is directly proportional to the mass flow rate and the nozzle exit velocity. The mass flow rate depends on the size of the engine, while the nozzle exit velocity is a function of engine performance.

The nozzle exit velocity, u_e, is directly proportional to u_{max} and E. In order to maximize design thrust, let us examine how u_e and E are affected by the propellant and rocket engine parameters.

The u_{max} velocity can be increased by reducing the molecular mass of the propellant, $\mathcal{M}$, and by increasing the combustion chamber temperature, T_{0c}. The temperature T_{0c} can be increased by using fuels with higher heating values. However, if T_{0c} is too high, dissociation will occur, which will restrict T_{0c}. In addition, if T_{0c} is high enough, the increase of u_e will reduce the residence time in the nozzle. Consequently, the recombination, which releases energy as the gas expands to lower temperatures in the nozzle, will not occur and therefore the engine efficiency will decrease. The dissociation, which is commonly seen as a loss because of lowering the combustion temperature, lowers the molecular mass of the propellant which in this case is beneficial.

For a given fuel–oxidizer propellant, the temperature, T_{0c}, and the molecular mass, $\mathcal{M}$, depend on the oxidizer/fuel ratio and on the combustion chamber stagnation temperature, p_{0c}. For a given rocket engine and pressure p_{0c}, there is an optimum oxidizer/fuel ratio that maximizes u_e. If the pressure p_{0c} is increased, the dissociation will decrease and therefore T_{0c} will increase. A higher pressure p_{0c} will, however, increase stresses in the chamber walls and will require a heavier chamber. A lower pressure p_{0c} will promote dissociation which leads to a lower T_{0c}, an undesirable outcome, and a lower $\mathcal{M}$, a beneficial outcome.

The pressure p_{0c} also affects expansion efficiency, E. If p_e is kept constant, an increase in p_{0c} leads to an increase of E. Figure 10.2 shows that the largest variation of E occurs at low values of T_{0c}/T_e that correspond to low values of p_{0c}/p_e. As the p_{0c} is further increased, the increase in E diminishes.

Rather than increasing p_{0c}, the p_e/p_{0c} ratio can be decreased by increasing the area ratio, $\epsilon = A_e/A_{cr}$. In this case, the increase in E is associated with an increase of the nozzle size, therefore an increase of the engine weight. If the pressure p_{0c} is increased, this requires an increase of the combustion chamber mass to compensate for the higher stresses. It should be apparent from this discussion that maximizing the nozzle exit velocity is a non-trivial optimization problem.

10.3.1 Dimensionless Mass Flow Function, $\Gamma(\gamma)$

The mass flow rate of the rocket engine can be calculated as a function of the Mach number, M

$$\dot{m} = \rho A V = \frac{pA\mathrm{M}}{\sqrt{T}}\sqrt{\frac{\gamma}{R}}$$

or using stagnation values

$$\dot{m} = \frac{p_{0c}}{\sqrt{T_{0c}}\left(1 + \frac{\gamma-1}{2}\mathrm{M}^2\right)^{\frac{\gamma+1}{2(\gamma-1)}}} A\mathrm{M}\sqrt{\frac{\gamma}{R}}. \tag{10.9}$$

It is useful to write the mass flow rate as a function of the nozzle throat critical conditions, so that (10.9) becomes

$$\dot{m} = \frac{p_{0c}}{\sqrt{T_{0c}}\left(\frac{\gamma+1}{2}\right)^{\frac{\gamma+1}{2(\gamma-1)}}} A_{cr}\sqrt{\frac{\gamma}{R}} \tag{10.10}$$

or, regrouping the terms:

$$\frac{\dot{m}\sqrt{RT_{0c}}}{p_{0c}A_{cr}} = \sqrt{\gamma}\left(\frac{2}{\gamma+1}\right)^{\frac{\gamma+1}{2(\gamma-1)}}. \tag{10.11}$$

The right-hand side of (10.11) is dimensionless, therefore the left-hand side must also be dimensionless. Consequently, one can define a *dimensionless mass flow function*, $\Gamma(\gamma)$, as

$$\boxed{\Gamma(\gamma) := \frac{\dot{m}\sqrt{RT_{0c}}}{p_{0c}A_{cr}}} \tag{10.12}$$

so that (10.11) yields

$$\boxed{\Gamma(\gamma) = \sqrt{\gamma}\left(\frac{2}{\gamma+1}\right)^{\frac{\gamma+1}{2(\gamma-1)}}.} \tag{10.13}$$

For γ varying between 1.15 and 1.66, $\Gamma(\gamma)$ varies between 0.6386 and 0.7262.

Figure 10.3 Variation of c^*/u_{max} vs. γ.

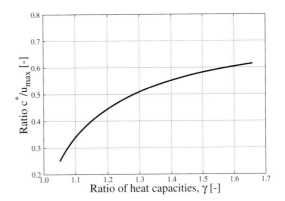

10.4 Characteristic Velocities

The terms of the dimensionless mass flow function definition (10.12) can be grouped so that they have the dimensions of a velocity:

$$c^* = \frac{\sqrt{RT_{0c}}}{\Gamma} = \frac{p_{0c}A_{cr}}{\dot{m}}. \tag{10.14}$$

This velocity c^* is called the *characteristic velocity* of the gas in the combustion (or stagnation) chamber. Although this velocity c^* has no physical equivalent, the value of c^* can be related to u_{max} defined in (10.5), so that

$$\frac{c^*}{u_{max}} = \frac{\sqrt{RT_{0c}}}{\Gamma}\frac{\sqrt{\gamma-1}}{\sqrt{2\gamma RT_{0c}}} = \frac{1}{\Gamma}\sqrt{\frac{\gamma-1}{2\gamma}}$$

or

$$c^* = u_{max}\frac{1}{\Gamma(\gamma)}\sqrt{\frac{\gamma-1}{2\gamma}}.$$

The variation of the ratio c^*/u_{max} is shown in Fig. 10.3. The characteristic velocity is roughly half of u_{max}, the ratio c^*/u_{max} varying between 0.4 and 0.615 when γ varies between 1.15 and 1.66.

It is important to note that the characteristic velocity c^* can be calculated in multiple ways, according to (10.14). Using the middle term of (10.14), c^* is dependent on the gas properties in the combustion chamber: T_{0c}, γ, and R (or $\mathcal{M}$, since $R = \mathcal{R}/\mathcal{M}$). This is called the *nominal characteristic velocity*, $c^*_{nom} = \sqrt{RT_{0c}}/\Gamma$, since it is a measure of the nominal performance of the gas in the combustion chamber.

Using the right-hand-side term of (10.14), the characteristic velocity c^* is a function of nozzle parameters A_{cr}, $\dot{m}$, and p_{0c}. Since these three parameters can be measured during rocket engine static ground tests, this is called the *characteristic experimental velocity*

$$c^*_{exp} = \frac{p_{0c}A_{cr}}{\dot{m}}.$$

Table 10.1 Characteristic velocities for different fuel–oxidizer compositions for vacuum expansion, $p = 6.89$ MPa and area ratio, $\epsilon = 40$.

Oxidizer	LOX[1]	LOX	LOX
Fuel	LH2[2]	RP-1[3] ($CH_{1.97}$)	Liquid CH_4
Oxidizer/fuel mass ratio	4.83	2.77	3.45
Temperature, T_{0c} [K]	3250	3700	3560
Average bulk density of propellant [kg/m^3]	320	1030	830
Specific impulse, I_{sp} [s]	455	358	369
Exit velocity, u_e [m/s]	4464	3512	3620
Characteristic velocity, c^* [m/s]	2386	1838	1783
Ratio c^*/u_e [-]	0.534	0.523	0.492

[1] liquid oxygen   [2] liquid hydrogen   [3] rocket propellant-1 or refined petroleum-1

By comparing the nominal characteristic velocity c^*_{nom} with the experimental characteristic velocity c^*_{exp}, one gets an indication of how good the rocket design is.

A third characteristic velocity can be defined for chemical rockets, where T_{0c}, γ, and R (or $\mathcal{M}$) can be calculated for a given propellant, so that

$$c^*_{chem} = \left(\frac{RT_{0c}}{\Gamma} \right)_{chem}.$$

The c^*_{chem} is determined from a theoretical chemical analysis of the combustion process.

Table 10.1 shows the variation of characteristic velocity as a function of the fuel–oxidizer composition. Three fuels are considered: liquid hydrogen, liquid methane, and RP-1, a liquid hydrocarbon. The oxidizer for all three cases is liquid oxygen. The oxidizer/fuel ratios were chosen to maximize the specific impulse. The combustion chamber temperature is the adiabatic combustion temperature calculated at a pressure of 6.89 MPa (1000 psia) assuming steady flow. The characteristic velocity, for all three fuels, is approximately 50% of the nozzle exit velocity. The ratio c^*/u_e is in line with the values shown in Fig. 10.3 for engines that have an expansion efficiency, E, of 0.9 or higher.

The ratio c^*_{exp}/c^*_{chem} is called the *performance index*, and its value is an indication of the design quality of the rocket engine. Values of the performance index exceeding 0.9 are expected for well-designed chemical rocket engines.

Example 10.4.1   A chemical rocket engine uses liquid oxygen and RP-1. The oxidizer-fuel mass ratio is 2.24 so that the molecular mass is $\mathcal{M} = 21.9$ kg/kmol, the ratio of specific heat capacities is $\gamma = 1.24$, and the chamber temperature is 3571 K. The experimental characteristic velocity is $c^*_{exp} = 1680$ m/s. Calculate the performance index for the rocket engine.

Solution

$$R = \frac{\mathcal{R}}{\mathcal{M}} = \frac{8314}{21.9} = 379.6 \text{ J/(kg K)}$$

$$\Gamma(\gamma) = \sqrt{\gamma} \left(\frac{2}{\gamma + 1} \right)^{\frac{\gamma+1}{2(\gamma-1)}} \rightarrow \Gamma(1.24) = 0.6562$$

$$c^*_{\text{chem}} = \frac{\sqrt{RT_{0c}}}{\Gamma} = \frac{\sqrt{379.6 \times 3571}}{0.6562} = 1774 \text{ m/s}$$

$$\frac{c^*_{\text{exp}}}{c^*_{\text{chem}}} = \frac{1680}{1774} = 0.947.$$

10.5 Thrust Coefficient, C_T

The expression of the rocket thrust

$$T = \dot{m}u_e + (p_e - p_a)A_e \tag{10.1}$$

can be written as a function of combustion chamber parameters by substituting the mass flow rate from (10.12)

$$\dot{m} = \frac{p_{0c}A_{cr}\Gamma(\gamma)}{\sqrt{RT_{0c}}} \tag{10.15}$$

and the exit velocity, u_e, from (10.4), so that the thrust becomes

$$T = p_{0c}A_{cr}\Gamma \sqrt{\frac{2\gamma}{\gamma - 1} \left[1 - \left(\frac{p_e}{p_{0c}} \right)^{\frac{\gamma-1}{\gamma}} \right]} + (p_e - p_a)A_e.$$

In this form, the thrust depends on the geometry (A_e and A_{cr}), the propellant (γ), and the pressure (p_{0c} and p_e).

To compare rocket engines of different sizes, it is convenient to define a dimensionless expression of the thrust. This dimensionless thrust is called the *thrust coefficient* and is defined as

$$\boxed{C_T := \frac{T}{p_{0c}A_{cr}}.} \tag{10.16}$$

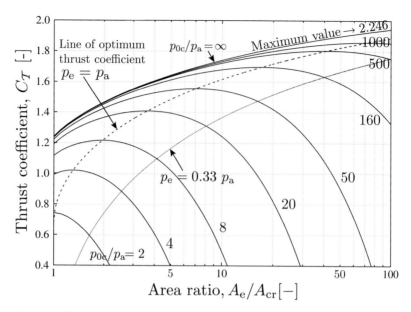

Figure 10.4 Thrust coefficient vs. area ratio for $\gamma = 1.2$.

Let us write the thrust coefficient using the thrust equation (10.1). Then, substituting u_e by (10.4). Then, $\dot{m}$ by (10.15) yields

$$C_T = \frac{\dot{m} u_e}{p_{0c} A_{cr}} + \frac{p_e - p_a}{p_{0c}} \frac{A_e}{A_{cr}} \overset{(10.4),(10.15)}{=} \Gamma \sqrt{\frac{2\gamma}{\gamma - 1}} \sqrt{1 - \left(\frac{p_e}{p_{0c}}\right)^{\frac{\gamma-1}{\gamma}}} + \frac{p_e - p_a}{p_{0c}} \frac{A_e}{A_{cr}} \overset{(10.7)}{=}$$

$$= \Gamma \sqrt{\frac{2\gamma}{\gamma - 1}} E + \frac{p_e - p_a}{p_{0c}} \frac{A_e}{A_{cr}}. \tag{10.17}$$

For a propellant with a given ratio of specific heat capacities, γ, Fig. 10.4 shows the variation of C_T as a function of A_e/A_{cr} for different pressure ratios p_{0c}/p_a. The figure also shows that, for every p_{0c}/p_a value, the largest thrust coefficient occurs when the nozzle exit pressure is equal to the atmospheric pressure, $p_e = p_a$. This is called the *design thrust coefficient*, C_{T_D}. One can prove that the maximum thrust coefficient corresponds to the case when $p_e = p_a$, by taking the derivative of (10.17) with respect to A_e/A_{cr} and setting it to zero. As a result,

$$C_{T_D} = \frac{\dot{m} u_e}{p_{0c} A_{cr}} = \Gamma \sqrt{\frac{2\gamma}{\gamma - 1}} E, \tag{10.18}$$

and using (10.14) yields

$$C_{T_D} = \frac{u_e}{c^*}. \tag{10.19}$$

Substituting C_{T_D} from (10.18) into (10.17) yields

$$C_T = C_{T_D} + \frac{p_e - p_a}{p_{0c}} \frac{A_e}{A_{cr}}. \tag{10.20}$$

The second term of the right-hand side of (10.20) is a measure of the deviation from the design condition where $p_e = p_a$. The design condition states that for a given chamber pressure p_{0c}, the maximum thrust (coefficient) at any atmospheric pressure p_a is obtained when $p_e = p_a$.

In addition, Fig. 10.4 shows the line corresponding to $p_e = p_a/3$ beyond which the flow will separate in the nozzle. As the ratio p_e/p_a drops further below the $1/3$ value, flow separation becomes unavoidable.

The C_T values given in Fig. 10.4 are ideal values that do not take into account the losses due to nonuniform or non-axial flow in the nozzle, viscous effects, separated flows and shock waves, or nonisentropic expansion.

10.5.1 Dimensionless Thrust Function, Γ_T

The term $\Gamma\sqrt{\frac{2\gamma}{\gamma-1}}$ in (10.17) is called the *dimensionless thrust function* and is denoted by Γ_T:

$$\Gamma_T := \Gamma\sqrt{\frac{2\gamma}{\gamma-1}} \tag{10.21}$$

so that the dimensionless thrust coefficient (10.17) can be written as

$$C_T = \Gamma_T E + \frac{p_e - p_a}{p_{0c}} \frac{A_e}{A_{cr}}. \tag{10.22}$$

For design conditions, (10.22) yields

$$C_{T_D} = \Gamma_T E. \tag{10.23}$$

Combining (10.5), (10.14), and (10.21) yields

$$\Gamma_T = u_{max}/c^*,$$

which shows that Γ_T is also a dimensionless velocity function.

Figure 10.5 shows the variation of the dimensionless mass function Γ and the dimensionless thrust function with γ. For γ varying between 1.15 and 1.67, Γ varies between 0.6386 and 0.7262 while Γ_T varies between 2.5008 and 1.6238, as shown in Table 10.2.

The thrust can be written using the thrust coefficient as

$$T = C_T p_{0c} A_{cr}, \tag{10.24}$$

or substituting $p_{0c} A_{cr} = \dot{m} c^*$ from (10.14) into (10.24) yields

$$T = C_T \dot{m} c^*.$$

The specific thrust is then

$$T/\dot{m} = C_T c^*. \tag{10.25}$$

Table 10.2 Dimensionless mass flow and thrust functions vs. γ.

γ	Γ	Γ_T
1.05	0.61767	4.00295
1.10	0.62836	2.94727
1.15	0.63864	2.50077
1.20	0.64853	2.24658
1.25	0.65806	2.08098
1.30	0.66726	1.96437
1.35	0.67614	1.87797
1.40	0.68473	1.81163
1.45	0.69304	1.75934
1.50	0.70108	1.71730
1.55	0.70888	1.68296
1.60	0.71644	1.65456
1.65	0.72379	1.63083
1.666667	0.72618	1.62380

Figure 10.5 Dimensionless mass flow and thrust functions vs. γ.

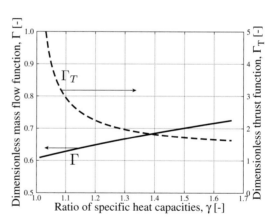

Example 10.5.1 A liquid rocket engine uses liquid oxygen and liquid methane. For an oxidizer/fuel ratio (by mass) of 3.2, the molecular mass of the propellant is $\mathcal{M} = 20.3$, the ratio of specific heat capacities is $\gamma = 1.2$, and the chamber temperature (at 68.95 bar) is $T_{0c} = 3526$ K. The area ratio is $\epsilon = A_e/A_{cr} = 77$ and the nozzle exit diameter is $D = 2.286$ m. The rocket engine operates at 10 km altitude. What are the nozzle exit velocity, thrust, and mass flow rate at design conditions?

Solution

The atmospheric pressure at 10 km is obtained from the ICAO Standard Atmosphere (Appendix A), $p_a = 2.65 \times 10^4$ Pa. Since the engine operates at design conditions, $p_e = p_a = 2.65 \times 10^4$ Pa.

The expansion efficiency is

$$E := \sqrt{1 - \left(\frac{p_e}{p_{0c}}\right)^{\frac{\gamma - 1}{\gamma}}} = \sqrt{1 - \left(\frac{2.65 \times 10^4}{68.95 \times 10^5}\right)^{\frac{0.2}{1.2}}} = 0.7773.$$

The dimensionless thrust function Γ_T is calculated using (10.13) and (10.21) or is read from Table 10.2, so that

$$\Gamma_T(\gamma = 1.2) = 2.2466.$$

The design thrust coefficient is obtained from (10.23):

$$C_{T_D} = \Gamma_T E = 2.2466 \times 0.7773 = 1.7463.$$

The gas constant is

$$R = \mathcal{R}/\mathcal{M} = 8314/20.3 = 409.6 \text{ J/(kg K)}.$$

The dimensionless mass flow function Γ is calculated using (10.13) or is read from Table 10.2, so that

$$\Gamma(\gamma = 1.2) = 0.6485.$$

The characteristic velocity is calculated using (10.14):

$$c^* = \frac{\sqrt{RT_{0c}}}{\Gamma} = \frac{\sqrt{409.6 \times 3526}}{0.6485} = 1853 \text{ m/s}.$$

The nozzle exit velocity is calculated using (10.19):

$$u_e = C_{T_D} c^* = 1.7463 \times 1853 = 3235.8 \text{ m/s}.$$

The nozzle exit area is

$$A_e = \frac{\pi}{4} D^2 = \frac{\pi}{4} 2.286^2 = 4.104 \text{ m}^2$$

and the critical (or throat) area of the nozzle is

$$A_{cr} = \frac{A_e}{\epsilon} = \frac{4.104}{77} = 0.0533 \text{ m}^2.$$

The engine thrust is

$$T_D = C_{T_D} p_{0c} A_{cr} = 1.7463 \times 68.95 \times 10^5 \times 0.0533 = 641,809 \text{ N}$$

and the mass flow rate is

$$\dot{m} = \frac{p_{0c} A_{cr}}{c^*} = \frac{68.95 \times 10^5 \times 0.0533}{1853} = 198.3 \text{ kg/s}.$$

10.6 Maximum Thrust

As shown in Fig. 10.4, the optimum (or design) thrust is not the maximum thrust the rocket engine can produce. The maximum thrust occurs when the atmospheric pressure p_a is zero, that is, the engine operates in a vacuum. When $p_a = 0$, (10.1) becomes

$$T_{max} = \dot{m} u_e + p_e A_e. \tag{10.26}$$

Substituting

$$T_D = \dot{m} u_e \tag{10.8}$$

in (10.26) yields

$$T_{max} = T_D + p_e A_e$$

and after dividing by $p_{0c} A_{cr}$ gives

$$C_{T_{max}} = C_{T_D} + \frac{p_e A_e}{p_{0c} A_{cr}}. \tag{10.27}$$

Combining the thrust (10.1) and the design thrust (10.8) definitions yields

$$T = T_D + (p_e - p_a) A_e$$

which nondimensionalized by $p_{0c} A_{cr}$ gives

$$C_T = C_{T_D} + \frac{p_e - p_a}{p_{0c}} \frac{A_e}{A_{cr}}. \tag{10.28}$$

Substituting C_{T_D} from (10.28) into (10.27) yields

$$C_{T_{max}} = C_T - \frac{p_a A_e}{p_{0c} A_{cr}}.$$

As shown in Fig. 10.4, the *maximum thrust coefficient* corresponds to the chamber pressure being much higher than the atmospheric pressure, that is, $p_{0c}/p_a \to \infty$.

For an engine operating in a vacuum, that is, in outer space, the maximum thrust will also be the design thrust if $p_e = p_a = 0$. Expanding the propellant in the nozzle until it reaches

$p_e = 0$ requires the nozzle to be infinitely long, which is not practical. The expansion ratio and therefore the area ratio of rocket engines operating in outer space are much higher that those of the low-altitude boosters. For example, the area ratio of the Space Shuttle solid rocket motors, which operate the first 123 seconds at liftoff, is $\epsilon = 7.72$, while the area ratio of the Space Shuttle main engines that operate for the majority[1] of their time in outer space is $\epsilon = 69$.

Let us determine the expression of $C_{T_{max}}$ as a function of γ and the pressure ratio p_e/p_{0c}. The Saint-Venant equation (3.60) applied at the nozzle exit gives

$$
\frac{\dot{m}}{A_e} = \sqrt{\frac{2\gamma}{\gamma - 1} p_{0c}\rho_{0c} \left(\frac{p_e}{p_{0c}}\right)^{\frac{2}{\gamma}} \left[1 - \left(\frac{p_e}{p_{0c}}\right)^{\frac{\gamma-1}{\gamma}}\right]}
$$

$$
= \frac{p_{0c}}{\sqrt{RT_{0c}}} \sqrt{\frac{2\gamma}{\gamma - 1} \left(\frac{p_e}{p_{0c}}\right)^{\frac{2}{\gamma}} \left[1 - \left(\frac{p_e}{p_{0c}}\right)^{\frac{\gamma-1}{\gamma}}\right]}. \tag{10.29}
$$

Substituting $\dot{m}$ from (10.12) into (10.29) yields

$$
\frac{A_e}{A_{cr}} = \frac{\Gamma}{\sqrt{\dfrac{2\gamma}{\gamma - 1} \left(\dfrac{p_e}{p_{0c}}\right)^{\frac{2}{\gamma}} \left[1 - \left(\dfrac{p_e}{p_{0c}}\right)^{\frac{\gamma-1}{\gamma}}\right]}} \tag{10.30}
$$

which replaced in (10.22) gives

$$
C_T = \Gamma_T E \left[1 + \frac{1}{E^2} \frac{\gamma - 1}{2\gamma} \left(\frac{p_{0c}}{p_e}\right)^{\frac{1}{\gamma}} \frac{p_e - p_a}{p_{0c}}\right]. \tag{10.31}
$$

The maximum thrust coefficient is obtained from (10.31) when $p_a = 0$:

$$
C_{T_{max}} = \Gamma_T E \left[1 + \frac{1}{E^2} \frac{\gamma - 1}{2\gamma} \left(\frac{p_e}{p_{0c}}\right)^{1-\frac{1}{\gamma}}\right]. \tag{10.32}
$$

The maximum maximorum values of $C_{T_{max}}$ are obtained when $p_e/p_{0c} \to 0$, so that

$$
\lim_{p_e/p_{0c} \to 0} C_{T_{max}} = \Gamma_T. \tag{10.33}
$$

As shown in Fig. 10.4, when the area ratio $\epsilon = A_e/A_{cr}$ goes to infinity, the maximum thrust coefficient asymptotically aproaches its maximum maximorum value, Γ_T. For $\gamma = 1.2$ this value is $\lim_{p_e/p_{0c} \to 0} C_{T_{max}} = \Gamma_T = 2.2466$.

---

[1] The Space Shuttle used both the solid rocket motors and the main engines at liftoff.

Example 10.6.1 Calculate the maximum thrust of the rocket engine from Example 10.5.1.

Solution
Using (10.33)

$$C_{T_{\max}} = \Gamma_T(\gamma = 1.2) = 2.246.$$

The maximum thrust is then obtained using (10.16):

$$T_{\max} = C_{T_{\max}} p_{0c} A_{cr} = 2.246 \times 68.95 \times 10^5 \times 0.0533 = 825{,}413 \text{ N.}$$

The design thrust at 10 km altitude was calculated in Example 10.5.1:

$$T_{D_{10\,\mathrm{km}}} = 641{,}809 \text{ N.}$$

Example 10.6.2 Calculate the pressure ratios p_e/p_{0c} for area ratios A_e/A_{cr} 5, 20, 50, 100, 200, 1000 and γ equal to 1.15, 1.2, 1.25, 1.3, 1.35.

Solution
Using the Newton–Raphson method applied to (10.30) yields the values tabulated in Table 10.3.

Table 10.3 Pressure ratio p_e/p_{0c} as a function of area ratio and γ.

γ \\ A_e/A_{cr}	5	20	50	100	200	1000
1.15	0.079265	0.006329	0.001934	0.000833	0.000361	0.000054
1.20	0.062093	0.005171	0.001573	0.000657	0.000277	0.000038
1.25	0.049618	0.004305	0.001280	0.000519	0.000212	0.000027
1.30	0.040357	0.003638	0.001043	0.000410	0.000163	0.000020
1.35	0.033345	0.003086	0.000849	0.000324	0.000125	0.000014

Figure 10.6 shows the variation of pressure ratio p_e/p_{0c} as a function of area ratio and γ.

The specific impulse was defined in Section 9.1.2 and was used to compare different rocket engines. Let us explore this parameter in more detail and find how it relates to other rocket parameters.

Typically the same notation, I_{sp}, is used for both definitions of specific impulse

$$I_{sp} = \frac{I_t}{\int_0^t \dot{m}_f \, dt} \tag{9.3}$$

$$I_{sp} = \frac{I_t}{g \int_0^t \dot{m}_f \, dt}, \tag{9.4}$$

Figure 10.6 Nozzle pressure ratio p_e/p_{0c} as a function of area ratio A_e/A_{cr} and the ratio of specific heat capacities, γ.

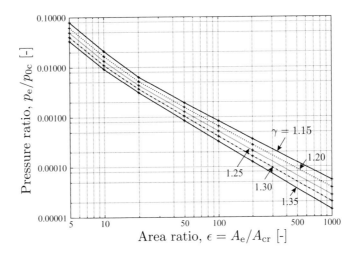

and the reader must differentiate them by looking at the units. To avoid confusion, the notation used for the value defined in (9.3) will be $I_{sp,m}$.

The specific impulses defined in (9.3) and (9.4) are time-averaged values. The instantaneous specific impulse is defined as

$$I_{sp} = \frac{T}{\dot{w}} \tag{10.34}$$

where $\dot{w}$ is the weight flow rate of propellant, $\dot{w} = \dot{m}g$. If the thrust and weight flow rate are constant, then

$$\Delta I = T \Delta t = I_{sp} \Delta w$$

so that

$$I_{sp} = \frac{\Delta I}{\Delta w} = \frac{\Delta I}{g \Delta m}.$$

Consequently, the specific impulse is the impulse per unit weight of propellant.

Substituting the thrust expression (10.1) into (10.34) yields

$$I_{sp} = \frac{u_e}{g} + \frac{(p_e - p_a)A_e}{\dot{w}}.$$

Using the definition of the *characteristic exhaust velocity*

$$c := \frac{T}{\dot{m}} \tag{10.35}$$

$$\stackrel{(10.1)}{=} u_e + \frac{(p_e - p_a)A_e}{\dot{m}},$$

the specific impulse becomes

$$I_{sp} = \frac{c}{g} \tag{10.36}$$

and

$$I_{sp,m} = c.$$

Combining (10.25) and (10.35) yields

$$c = C_T c^*$$

so that

$$I_{sp} = \frac{c^*}{g} C_T \quad \text{and} \quad I_{sp,m} = c^* C_T.$$

Following the same arguments used for determining the design (or optimal) thrust and maximum thrust, the optimum specific impulse corresponds to $p_a = p_e$

$$I_{sp_{opt}} = \frac{u_e}{g},$$

and the maximum specific impulse corresponds to $p_a = 0$

$$I_{sp_{max}} = \frac{u_e}{g} + \frac{p_e A_e}{\dot{w}}.$$

The specific impulse can be interpreted in two ways [Archer and Saarlas, 1996, p. 449]: (1) the amount of thrust produced by a unit of weight flow rate or (2) the time during which a unit weight of propellant produces one unit of thrust. As an example, a rocket engine with $I_{sp} = 300$ s will: (1) produce 300 N for each 1 N/s of propellant, and (2) enable 1 N/s of propellant to produce 1 N of thrust for 300 s.

10.7 Conical Nozzle Thrust

The expressions of the ideal thrust (10.1) and of the dimensionless thrust coefficient (10.21) were obtained assuming that the flow in the rocket nozzle is one-dimensional and isentropic. As a result, the only geometric variable included in the thrust expression is the area ratio, A_e/A_{cr}. The flow in the nozzle, however, is neither one-dimensional nor isentropic. There are several types of losses in a real nozzle, including those due to shock waves, viscous effects, momentum loss due to nozzle angularity, heat transfer, and manufacturing geometric imperfections.

The detailed design and analysis of the nozzle must take into account all these losses, especially on the divergent part of the nozzle. The convergent part of the nozzle, where the subsonic flow has a favorable pressure gradient, is less problematic and challenging than the divergent part. A smooth nozzle variation is usually sufficient to provide a good subsonic flow in the convergent part of the nozzle.

The shape of the diverging part of the nozzle, where the flow is supersonic, is critical since incorrect area variation can lead to shock formation and performance loss. In the absence of

Figure 10.7 Conical nozzle.

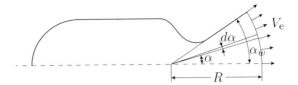

strong boundary-layer/shock-wave interaction, the most important losses in the divergence nozzle are due to friction and divergence of the exhaust stream, that is, flow angularity.

The expressions of the ideal thrust and dimensionless thrust coefficient can be improved to account for the effect of nonaxial exhaust velocities, while still assuming constant composition flow. Let us consider the nozzle flow in Fig. 10.7, where the nozzle cone angle is α_w. The exit boundary is a spherical cone, defined by the nozzle cone angle α_w and the cone radius R. Consequently, the momentum term $\dot{m}u_e$ of the ideal thrust equation (10.1) becomes

$$\int_{A_e} V_{e_x} d\dot{m},$$

where $V_{e_x} = V_e \cos \alpha$ for an arbitrary α angle. The mass flow rate $d\dot{m}$ is

$$d\dot{m} = \rho_e V_e dA = \rho_e V_e 2\pi R \sin \alpha \, R d\alpha.$$

The momentum term becomes

$$\int_{A_e} V_{e_x} d\dot{m} = \int_0^{\alpha_w} V_e \cos \alpha \rho_e V_e 2\pi R^2 \sin \alpha \, d\alpha = 2\pi \rho_e V_e^2 R^2 \left(1 - \cos^2 \alpha_w\right) \frac{1}{2}.$$

Since the area of the spherical cone is $A_e = 2\pi R^2 \left(1 - \cos \alpha_w\right)$, the momentum term is

$$\int_{A_e} V_{e_x} d\dot{m} = \rho_e V_e^2 A_e \frac{1 + \cos \alpha_w}{2} = \dot{m} V_e \frac{1 + \cos \alpha_w}{2}$$

so that the thrust becomes

$$T = \frac{1 + \cos \alpha_w}{2} \dot{m} V_e + (p_e - p_a) A_e = \lambda \dot{m} V_e + (p_e - p_a) A_e,$$

where $\lambda = (1 + \cos \alpha_w)/2$ is called the *nozzle divergence correction factor*.

It is apparent that as the cone angle α_w increases, the thrust decreases. For example, if α_w increases from 15° to 30°, λ decreases from 0.983 to 0.933. For a given nozzle exit area A_e, a smaller cone angle yields a longer, therefore heavier, nozzle. For this reason, only small rocket motors have conical nozzles. Larger nozzles are typically contoured so that the nozzle angle varies from α_w at throat to a smaller α_e at exit. In this case, an approximate value of the nozzle divergence correction factor is [Archer and Saarlas, 1996, p. 459]

$$\lambda = \left(1 + \cos \frac{\alpha_w + \alpha_e}{2}\right)/2.$$

The purpose of nozzle contouring is to minimize divergence losses and to obtain a near-parallel exhaust jet for a wide range of operating conditions. Nozzle contouring is done by using the method of characteristics [Shapiro, 1953, Chapter 15] followed by a computational fluid dynamics (CFD) analysis.

10.8 Vertical Rocket Trajectory

Let us determine the velocity of a rocket that has a purely vertical ascent. The force balance gives

$$T - W - D = M\frac{dV}{dt} \tag{10.37}$$

where the thrust is $T = \dot{m}c$, the rocket weight is $W = Mg$, and $\frac{dV}{dt}$ is the acceleration. The drag is $D = \frac{1}{2}\rho V^2 A C_D$ where ρ is the density, A is the frontal area of the rocket, and C_D is the drag coefficient. The mass of the rocket varies in time, $M(t)$. At liftoff the mass is $M(t = 0) = M_0$. Assuming the mass flow rate of propellant is constant, the mass at time t is $M(t) = M_0 - \dot{m}t = M_0 - M_p(t)$, where M_p is the mass of propellant used from liftoff until time t.

Dividing (10.37) by Mg and substituting T and W yields

$$\frac{\dot{m}c}{Mg} - 1 - \frac{D}{W} = \frac{1}{g}\frac{dV}{dt}. \tag{10.38}$$

The mass flow rate of propellant leaving the rocket leads to the decrease of rocket mass, that is, $\dot{m} = -dM/dt$, so that (10.38) yields

$$\frac{dV}{c} = -\frac{dM}{M} - \frac{g}{c}\left(1 + \frac{D}{W}\right)dt. \tag{10.39}$$

Substituting $g/c \overset{(10.36)}{=} 1/I_{sp}$ in (10.39) and integrating from liftoff to time t yields the rocket velocity at time t:

$$\frac{V}{c} = \ln\frac{M_0}{M} - \int_0^t \frac{1}{I_{sp}}\left(1 + \frac{D}{W}\right)dt. \tag{10.40}$$

The ambient pressure p_a becomes negligible at altitudes above 30 km so that the thrust coefficient C_T is constant, and therefore $I_{sp} = c^* C_T/g$ is also constant. In addition, at these high altitudes the drag is negligible, so that (10.40) yields

$$\frac{V}{c} = \ln\frac{M_0}{M} - \frac{t}{I_{sp}} \tag{10.41}$$

and using (10.36) gives

$$V = c\ln\frac{M_0}{M} - gt. \tag{10.42}$$

If the rocket is far away from Earth so that gravity can be neglected, (10.42) yields

$$V(t) = c\ln\frac{M_0}{M(t)} = I_{sp}g\ln\frac{M_0}{M(t)}, \tag{10.43}$$

which is called the *Tsiolkovsky equation*. This equation shows that the rocket velocity is linearly proportional to the specific impulse I_{sp} and the logarithm of the *mass ratio*, M_0/M.

Table 10.4 Mass fractions of different launch vehicles.

Vehicle	Mass fraction
Ariane 5 GS booster	0.8791
Ariane 5 GS core	0.9284
Ariane 5 GS second stage	0.8929
Delta II	0.9147
Delta III	0.9054
Shuttle solid rocket booster	0.8599
V-2	0.7335

The maximum velocity is reached when $M(t)$ is minimum, that is, at burnout. At burnout, $t = t_b$ and $M_b = M(t_b)$, so that

$$M_b = M_0 - M_p \tag{10.44}$$

where M_p is the mass of propellant loaded on the rocket and completely used at burnout. The velocity at burnout is

$$V(t_b) \overset{(10.43)}{=} I_{sp}g \ln \frac{M_0}{M_b} \overset{(10.44)}{=} I_{sp}g \ln \frac{M_0}{M_0 - M_p} = I_{sp}g \ln \frac{1}{1 - M_p/M_0} \tag{10.45}$$

where M_p/M_0 is called the *propellant mass fraction*. Typical values of propellant mass fractions are in the range of 0.75 to 0.95, as shown in Table 10.4.

Example 10.8.1 The engine of a rocket that uses liquid oxygen and liquid hydrogen propellant has a specific impulse $I_{sp} = 455$ m/s. What should the propellant mass fraction be so that the rocket achieves the orbital speed of 8 km/s?

Solution
Using (10.45)

$$\frac{V}{I_{sp}g} = -\ln\left(1 - M_p/M_0\right)$$

yields

$$\exp\left(-\frac{V}{I_{sp}g}\right) = 1 - M_p/M_0$$

so that

$$M_p/M_0 = 1 - \exp\left(-\frac{V}{I_{sp}g}\right) = 1 - \exp\left(-\frac{8000}{455 \times 9.8}\right) = 0.834.$$

This example shows that single stage to orbit is possible if the propellant mass fraction exceeds 83.4%. This estimate is based on the assumptions that air drag and gravity can be neglected. Although the gravity and air drag effects are small, taking them into account requires increasing the propellant mass fraction to values around 90% [Archer and Saarlas, 1996, p. 453].

Let us consider the impact of burn time, t_b, on rocket velocity, V. For constant, unthrottled mass flow $\dot{m}$, the ratio M_0/M is

$$\frac{M_0}{M(t)} = \frac{M_0}{M_0 - \dot{m}t}.$$

At burnout, this ratio becomes

$$\frac{M_0}{M(t_b)} = \frac{M_0}{M_0 - \dot{m}t_b}$$

so that the larger t_b is, the larger the ratio $M_0/M(t_b)$ becomes. Consequently, the first term of the right-hand side of (10.42) increases, therefore increasing V. The increase of t_b, however, increases the second term of the right-hand side of (10.42), therefore decreasing V. Consequently, the variation of burn time t_b affects the two terms of the right-hand side of (10.42) in an antagonistic way. Reducing t_b, that is, having a short impulsive burn, results in a high rocket acceleration, which requires a stronger and heavier structure.

Let us determine the value of the burn time for a constant, unthrottled mass flow $\dot{m}$:

$$t_b = \frac{M_p}{\dot{m}}.$$

Using $\dot{m} \overset{(10.35)}{=} T/c$ yields

$$t_b = \frac{M_p g}{T} \frac{c}{g} \overset{(10.36)}{=} I_{sp} \frac{M_p g}{T}. \tag{10.46}$$

Since the thrust must exceed the weight at liftoff, $M_0 g$, and $M_0 > M_p$, then $t_b < I_{sp}$.

An alternate expression of (10.46)

$$t_b = I_{sp} \frac{M_0 g}{T} \frac{M_p}{M_0} \tag{10.47}$$

can be used to calculate t_b since the thrust to gross weight $T/(M_0 g)$, and the propellant mass fraction M_p/M_0 are typically specified.

Example 10.8.2 Calculate the burn time of a rocket engine that has a specific impulse $I_{sp} = 455$ s, a thrust to gross weight $T/(M_0 g) = 1.54$, and a propellant mass fraction $M_p/M_0 = 0.88$.

Solution
Using (10.47) yields

$$t_b = I_{sp} \frac{M_0 g}{T} \frac{M_p}{M_0} = 455 \times \frac{1}{1.54} \times 0.88 = 260 \text{ s}.$$

If the engine can be throttled, the burn time can be higher than the value predicted using (10.47). For example, the Space Shuttle main engine can be throttled between 67% and 109% of the rated power level. The specific impulse is $I_{sp} = 455$ s but, because of throttling, the burn time is $t_b = 480$ s.

Problems

1. A liquid rocket engine uses liquid oxygen and liquid hydrogen. For an oxidizer/fuel ratio (by mass) of 4.02, the molecular mass of the propellant is $\mathcal{M} = 10$, the ratio of specific heat capacities is $\gamma = 1.26$, and the chamber temperature (at 68.95 bar) is $T_{0c} = 2999$ K. The area ratio is $\epsilon = A_e/A_{cr} = 69$ and the nozzle exit diameter is $D_e = 2.286$ m. The rocket engine operates at 10 km altitude. What are the nozzle exit velocity, thrust, and mass flow rate at design conditions?

2. A rocket engine has a throat area of $A_{cr} = 0.01$ m^2. The propellant has a molecular mass $\mathcal{M} = 10$ kg/kmol and $\gamma = 1.26$. The chamber stagnation temperature is $T_{0c} = 3000$ K and the stagnation pressure is $p_{0c} = 100$ bar.

   1. What is the mass flow rate of propellant in the rocket nozzle?
   2. What is the amount of propellant consumed in a 100-second burn?

3. Fluorine–hydrazine is the oxidizer–fuel mix used in a liquid rocket engine. The chamber stagnation temperature is $T_{0c} = 4713$ K and the propellant has a molecular mass $\mathcal{M} = 19.4$ kg/kmol and $\gamma = 1.33$.

   1. What is the characteristic velocity c^*_{chem} of this engine?
   2. What are the characteristic velocity c^*_{exp} and the performance index if the throat nozzle area is $A_{cr} = 0.04$ m^2 and the stand measurements give a mass flow rate of $\dot{m} = 200$ kg/s and a chamber stagnation pressure $p_{0c} = 100$ bar?

4. A rocket engine uses unsymmetrical dimethyl hydrazine (UDMH) as the fuel and oxygen as the oxidizer. The molecular mass of the propellant is $\mathcal{M} = 21.3$ kg/kmol and $\gamma = 1.25$. The area ratio is $\epsilon = A_e/A_{cr} = 5$.

   1. What is the specific impulse I_{sp} if the rocket engine is operating at sea level?
   2. What is the specific impulse in a vacuum?

5. Liquid oxygen and RP-1 fuel are in a liquid rocket engine that has a chamber stagnation temperature $T_{0c} = 3677$ K and a chamber stagnation pressure $p_{0c} = 68.95$ bar. The propellant has a molecular mass $\mathcal{M} = 23.3$ kg/kmol and $\gamma = 1.24$. The nozzle expansion ratio is $p_{0c}/p_e = 120$.

   1. What is the maximum velocity of the propellant at nozzle exit?
   2. What is the actual velocity of the propellant at nozzle exit?
   3. What is the design altitude?
   4. What are the nozzle throat area and mass flow rate needed to produce a thrust of 50,000 N?

6. The chamber stagnation temperature in a solid propellant rocket motor is $T_{0c} = 3200$ K. The propellant has a molecular mass $\mathcal{M} = 26$ kg/kmol and $\gamma = 1.2$. The rocket motor is designed for sea-level exhaust conditions. What are the design thrust coefficient and the sea-level specific impulse if the nozzle pressure ratio is $p_{0c}/p_e = 40$?

7. The propellant of a rocket motor has a molecular mass $\mathcal{M} = 27.5$ kg/kmol, and $\gamma = 1.18$. The chamber stagnation temperature and stagnation pressure are $T_{0c} = 2850$ K and $p_{0c} = 60$ bar, respectively. The nozzle exit pressure is $p_e = 1.013$ bar. What are the maximum nozzle exit velocity, u_{max}, the nozzle exit velocity, u_e, the characteristic velocity, c^*, and the sea-level specific impulse, I_{sp}?

8. A rocket engine has a chamber stagnation pressure $p_{0c} = 30$ bar. The propellant has $\gamma = 1.25$. The expansion ratio in the nozzle is $p_{0c}/p_e = 10$.
   1. What is the ratio between the sea-level thrust and the thrust at 20 km altitude?
   2. What is the design altitude?

9. A rocket engine has the following specifications: chamber stagnation temperature is $T_{0c} = 3500$ K, chamber stagnation pressure $p_{0c} = 60$ bar, expansion pressure ratio $p_{0c}/p_e = 69$, nozzle throat area $A_{cr} = 0.25$ m^2. The propellant has a molecular mass $\mathcal{M} = 24$ kg/kmol, and $\gamma = 1.22$. Calculate the following:
   1. thrust coefficient at sea level and in vacuum;
   2. thrust at sea level and in vacuum;
   3. specific impulse at sea level and in vacuum;
   4. propellant exit velocity.

10. The solid propellant motor of the first-stage of the Minuteman missile has a chamber stagnation temperature is $T_{0c} = 3200$ K and a chamber stagnation pressure $p_{0c} = 53$ bar. The area ratio is $\epsilon = A_e/A_{cr} = 10$ and the throat nozzle area is $A_{cr} = 0.106$ m^2. Assuming that the molecular mass $\mathcal{M} = 23.8$ kg/kmol and $\gamma = 1.31$, calculate:
   1. takeoff thrust;
   2. thrust at 10 km altitude;
   3. maximum thrust;
   4. specific impulse at takeoff and in vacuum.

Bibliography

R. D. Archer and M. Saarlas. *An Introduction to Aerospace Propulsion*. Prentice Hall, 1996. 425, 440, 441, 444

A. H. Shapiro. *The Dynamics and Thermodynamics of Compressible Fluid Flow*, volume I. Ronald Press Company, 1953. 441

APPENDIX A

ICAO Standard Atmosphere

H [m]	T [K]	p [Pa]	ρ [kg/m^3]	a [kg/m^3]	ν [m^2/s]
0	288.150	1.01325×10^5	1.2250	340.294	1.4607×10^{-5}
1000	281.651	8.9876×10^4	1.1117	336.435	1.5813×10^{-5}
2000	275.154	7.9501×10^4	1.0066	332.532	1.7147×10^{-5}
3000	268.659	7.0121×10^4	9.0925×10^{-1}	328.583	1.8628×10^{-5}
4000	262.166	6.1660×10^4	8.1935×10^{-1}	324.589	2.0275×10^{-5}
5000	255.676	5.4048×10^4	7.3643×10^{-1}	320.545	2.2110×10^{-5}
6000	249.187	4.7217×10^4	6.6011×10^{-1}	316.452	2.4162×10^{-5}
7000	242.700	4.1105×10^4	5.9002×10^{-1}	312.306	2.6461×10^{-5}
8000	236.215	3.5651×10^4	5.2579×10^{-1}	308.105	2.9044×10^{-5}
9000	229.733	3.0800×10^4	4.6706×10^{-1}	303.848	3.1957×10^{-5}
10000	223.252	2.6500×10^4	4.1351×10^{-1}	299.532	3.5251×10^{-5}
11000	216.774	2.2700×10^4	3.6480×10^{-1}	295.154	3.8988×10^{-5}
12000	216.650	1.9399×10^4	3.1194×10^{-1}	295.069	4.5574×10^{-5}
13000	216.650	1.4170×10^4	2.6660×10^{-1}	295.069	5.3325×10^{-5}
14000	216.650	1.4170×10^4	2.2786×10^{-1}	295.069	6.2391×10^{-5}
15000	216.650	1.2112×10^4	1.9475×10^{-1}	295.069	7.2995×10^{-5}
16000	216.650	1.0353×10^4	1.6647×10^{-1}	295.069	8.5397×10^{-5}
17000	216.650	8.8496×10^3	1.4230×10^{-1}	295.069	9.9902×10^{-5}
18000	216.650	7.5652×10^3	1.2165×10^{-1}	295.069	1.1686×10^{-4}
19000	216.650	6.4674×10^3	1.0400×10^{-1}	295.069	1.3670×10^{-4}
20000	216.650	5.5293×10^3	8.8910×10^{-2}	295.069	1.5989×10^{-4}
21000	217.581	4.7274×10^3	7.5715×10^{-2}	295.703	1.8843×10^{-4}
22000	218.574	4.0420×10^3	6.4510×10^{-2}	296.377	2.2201×10^{-4}
23000	219.567	3.4562×10^3	5.5006×10^{-2}	297.049	2.6135×10^{-4}
24000	220.560	2.9554×10^3	4.6938×10^{-2}	297.720	3.0743×10^{-4}

(cont.)

H [m]	T [K]	p [Pa]	ρ [kg/m³]	a [kg/m³]	ν [m²/s]
25000	221.552	2.6077×10^3	4.0084×10^{-2}	298.389	3.6135×10^{-4}
26000	222.544	2.1632×10^3	3.4257×10^{-2}	299.056	4.2439×10^{-4}
27000	223.536	1.8555×10^3	2.9298×10^{-2}	299.722	4.9805×10^{-4}
28000	224.527	1.5949×10^3	2.5076×10^{-2}	300.386	5.8405×10^{-4}
29000	225.518	1.3737×10^3	2.1478×10^{-2}	301.048	6.8438×10^{-4}
30000	226.509	1.1855×10^3	1.8410×10^{-2}	301.709	8.0134×10^{-4}

https://www.digitaldutch.com/atmoscalc/

APPENDIX B

Thermodynamic Properties Tables for Air

T [K]	h [kJ/kg]	u [kJ/kg]	s [kJ/(kg K)]	a [m/s]
213.16	213.1	151.9	6.3629	292.9
223.16	223.1	159.0	6.4055	299.7
233.16	233.1	166.2	6.4481	306.4
243.16	243.1	173.3	6.4906	312.8
253.16	253.2	180.5	6.5311	319.7
263.16	263.2	187.6	6.5696	325.4
273.16	273.2	194.8	6.6081	331.5
283.16	283.3	201.9	6.6432	337.5
288.16	288.3	205.5	6.6608	340.4
293.16	293.3	209.1	6.6784	343.3
303.16	303.3	216.3	6.7121	349.1
313.16	313.4	223.4	6.7444	354.7
323.16	323.4	230.6	6.7769	360.3
333.16	333.5	237.8	6.8069	365.8
343.16	343.6	245.1	6.8369	371.2
353.16	353.6	252.2	6.8659	376.5
363.16	363.7	259.4	6.8938	381.7
373.16	373.8	266.6	6.9218	386.9
383.16	383.9	273.9	6.9480	392.0
393.16	394.0	281.1	6.9743	397.0
403.16	404.1	288.4	6.9996	401.9
413.16	414.3	295.6	7.0243	406.8
423.16	424.4	302.9	7.0491	411.6
433.16	434.6	310.2	7.0725	416.2
443.16	444.8	317.5	7.0959	420.9
453.16	454.9	324.8	7.1187	425.4
463.16	465.2	332.2	7.1409	429.9
473.16	475.4	339.5	7.1630	434.4
483.16	485.7	346.9	7.1841	438.8
493.16	495.9	354.3	7.2053	443.2
503.16	506.2	361.7	7.2260	447.5

(*cont.*)

T [K]	h [kJ/kg]	u [kJ/kg]	s [kJ/(kg K)]	a [m/s]
513.16	516.5	369.2	7.2462	451.7
523.16	526.9	376.6	7.2664	456.0
533.16	537.2	384.1	7.2857	460.1
543.16	547.6	391.6	7.3051	464.2
553.16	558.0	399.1	7.3241	468.3
563.16	568.4	406.7	7.3427	472.4
573.16	578.8	414.2	7.3612	476.4
583.16	589.3	421.8	7.3791	480.3
593.16	599.8	429.4	7.3970	484.2
603.16	610.3	437.0	7.4145	488.1
613.16	620.8	444.7	7.4318	491.9
623.16	631.3	452.4	7.4491	495.7
633.16	641.9	460.1	7.4658	499.4
643.16	652.5	467.8	7.4824	503.2
653.16	663.1	475.6	7.4988	506.9
663.16	673.8	483.4	7.5149	510.7
673.16	684.4	491.1	7.5310	514.2
683.16	695.2	499.0	7.5466	517.8
693.16	705.9	506.8	7.5622	521.3
703.16	716.7	514.7	7.5777	524.9
713.16	727.4	522.5	7.5928	528.3
723.16	738.1	530.5	7.6080	531.8
733.16	749.0	538.4	7.6227	535.3
743.16	759.8	546.4	7.6370	538.7
753.16	770.7	554.4	7.6520	542.1
763.16	781.5	562.4	7.6663	545.5
773.16	792.4	570.4	7.6806	548.9
783.16	803.4	578.5	7.6945	552.2
793.16	814.3	586.6	7.7085	555.5
803.16	825.3	594.7	7.7228	558.8
813.16	836.3	602.8	7.7358	562.1
823.16	847.3	610.9	7.7494	565.3
833.16	858.4	619.1	7.7626	568.5
843.16	869.5	627.3	7.7759	571.7
853.16	880.6	635.6	7.7889	574.9
863.16	891.7	643.8	7.8019	578.0
873.16	902.8	652.1	7.8148	581.1
883.16	913.9	660.4	7.8274	584.2

(*cont.*)

T [K]	h [kJ/kg]	u [kJ/kg]	s [kJ/(kg K)]	a [m/s]
893.16	925.1	668.7	7.8401	587.3
903.16	936.3	677.0	7.8526	590.4
913.16	947.6	685.3	7.8649	593.4
923.16	958.8	693.7	7.8772	596.4
933.16	970.1	702.1	7.8893	599.5
943.16	981.4	710.5	7.9013	602.5
953.16	992.7	718.9	7.9132	605.5
963.16	1004.0	727.4	7.9250	608.5
973.16	1015.3	735.9	7.9368	611.5
983.16	1026.7	744.4	7.9483	614.4
993.16	1038.1	752.9	7.9599	617.4
1003.16	1049.5	761.4	7.9713	620.3
1013.16	1060.9	770.0	7.9826	623.2
1023.16	1072.4	778.5	7.9939	626.1
1033.16	1083.8	787.1	8.0049	629.0
1043.16	1095.3	795.7	8.0160	631.8
1053.16	1106.8	804.4	8.0270	634.7
1063.16	1118.3	813.0	8.0379	637.6
1073.16	1129.8	821.7	8.0488	640.4
1083.16	1141.4	830.4	8.0594	643.2
1093.16	1153.0	839.1	8.0701	646.0
1103.16	1164.6	847.8	8.0806	648.7
1113.16	1176.2	856.5	8.0911	651.5
1123.16	1187.8	865.3	8.1015	654.2
1133.16	1199.4	874.0	8.1118	657.0
1143.16	1211.1	882.8	8.1220	659.8
1153.16	1222.8	891.6	8.1322	662.9
1163.16	1234.5	900.4	8.1423	665.3
1173.16	1246.2	909.3	8.1524	668.1
1183.16	1257.9	918.1	8.1623	670.8
1193.16	1269.6	927.0	8.1722	673.5
1203.16	1281.3	935.8	8.1821	676.1
1213.16	1293.1	944.7	8.1918	678.8
1223.16	1304.8	953.6	8.2014	681.4
1233.16	1316.6	962.5	8.2109	684.1
1243.16	1328.4	971.4	8.2205	686.8
1253.16	1340.2	980.4	8.2300	689.4
1263.16	1352.1	989.3	8.2394	692.1

(cont.)

T [K]	h [kJ/kg]	u [kJ/kg]	s [kJ/(kg K)]	a [m/s]
1273.16	1363.9	998.3	8.2487	694.7
1283.16	1375.8	1007.3	8.2579	697.3
1293.16	1387.6	1016.3	8.2672	699.8
1303.16	1399.5	1025.3	8.2763	702.3
1313.16	1411.4	1034.3	8.2854	704.9
1323.16	1423.3	1043.3	8.2945	707.4
1333.16	1435.2	1052.4	8.3034	710.0
1343.16	1447.1	1061.4	8.3123	712.5
1353.16	1459.1	1070.5	8.3212	715.0
1363.16	1471.0	1079.6	8.3300	717.6
1373.16	1483.0	1088.7	8.3388	720.1
1383.16	1495.0	1097.8	8.3475	722.7
1393.16	1507.0	1106.9	8.3561	725.2
1403.16	1519.0	1116.0	8.3646	727.7
1413.16	1531.0	1125.2	8.3732	730.3
1423.16	1543.0	1134.3	8.3817	732.8
1433.16	1555.0	1143.5	8.3901	735.3
1443.16	1567.1	1152.7	8.3985	737.7
1453.16	1579.1	1161.8	8.4068	740.1
1463.16	1591.2	1171.0	8.4151	742.6
1473.16	1603.3	1180.3	8.4233	745.0
1483.16	1615.4	1189.5	8.4314	747.4
1493.16	1627.5	1198.7	8.4396	749.8
1503.16	1639.6	1207.9	8.4476	752.2
1513.16	1651.7	1217.1	8.4556	754.6
1523.16	1663.8	1226.4	8.4637	757.0
1533.16	1675.9	1235.7	8.4716	759.3
1543.16	1688.1	1244.9	8.4795	761.7
1553.16	1700.2	1254.2	8.4874	764.0
1563.16	1712.4	1263.3	8.4953	766.4
1573.16	1724.6	1272.8	8.5030	768.7
1583.16	1736.8	1282.1	8.5107	771.1
1593.16	1749.0	1291.5	8.5184	773.4
1603.16	1761.2	1300.8	8.5260	775.8
1613.16	1773.4	1310.1	8.5335	778.2
1623.16	1785.6	1319.5	8.5411	780.5
1633.16	1797.8	1328.8	8.5486	782.8
1643.16	1810.1	1338.2	8.5560	785.1

(*cont.*)

T [K]	*h* [kJ/kg]	*u* [kJ/kg]	*s* [kJ/(kg K)]	*a* [m/s]
1653.16	1822.4	1347.8	8.5635	787.3
1663.16	1834.6	1357.6	8.5709	789.6
1673.16	1846.8	1367.3	8.5783	791.8
1683.16	1859.1	1376.3	8.5856	794.1
1693.16	1871.4	1385.4	8.5929	796.4
1703.16	1883.7	1394.6	8.6001	798.7
1713.16	1896.0	1404.0	8.6073	801.0
1723.16	1908.3	1413.4	8.6145	803.8
1733.16	1920.6	1422.9	8.6216	805.5
1743.16	1932.9	1432.3	8.6287	807.8
1753.16	1945.2	1441.8	8.6357	810.0
1763.16	1957.6	1451.2	8.6427	812.3
1773.16	1969.9	1460.7	8.6497	814.5

APPENDIX C

Thermodynamic Properties of Stoichiometric Combustion Products

T [K]	h [kJ/kg]	u [kJ/kg]	s [kJ/(kg K)]	a [m/s]
273.16	280.9	202.5	6.8119	327.8
283.16	291.4	210.1	6.8493	333.5
288.16	296.7	214.0	6.8680	336.5
293.16	302.0	217.8	6.8867	339.2
303.16	312.6	225.5	6.9222	344.9
313.16	323.2	233.2	6.9560	350.2
323.16	333.7	240.9	6.9899	355.5
333.16	344.4	248.7	7.0219	360.7
343.16	355.0	256.4	7.0539	365.9
353.16	365.7	264.3	7.0846	371.1
363.16	376.4	272.1	7.1142	376.2
373.16	387.2	280.0	7.1439	381.0
383.16	398.0	288.0	7.1721	386.0
393.16	408.9	296.0	7.2003	390.7
403.16	419.7	304.0	7.2274	395.5
413.16	430.6	311.9	7.2537	400.2
423.16	441.4	319.9	7.2800	404.7
433.16	452.4	328.0	7.3051	409.3
443.16	463.3	336.1	7.3302	413.7
453.16	474.3	344.2	7.3547	418.2
463.16	485.3	352.3	7.3785	422.8
473.16	496.3	360.4	7.4023	427.0
483.16	507.4	368.6	7.4251	431.3
493.16	518.4	376.8	7.4479	435.3
503.16	529.6	385.1	7.4701	439.5
513.16	540.7	393.3	7.4919	443.7
523.16	551.9	401.6	7.5136	447.7
533.16	563.1	410.0	7.5345	451.7
543.16	574.3	418.3	7.5554	455.6
553.16	585.5	426.7	7.5759	459.6
563.16	596.8	435.1	7.5960	463.6

(*cont.*)

T [K]	h [kJ/kg]	u [kJ/kg]	s [kJ/(kg K)]	a [m/s]
573.16	608.1	443.5	7.6162	467.4
583.16	619.5	452.1	7.6356	471.3
593.16	630.9	460.6	7.6550	475.1
603.16	642.3	469.1	7.6741	478.7
613.16	653.8	477.7	7.6927	482.5
623.16	665.2	486.2	7.7112	486.3
633.16	676.7	494.9	7.7295	490.0
643.16	688.2	503.3	7.7478	493.6
653.16	699.8	512.2	7.7657	497.2
663.16	711.4	521.0	7.7833	500.7
673.16	723.1	529.7	7.8008	504.3
683.16	734.8	538.6	7.8179	507.7
693.16	746.5	547.4	7.8350	511.2
703.16	758.2	556.3	7.8518	514.6
713.16	770.0	565.2	7.8684	517.9
723.16	781.8	574.1	7.8850	521.4
733.16	793.6	583.1	7.9011	524.7
743.16	805.5	592.1	7.9172	528.1
753.16	817.4	601.1	7.9330	531.4
763.16	829.3	610.2	7.9487	534.6
773.16	841.3	619.3	7.9645	537.9
783.16	853.3	628.4	7.9796	541.1
793.16	865.3	637.5	7.9947	544.4
803.16	877.3	646.7	8.0096	547.7
813.16	889.4	655.9	8.0245	550.9
823.16	901.5	665.2	8.0395	554.0
833.16	913.7	674.4	8.0541	557.2
843.16	925.9	683.7	8.0687	560.3
853.16	938.1	693.1	8.0829	563.4
863.16	950.3	702.4	8.0970	566.6
873.16	962.5	711.8	8.1111	569.7
883.16	974.8	721.2	8.1251	572.8
893.16	987.2	730.7	8.1391	575.8
903.16	999.5	740.2	8.1529	578.7
913.16	1011.9	749.7	8.1663	581.8
923.16	1024.3	759.2	8.1797	584.7
933.16	1036.8	768.8	8.1931	587.7

(*cont.*)

T [K]	h [kJ/kg]	u [kJ/kg]	s [kJ/(kg K)]	a [m/s]
943.16	1049.2	778.4	8.2065	590.6
953.16	1061.7	788.0	8.2197	593.5
963.16	1074.3	797.7	8.2328	596.4
973.16	1086.8	807.3	8.2459	599.4
983.16	1099.4	817.0	8.2585	602.2
993.16	1111.9	826.7	8.2711	605.1
1003.16	1124.5	836.4	8.2836	607.9
1013.16	1137.1	846.2	8.2963	610.8
1023.16	1149.8	856.0	8.3091	613.6
1033.16	1162.5	865.8	8.3212	616.5
1043.16	1175.2	875.6	8.3333	619.2
1053.16	1187.9	885.5	8.3453	622.0
1063.16	1200.6	895.4	8.3576	624.9
1073.16	1213.4	905.2	8.3698	627.7
1083.16	1226.2	915.2	8.3815	630.3
1093.16	1239.0	925.1	8.3932	633.1
1103.16	1251.9	935.1	8.4048	635.9
1113.16	1264.8	945.1	8.4163	638.7
1123.16	1277.6	955.1	8.4280	641.4
1133.16	1290.6	965.2	8.4394	644.1
1143.16	1303.5	975.2	8.4508	646.7
1153.16	1316.4	985.3	8.4621	649.5
1163.16	1329.4	995.4	8.4733	652.1
1173.16	1342.4	1005.5	8.4845	654.7
1183.16	1355.5	1015.7	8.4956	657.4
1193.16	1368.4	1025.9	8.5067	659.9
1203.16	1381.6	1036.1	8.5174	662.6
1213.16	1394.7	1046.3	8.5278	665.2
1223.16	1407.8	1056.5	8.5389	667.8
1233.16	1420.9	1066.8	8.5496	670.4
1243.16	1434.0	1077.0	8.5603	673.1
1253.16	1447.2	1087.3	8.5710	675.5
1263.16	1460.4	1097.7	8.5814	678.2
1273.16	1473.6	1108.0	8.5917	680.7
1283.16	1486.8	1118.4	8.6021	683.3
1293.16	1500.1	1128.8	8.6125	685.9
1303.16	1513.4	1139.2	8.6227	688.3

(*cont.*)

T [K]	h [kJ/kg]	u [kJ/kg]	s [kJ/(kg K)]	a [m/s]
1313.16	1526.7	1149.6	8.6327	690.9
1323.16	1539.9	1160.0	8.6428	693.4
1333.16	1553.3	1170.5	8.6528	695.9
1343.16	1566.6	1180.9	8.6628	698.5
1353.16	1580.0	1191.4	8.6727	700.8
1363.16	1593.4	1201.9	8.6824	703.4
1373.16	1606.7	1212.4	8.6922	705.8
1383.16	1620.1	1222.9	8.7021	708.3
1393.16	1633.6	1233.5	8.7120	710.8
1403.16	1647.0	1244.0	8.7216	713.1
1413.16	1660.4	1254.6	8.7310	715.6
1423.16	1673.9	1265.2	8.7404	718.0
1433.16	1687.4	1275.8	8.7499	720.4
1443.16	1700.9	1286.5	8.7594	722.9
1453.16	1714.4	1297.1	8.7688	725.1
1463.16	1727.9	1307.8	8.7780	727.6
1473.16	1741.4	1318.4	8.7872	730.0
1483.16	1755.0	1329.1	8.7964	732.3
1493.16	1768.6	1339.8	8.8056	734.8
1503.16	1782.2	1350.5	8.8144	736.9
1513.16	1795.8	1361.3	8.8235	739.3
1523.16	1809.4	1372.0	8.8325	741.7
1533.16	1823.0	1382.8	8.8415	743.9
1543.16	1836.7	1393.5	8.8505	746.3
1553.16	1850.3	1404.3	8.8591	748.8
1563.16	1864.1	1415.2	8.8680	751.1
1573.16	1877.7	1426.0	8.8764	753.4
1583.16	1891.4	1436.8	8.8851	755.6
1593.16	1905.2	1447.7	8.8938	758.0
1603.16	1918.9	1458.5	8.9023	760.4
1613.16	1932.6	1469.4	8.9109	762.8
1623.16	1946.4	1480.3	8.9196	765.0
1633.16	1960.2	1491.2	8.9282	767.2
1643.16	1973.9	1502.1	8.9368	769.5
1653.16	1987.7	1513.0	8.9451	771.6
1663.16	2001.5	1523.9	8.9536	773.9
1673.16	2015.3	1534.8	8.9620	776.1

(cont.)

T [K]	h [kJ/kg]	u [kJ/kg]	s [kJ/(kg K)]	a [m/s]
1683.16	2029.1	1545.8	8.9703	778.2
1693.16	2043.0	1556.8	8.9786	780.5
1703.16	2056.8	1567.7	8.9865	782.8
1713.16	2070.7	1578.7	8.9944	785.1
1723.16	2084.5	1589.7	9.0024	787.3
1733.16	2098.4	1600.7	9.0104	789.4
1743.16	2112.3	1611.7	9.0184	791.7
1753.16	2126.2	1622.7	9.0264	793.9
1763.16	2140.1	1633.7	9.0343	796.2
1773.16	2153.9	1644.8	9.0422	798.1

APPENDIX D

Reynolds' Transport Theorem

The Reynolds Transport Theorem states

$$\frac{d}{dt} \int_{\tau^*} \chi \, d\tau = \int_{\tau^*} \frac{\partial \chi}{\partial t} \, d\tau + \int_{\sigma^*} \chi \vec{V}^* \cdot \vec{n} \, d\sigma \tag{D.1}$$

where $\chi = \chi(x, y, z, t)$ can be any scalar-, vector- or tensor-valued function.

To prove (D.1) let us consider that the initial volume (at time t) is $\tau^* = \tau_1 + \tau_2$. The volume at time $t + \Delta t$ is $\tau_1 + \tau_3$, as shown in Fig. D.1 [Constantinescu and Galetuse, 1983, p. 82].

Let's define $I(t) = \int_{\tau_1 + \tau_2} \chi(x, y, z, t) \, d\tau$. By definition, the derivative of $I(t)$ is

$$\frac{d}{dt} \int_{\tau^*} \chi(x, y, z, t) d\tau \equiv \lim_{\Delta t \to 0} \frac{1}{\Delta t} \left\{ \int_{\tau_1 + \tau_3} \chi(x, y, z, t + \Delta t) \, d\tau - \int_{\tau_1 + \tau_2} \chi(x, y, z, t) \, d\tau \right\} \tag{D.2}$$

and

$$\int_{\tau^*} \frac{\partial \chi}{\partial t} \, d\tau = \lim_{\Delta t \to 0} \frac{1}{\Delta t} \left\{ \int_{\tau_1 + \tau_2} \chi(x, y, z, t + \delta t) \, d\tau - \int_{\tau_1 + \tau_2} \chi(x, y, z, t) \, d\tau \right\}. \tag{D.3}$$

(D.2) can also be written as:

$$\frac{d}{dt} \int_{\tau^*} \chi(x, y, z, t) d\tau = \lim_{\Delta t \to 0} \frac{1}{\Delta t} \left\{ \int_{\tau_1 + \tau_2} \chi(x, y, z, t + \Delta t) \, d\tau + \right.$$

$$\left. + \int_{\tau_3} \chi(x, y, z, t + \Delta t) \, d\tau - \int_{\tau_2} \chi(x, y, z, t + \Delta t) \, d\tau - \int_{\tau_1 + \tau_2} \chi(x, y, z, t) \, d\tau \right\} \tag{D.4}$$

where $\int_{\tau_2} \chi(x, y, z, t + \Delta t) \, d\tau$ was added and subtracted.

Taking into account (D.3), (D.4) becomes

$$\frac{d}{dt} \int_{\tau^*} \chi(x, y, z, t) d\tau = \int_{\tau^*} \frac{\partial \chi}{\partial t} \, d\tau + \lim_{\Delta t \to 0} \frac{1}{\Delta t} \int_{\tau_3} \chi(x, y, z, t + \Delta t) \, d\tau$$

$$- \lim_{\Delta t \to 0} \frac{1}{\Delta t} \int_{\tau_2} \chi(x, y, z, t + \Delta t) \, d\tau. \tag{D.5}$$

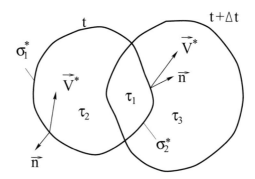

Figure D.1 Control volume at times t and $t + \Delta t$.

The last two integrals of the RHS of (D.5) can be written by taking into account that the volume $d\tau$ is

$$d\tau = \vec{n}\vec{V}^* \, d\sigma \, dt \quad \text{for } \tau_3 \tag{D.6}$$

and

$$d\tau = -\vec{n}\vec{V}^* \, d\sigma \, dt \quad \text{for } \tau_2 \tag{D.7}$$

such that both volumes of (D.6) and (D.7) are positive, given the relative directions of $\vec{n}$ and $\vec{V}^*$ shown in Fig. D.1. Equation (D.5) becomes

$$\frac{d}{dt} \int_{\tau^*} \chi(x,y,z,t) \, d\tau = \int_{\tau^*} \frac{\partial \chi}{\partial t} \, d\tau + \lim_{\Delta t \to 0} \frac{1}{\Delta t} \int_{\sigma_2^*} \chi \vec{n} \vec{V}^* \, d\sigma \, \Delta t + \lim_{\Delta t \to 0} \frac{1}{\Delta t} \int_{\sigma_1^*} \chi \vec{n} \vec{V}^* \, d\sigma \, \Delta t \tag{D.8}$$

or

$$\frac{d}{dt} \int_{\tau^*} \chi(x,y,z,t) \, d\tau = \int_{\tau^*} \frac{\partial \chi}{\partial t} \, d\tau + \int_{\sigma^*} \chi \vec{n} \vec{V}^* \, d\sigma \tag{D.9}$$

where $\sigma^* = \sigma_1^* + \sigma_2^*$.

Bibliography

V. N. Constantinescu and S. Galetuse. *Mecanica Fluidelor si Elemente de Aerodinamics*. Editura Didactica si Pedagogica, Bucuresti, 1983. in Romanian. 459

Index